Bridging the Communication Gap in Science and Technology

Pallava Bagla · V.V. Binoy
Editors

Bridging the Communication Gap in Science and Technology

Lessons from India

Springer

Editors
Pallava Bagla
New Delhi Television (NDTV)
Delhi
India

V.V. Binoy
School of Natural Sciences and Engineering
National Institute of Advanced Studies
Bengaluru, Karnataka
India

ISBN 978-981-10-1024-8 ISBN 978-981-10-1025-5 (eBook)
DOI 10.1007/978-981-10-1025-5

Library of Congress Control Number: 2016942524

Printed on acid-free paper

This Springer imprint is published by Springer Nature
The registered company is Springer Science+Business Media Singapore Pte Ltd.

Foreword

India is one of the few countries in the world that chose the path of science and technology for development almost immediately after coming out of centuries of colonial rule. The early decades were indeed tough. While on one hand the national science and technology infrastructure (education, research and capabilities for technology industries) was poor, scientific literacy among the majority of the population living in our villages was indeed nonexistent. Not that they were ignorant but simply not up to date on 20th century developments in science and technology. Across the world, high-technology products and services were indeed the luxury of the rich and had very little to do with the life of the common man. Science communication in those decades was synonymous with inculcation of scientific temper among the population and removal of superstitions.

Things changed rather dramatically over the next few decades. Technology products were making deep inroads into the life of Indians, rich or poor. With poor scientific literacy, the common man had difficulties in understanding the relevance of technology to his day-to-day life. One may recall the early resistance to the introduction of computers in government offices since it was seen as an anti-labour measure. I still recall the early discussions in our government corridors that satellite communication was a luxury of the rich and had no relevance to a developing country like India. The need of the hour was clearly science communication that would let people know how science and technology were progressing across the world, and what this progress could do to improve their day-to-day lives.

The 21st century is indeed very different. Today, there is no aspect of human life that is untouched by science and technology in some form or the other cutting across economic and social boundaries. True to the dictum 'the consumer is the king' the consumers also have a say, often the final say, in the choice of technologies. In a democratic environment, the citizens also have the final say in regulatory matters. The debates on nuclear energy, genetically modified food and so on clearly demonstrate that taking public into confidence is an important component of technology options. It is only science communication that can ensure informed technology choices. Nevertheless, science communication today has to be bidirectional. Although it is important to communicate science and technology to the common man, it is also important to communicate the public perceptions to the scientists and other policy makers. This is a new challenge of the 21st century.

Are we geared to face the challenge? In my view, we are not. We do not have enough science communicators; the channels of communication are weak, the India-specific challenge of language multiplicity persists and, last but not the least, we are not doing enough. This project was an effort to identify the problems and the available options, share experiences, evolve a road map to strengthen science and technology communications and enable informed decision making in a democratic environment.

V.S. Ramamurthy
Former Director, National Institute of Advanced Studies, Bengaluru
Former Secretary, Department of Science and Technology
Government of India

Acknowledgements

We are indebted to Prof. V.S. Ramamurthy, former Director, National Institute of Advanced Studies (NIAS), Bengaluru and former Secretary, Department of Science and Technology, Government of India, for his visionary support that made this book possible. His immeasurable passion and matchless experience in reaching science and technology to all strata of Indian society, the pinnacle of diversity of opinions and preferences, was the guiding force behind the selection of themes and getting support of luminaries who contributed the chapters for this book.

Dr. Baldev Raj, Director, NIAS, another well-known figure in the field of science as well as science communication, provided the needed momentum for the completion of the volume. We use this opportunity to express our gratitude to him for his support, encouragement and timely interventions, which made sure that this book was inked in the scheduled period. The authors also thank Prof. K. VijayRaghavan, Secretary, Department of Biotechnology for his help.

Professor Anindya Sinha, Dean (Academic Affairs), NIAS, is another person whom we are thankful to. His active involvement in the discussions contributed significantly to the structural design of the book. Professor Sangeetha Menon at NIAS provided sound advice to make sure the project remained on track.

Pallav Bagla thanks Mr. Mohandas Pai for having instituted Shri T. V. Raman Pai Chair Visiting Professorship at NIAS which Bagla was honored to hold in 2014.

It is uncontested that managing the compilation of the experiences of experts who contributed to divergent dimensions of science communication in India is not a trivial task. However, the presence of a skilled and dedicated person like Ms. Mariyammal, Secretary to Director, NIAS, in the team can make such a colossal job effortless. Her support during various stages of the preparation of the manuscript is acknowledged with gratitude.

Thanks are due to Mr. Srinivas Aithal, Head Administration, NIAS and his team for their selfless and timely supports for materialising this volume.

The team at Springer led by Ms. Shinjini Chatterjee and Ms. Shruti Raj worked tirelessly to make the book a reality.

We also thank our families who stood by us even as we spent endless hours editing the manuscripts.

Last but not least, contributors of the chapters are appreciated for their enthusiasm, kindness to share their experience and the timely submission of the manuscripts.

Readers should note that the contents of the chapters have been updated till 2014.

Delhi, India Pallava Bagla
Bengaluru, India V.V. Binoy

Contents

Editors and Contributors

About the Editors

Pallava Bagla is an Indian science journalist, communicator and a photojournalist for over 25 years. During this time he has come to be respected for his analysis and breaking-news stories and for having showcased Indian science and technology to the world in his almost two decades of writing for *Science*, the prestigious weekly magazine of the American Association for the Advancement of Science (AAAS), and as Science Editor for the NDTV (New Delhi Television) news channel. He is author of several books: *Destination Moon: India's Quest for Moon, Mars and Beyond* (HarperCollins, 2008), *Reaching for the Stars: India's Journey to Mars and Beyond* (Bloomsbury, 2014). A few of the awards he has been bestowed with include the 2006 National Award for Outstanding Effort in Science and Technology Communication in Print Medium, which is the highest honour of its kind for science journalism in India and is given by the Ministry of Science and Technology, Government of India, in 2016 he won the same award for his work in the Electronic Medium. In 2010, he was awarded the David Perlman Award for Excellence in Science Journalism, given by the American Geophysical Union (AGU), for his landmark writings which exposed the Himalayan glacier blunder by the United Nation's Intergovernmental Panel on Climate Change (IPCC). He was Shri T. V. Raman Pai Chair Visiting Professor of Science Communication at National Institute of Advanced Studies (NIAS), Bangalore in 2014.

V.V. Binoy is an Assistant Professor at the School of Natural Sciences and Engineering, National Institute of Advanced Studies (NIAS), Bengaluru. He is interested in understanding the biological and environmental basis of social cognition in both animals and human beings. His research focuses on the social decision making and personality traits (coping style or behavioural syndrome) in animals and the cultural variation in the attitude towards biotic and abiotic natural resources, autobiographical memory and environmental decision making.

Binoy's interests include science education and communication. He coordinates a citizen science initiative named Student Network (http://www.nias.res.in/wash/), which aims to enhance the student–scientist interaction and joint knowledge production in various states of India. He leads the biology education team of the Connected Learning Initiative (CLIx), a joint venture of the Massachusetts Institute of Technology, USA and Tata Institute of Social Science (TISS), Mumbai. Binoy is a research affiliate of the Krasnow Institute for Advanced Study and the Centre for the Study of Neuroeconomics, George Mason University, USA.

Contributors

D. Balasubramanian joined the Professor Brien Holden Eye Research Centre of L.V. Prasad Eye Institute (LVPEI) after an illustrious scientific career, which included positions such as Director, Centre for Cellular and Molecular Biology and Professor and Dean, University of Hyderabad. He was honoured by the Government of India with the Padma Shri award in 2002, and has received several other awards from professional and institutional bodies. Notable among them are the Kalinga Prize for the Popularization of Science from UNESCO, the Third World Academy of Sciences Award for Basic Medical Science and the Chevalier de l'Ordre National du Merite from the President of France. Balasubramanian's research focuses on the biology of eye diseases. He has to his credit over 160 publications in national and international journals and has authored three books. He is a member of several editorial boards and serves on a range of national and international scientific committees. Currently, he is President of the Indian Academy of Sciences, and Secretary General of the Academy of Sciences for the Developing World (TWAS), Trieste, Italy.

Archita Bhatta is the Chief Editor of the Communications Cell, Department of Biotechnology, Government of India, and a contributor to London-based science, development and climate change portals, www.scidev.net and Thomson Reuters Foundation's AlertNet.org. She is J.N. Tata fellow and a winner of the Sustainability Research Bursary of the Leeds University, UK and has several national and international publications as well as popular science articles to her credit. An alumni of Leeds University (School of Earth and Environment), Sikkim Manipal University and the University of Calcutta, she is pursuing a Ph.D. in climate change communication at Madurai Kamraj University, Tamil Nadu.

R.K. Chadha is a Ph.D. in applied geophysics and is presently the Chief Scientist at the National Geophysical Research Institute, Hyderabad. He is involved in earthquake research, especially earthquakes occurring around critical structures like dams and nuclear power plants. He has more than 150 scientific contributions to his credit, which include 83 research articles in peer reviewed national and international journals of repute such as Nature and Tectonics. He holds one US Patent, has guest-edited three international journals and submitted 43 technical reports. Chadha is the present Secretary General of the Asian Seismological Commission and has served as the President of the Natural Hazard Society. He provided leadership to several national initiatives of the Government of India after the Latur earthquake of 1993 by setting up the digital network of broadband seismic stations in the country. Chadha is a recipient of the National Mineral Award in Geophysics given by the Ministry of Mines, Government of India, Decennial Award of the International Gas Union (IGU) and is a Fellow of the National Academy of Sciences, India, Andhra Pradesh Akademi of Sciences (APAS), International Gas Union (IGU) and International Growth Centre (IGC). He has since retired from NGRI.

Ravi Kumar Gupta joined the Defence Research and Development Organisation (DRDO) as Senior Scientific Officer II in 1979 after obtaining a postgraduate degree in chemistry from the University of Delhi and DIIT in high polymer and rubber technology from IIT, Kharagpur. He has worked in the area of special materials, including advanced polymers and composites for high-temperature applications, organic crystals and polymers for laser applications and polycrystalline laser materials. He also worked on chemical lasers, hard anodisation and special-purpose electroless plating. In June 2008, he joined the newly formed Directorate of Public Interface and took over as its Director in March 2010. He has since retired from DRDO.

Arvind Gupta graduated from Indian Institute of Technology, Kanpur (1975) with a degree in electrical engineering. He has written 20 books on science activities, translated 150 books in Hindi and presented 125 films on science activities on Doordarshan, the India's national television channel. His first book *Matchstick Models and Other Science Experiments* was translated in 12 Indian languages. He has received several honours including the inaugural *National Award for Science Popularization amongst Children* (1988), *Distinguished Alumnus Award of IIT, Kanpur* (2000), *Indira Gandhi Award for Science Popularization* (2008) and the *Third World Academy of Science Award* (2010). He shares his passion for books and science through his web site: http://arvindguptatoys.com.

T.V. Jayan has a Bachelor of Science Communication degree from Madurai Kamaraj University. Since graduation, he has been writing on diverse topics of science, technology and medicine. Currently, he is a science editor with *The Telegraph*, a daily newspaper published from Kolkata. Jayan has earlier worked with the Press Trust of India (PTI), *Deccan Herald*, Vigyan Prasar and *Down To Earth* magazine.

Vinay B. Kamble is a nuclear physicist by training and has been a science communicator since 1980s. He worked as a scientist at the Vikram A. Sarabhai Community Science Centre, Ahmedabad, during 1978–1989, where he was engaged in research and development in nonformal science education and popularisation of science and technology (S&T). He joined the Department of Science and Technology (DST), Government of India, in 1989 and was instrumental in initiating and implementing major science communication projects such as Amateur Radio, Bharat Jan Gyan Vigyan Jatha (1992), Total Solar Eclipses of 1995 and 1999 and Emergence of Modern Science (2000). Since 2000, in addition to his duties at the DST, Kamble has served as the Director of Vigyan Prasar. During 2003–2004, he conceptualised and implemented the project 'Vigyan Rail—Science Exhibition on Wheels'. Kamble has been responsible for launching science programmes on Doordarshan in collaboration with Indian Space Research Organisation, and for the establishment of a network of Edusat satellite interactive terminals exclusively for S&T communication in the country.

Nimish Kapoor is a scientist and science communicator working at Vigyan Prasar, Department of Science and Technology, Government of India since early 2000s. At present he is the scientist in-charge of Science Film Festival Division at Vigyan Prasar. His work profile includes development of resource material and conceptualisation and coordination of campaigns (outreach) on science communication including science film festivals and orientation programmes for film makers and journalists. Earlier, he had been associated as science communicator and broadcaster with Doordarshan and All India Radio, Principal Researcher for Science and Society based ZEE TV's programme *Dastak* aired by Zee TV and Zee International and Programme Executive at Eenadu Television of Ramoji Film City, Hyderabad. He has contributed popular science articles, papers and book chapters on science communication, radio talks and science drama scripts for Vividh Bharti (national radio) and directed nonfiction television programmes and documentaries.

K. Kasturirangan is an Indian space scientist who headed the Indian Space Research Organisation (ISRO) from 1994 to 2003. He was a member of the Upper House (Rajya Sabha) of the Indian Parliament and a member of the Planning Commission of Government of India, Chancellor of Jawaharlal Nehru University, Chairman of Karnataka Knowledge Commission and Director of the National Institute of Advanced Studies, Bengaluru. Earlier, he was the Director of ISRO Satellite Centre, overseeing the development of new-generation spacecraft, the Indian National Satellite (INSAT-2) and the Indian Remote Sensing Satellites (IRS-1A and 1B) as well as scientific satellites. He was also the Project Director for India's first two experimental earth observation satellites, *Bhaskara I* and *II*. Under his leadership, ISRO witnessed several major milestones including the successful launching and operationalisation of India's prestigious launch vehicles, the Polar Satellite Launch Vehicle (PSLV) and the Geosynchronous Satellite Launch Vehicle (GSLV). He is a recipient of the three major civilian awards from the Government of India: Padma Shri (1982), Padma Bhushan (1992) and Padma Vibhushan (2000).

Hasan Jawaid Khan is Editor of *Science Reporter*, the popular science monthly magazine published by the National Institute of Science Communication And Information Resources (NISCAIR), Council of Scientific and Industrial Research (CSIR), New Delhi. He is also Editor of the monthly CSIR newsletter *CSIR News*. He has to his credit over 500 articles, news reports and book reviews in English and Hindi published in various magazines and newspapers. He has delivered about 100 radio talks on science topics in English, Hindi and Urdu and has also authored radio scripts. He is the author of four books, coauthor of 10 books and editor of 12 books on various scientific topics. He was awarded the National Science Popularization Award by the Indian Science Writers' Association (ISWA) in 2000.

Anitha Kurup is a Professor at the School of Social Sciences and Anchor of the Education Programme at the National Institute of Advanced Studies (NIAS). She is currently leading the National Gifted Education Programme in India anchored at NIAS. Her research interests span the broad disciplines of education and gender. Her doctoral work on quality of primary education in rural India is one of the earliest often-cited work on grounded research in classroom processes and school–community relationship in rural India. Kurup's expertise in the area of gender covers a wide spectrum, from examining conceptual and methodological strands of gender relations to political participation and decentralised governance as well as understanding violence against women in rural areas. Kurup was awarded the Fulbright Nehru Senior Research Fellowship for the year 2011–2012 and was hosted at the University of California, Davis.

Swapnesh Kumar Malhotra after post-graduation in chemistry from Agra University in 1976, joined the prestigious Bhabha Atomic Research Centre (BARC) in Mumbai. After training in nuclear science and technology, he joined BARC as scientific officer. For more than two decades, he was engaged in research and development on heavy water—an important material for the Indian nuclear power programme. He was involved in the technology transfer activities of BARC for a couple of years. Since 1999, as Head, Public Awareness Division of the Department of Atomic Energy (DAE), he is responsible for planning, overseeing and assessing the entire outreach programme of the DAE. He has served as an expert with the International Atomic Energy Agency (IAEA) for assessing and advising them on their public information programme. In recognition of his contributions, S.K. Malhotra has been bestowed with many awards by the Department of Atomic Energy, Indian Nuclear Society and many other professional bodies. He has since retired from DAE.

S.S. Meenakshisundaram is a postgraduate in mathematics from Loyola College, Madras and holds a Ph.D. in development economics from Jamia Milia Islamia National University, New Delhi. He joined the Karnataka cadre of the Indian Administrative Service in 1968. During his administrative career he had held a number of assignments including Secretary, Rural Development and Panchayati Raj (decentralisation) department when the state of Karnataka launched its new system of Panchayati Raj institutions. After working as Principal Secretary to the Chief Minister of Karnataka and Joint Secretary to three prime ministers of India, he also served as Advisor in the Planning Commission and Secretary in the Ministry of Rural Development. He was a Visiting Fellow at the Institute of Development Studies at the University of Sussex, UK. Besides several articles and papers on Panchayati Raj and rural development, he has also published a book, *Decentralization in Developing Countries* (1994), with special reference to Nigeria, China and India. Currently, he is the Executive Vice Chairman of MYRADA (MYsore Resettlement And Development Agency), a voluntary organisation in the field of building people's institutions for development, and a visiting professor at the Institute of Social and Economic Change, Bengaluru.

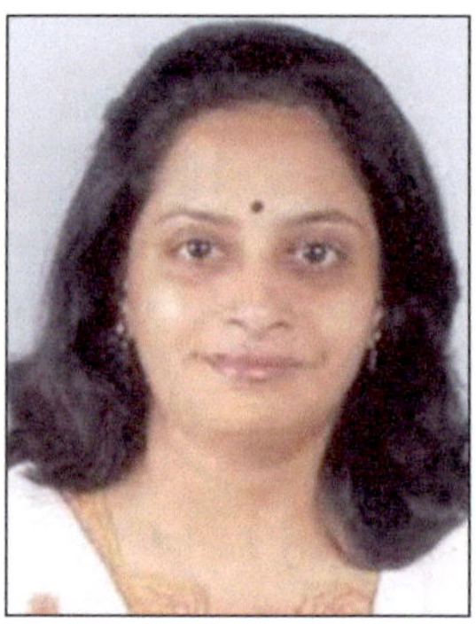

T.V. Padma is a Delhi-based science journalist. Her work has been published in *Nature, Nature India, Physics World, BioWorld, New Scientist, India Climate Dialogue* and *Third Pole Net*. Padma began her career as a science correspondent at the Press Trust of India (PTI), where she covered science regularly and was also responsible for the production of *PTI Science Service*, a fortnightly bulletin on national and international science developments. She later ran development communication projects at the South Asia office of the Panos Institute. In 2005, she joined SciDev.Net as its first South Asia Regional Coordinator and set up the first South Asian network of science writers. She reports on science, including key national and international conferences; she also commissioned and edited news reports from contributors for SciDev.Net. Padma holds an M.Sc. degree and a postgraduate diploma in bioethics, and was part of SciDev.Net's award-winning team in 2005 for best science reporting on the web, given by the Association of British Science Writers (ABSW), in recognition of SciDev.Net's coverage of the great Asian tsunami.

K.S. Parthasarathy is a consultant to the Department of Atomic Energy (DAE). He contributed significantly towards development of regulatory policies related to radiation equipment, protection of workers and members of public. He was a member of the Advisory Group of the International Atomic Energy Agency (IAEA), for which he issued a report on communication planning. He contributes science articles regularly to leading dailies such as The Hindu and to the Manorama Yearbook. He is a registered science writer of EurekAlert! of the American Association for Advancement of Science, British Medical Association, American Medical Association and the NucNet Agency.

Prabir Purkayastha is the Founder Editor of NewsClick India (Newsclick.in), a web site that covers and analyses key issues in today's world from points of view that are virtually missing from the coverage of the dominant corporate media. He is the President of the Centre of Technology and Development, an institution working on technology applications and policy issues, and the Vice-President of Free Software Movement of India. Purkayastha is also a founder member of Delhi Science Forum and the Chairperson of Knowledge Commons. He is the co-author of two books, 'Uncle Sam's Nuclear Cabin' and 'Enron Blowout: Corporate Capitalism and Theft of the Global Commons' and has written extensively on science, technology and policy issues in Economic and Political Weekly, Frontline and various newspapers. He has anchored various science programme on Rajya Sabha Television. Purkayastha holds a B.E. degree from the University of Calcutta and M.E. from Allahabad University.

Sindhu Radhakrishna is a Professor at the School of Natural Sciences and Engineering, National Institute of Advanced Studies, Bengaluru. She holds a master's degree in psychology and a doctorate in animal behaviour. Her research interests are in the fields of primatology, behavioural ecology and conservation biology, and her work has focused on gaining a better understanding of social behaviour and communication in nocturnal primates. She was awarded the Indian National Science Academy (INSA) Young Scientists Award in recognition of the significance of her doctoral research on the behavioural ecology of the slender loris, a nocturnal primate found in southern India.

S. Ramadorai is an Information Technology (IT) industry leader, a technocrat and a recognised global technology leader. He is associated with Tata Consultancy Services (TCS) Limited, India's best-known IT company, where he spent more than four decades. In his tenure as the CEO, before he stepped down in 2009, TCS had grown to be a US $6 billion global company employing 3,00,000 skilled workers in 46 countries. With his special gift in nurturing talent duly recognised by the Government of India, he has been piloting the National Skill Development Agency since 2011 as its chairman with the rank of a cabinet minister. An alumnus of the Indian Institute of Science, he is considered as one of the pioneers who transformed the Indian IT industry. Ramadorai is author of an acclaimed book, *The TCS Story ... and Beyond* (2011) and is also passionate about photography and Indian classical music.

Jairam Ramesh was the Minister of Environment and Forests, Government of India, from May 2009 to July 2011. Formerly, he was the Minister of Rural Development and had the additional charge of the Ministry of Drinking Water and Sanitation. He is a member of the International Council of the New York-based Asia Society and Visiting Fellow and Affiliated Researcher of the Institute of Chinese Studies, New Delhi. He anchored a number of popular television programmes (Business Breakfast and Crossfire) and has been a columnist for the Business Standard, Business Today, The Telegraph, Times of India and India Today. He authored the book *Making Sense of Chindia: Reflections on China and India* and co-edited another book titled *Mobilizing Technology for World Development*. He holds a B.Tech. in mechanical engineering from Indian Institute of Technology Bombay, Mumbai and M.Sc. in public policy and public management from Carnegie Mellon University's Heinz College.

Gauhar Raza was Head, Department of Science Communication through Multi-Media (SCM), National Institute of Science Communication and Information Resources (CSIR-NISCAIR), International Fellow, Mapungubwe Institute of Strategic Reflection, South Africa and Editor, *Journal of Scientific Temper*.

Sandhya Sekar is with Gubbi Labs, a research collective, which has started working in the area of communicating science with a larger audience. She has a Ph.D. from Centre for Ecological Sciences, Indian Institute of Science (IISc) and a master in science journalism from the University of Lincoln, UK. She has been freelancing for popular science outlets like *New Scientist*, *The Scientist* and *Nature India*.

A.S.K.V.S. Sharma is a popular science writer and well known to readers of Kannada as Kollegala Sharma. A Senior Principal Scientist in the Central Food Technological Research Centre (CFTRC), Mysore, he has been experimenting in science communication since 1980s. He is known as the second long-standing science columnist in Kannada, contributing a weekly column for 16 years at a stretch. He has published over 1600 science articles in Kannada and other 120 in English, and scripted 71 science dramas for radio. He is the recipient of the Best Science Communicator Award (2011) by the Vision Group on Science and Technology, Government of Karnataka, India.

Rajendra Singh popularly known as 'waterman of India', is a well-known water conservationist from Alwar district of Rajasthan, India. He is the founder–leader of Tarun Bharat Sangh (TBS), an organisation instrumental in empowering villagers from semiarid areas near the Thar Desert to acquire water security through community-based water resource management. Under the leadership of Singh, TBS built over 8600 *johad*s (rainwater tanks) and other water-conservation structures to collect rainwater for the dry seasons, brought water back to over 1000 villages and revived five rivers–Arvari, Ruparel, Sarsa, Bhagani and Jahajwali–in Rajasthan. He received the Ramon Magsaysay Award in 2001 for his leadership in water harvesting and water management, and *The Guardian* chose him as one among the '50 people who could save the planet'. He is a member of the National Ganga River Basin Authority (NGRBA) under the Ministry of Environment, Government of India.

Anuj Sinha is Honorary Chairman, Network for Indian Science Communicator Organizations (NISCO), Delhi. He is a highly decorated science communicator who led Vigyan Prasar, New Delhi for two years and was Head and Adviser, Science Communication Division, Department of Science and Technology for ten years. He has over ten years of field experience as a chemical engineer in a process industry and also served with distinction at the Bureau of Industrial Costs and Prices for three years. He piloted many policy and programme initiatives of applying science to society and has been involved in all major science outreach campaigns in the country for the last two decades. He holds B.Tech. (Nagpur University), MBA (University of Delhi) and M.Sc. (University of London) degrees.

H.S. Sudhira obtained his Ph.D. from the Indian Institute of Science (IISc), and is now with Gubbi Labs, studying the evolution of towns and cities using transdisciplinary approaches. For a while, he has been interested in communicating science to general public and is now working with researchers at Gubbi Labs to raise the bar of scientific literacy.

R.S. Sundar is a mechanical engineer from the Coimbatore Institute of Technology, Coimbatore. He joined Department of Atomic Energy in 1980 and served at Tarapur Atomic Power Station in various capacities and gained experience in the operation of nuclear power plants. He was the team leader for training on the Water-Water Energetic Reactor (VVER) class reactor of the Kudankulam Nuclear Power Project (KKNPP) at the Russian Federation. In May 2012, he took over as Site Director of the KKNPP. During his tenure he has increased awareness among the community about the safety of the nuclear power. He has also initiated welfare measures on health and education at and around the KKNPP site as part of corporate social responsibility (CSR). Elevated in the year 2012, he currently holds the position of Outstanding Scientist in the Department of Atomic Energy.

Ajit Tyagi is currently Koteswaram Chair Professor with Ministry of Earth Sciences and Advisor to North East Space Application Centre on Disaster Risk Reduction. He served as Director General of Meteorology, India Meteorological Department for a period of four years till January 2012. He was a Permanent Representative of India with World Meteorological Organization and member of its Executive Council. Prior to taking over as Director General, Dr. Tyagi was Air Vice Marshal in Indian Air Force and served as Principal Director of Information Technology and Assistant Chief of Air Staff (Meteorology). He was awarded with the Vishisht Seva Medal by the President of India for his distinguished services. Tyagi holds a Ph.D. in numerical weather prediction and has about 100 research articles and technical reports in the field of operational meteorology to his credit. Dr. Tyagi has served on advisory boards of National Atmospheric Research Laboratory, Indian Institute of Tropical Meteorology, United Nations Centre for Space Science and Technology Education in Asia and Pacific and National Knowledge Mission on Climate Change and on many Expert Committees of Department of Science and Technology.

T.V. Venkateswaran is a scientist with Vigyan Prasar. He has worked as the course director of the postgraduate programme on science and development communication at Centre for Development of Imaging Technology (C-DIT), Thiruvananthapuram, and has been a science writer with contributions to several periodicals including *Frontline, Down to Earth, Sandarbh* (in Hindi)*, Thulir* (in Tamil) and so on. He has many science television documentaries to his credit. His book, *A Revolutionary Scientist: Einstein and His Ideas*, published by the National Book Trust, India has had many editions. He delivers popular science lectures to public and conducts training for media students on science communication. His research interests include history of science popularisation, science in Indian media and risk communication.

Milind Watve is a science teacher who worked at Abasaheb Garware College in Pune from 1979 to 2009 and is currently with Indian Institute of Science Education and Research, Pune since 2009. He experimented on science education, particularly using research as a tool in undergraduate education. Most of his research publications are an output from work involving undergraduate creativity and initiative. His interests range widely and include principles of evolution, evolution of behaviour, behaviour–physiology interface, ecology of diseases, origin and stability of social systems, biodiversity and conservation biology.

Introduction: When Science Meets the Public—Bridging the Gap

V.V. Binoy

Abstract Reaching advancement in science and technology down to the lower-most stratum of the social pyramid and encouraging bidirectional communication between scientists and public are vital for 'scientising' the policies, attaining social and environmental sustainability and economic growth. Inefficacy or absence of a mechanism for reducing the chasm between science and public can kindle unnecessary clash between stakeholders and retard the development of the nation. However, in a society where stakeholders vary significantly in the value endorsed to science and technology, gaining trust of the public and promoting scientific temper is not a cakewalk. India being the zenith of the diversity of beliefs and value systems, plurality —hallmark of this nation—can be seen reflecting in the science communication and popularisation strategies as well. This book is a compilation of the views and experiences of the successful science communicators from India, whose interventions have led to meaningful connections between various branches of science and the society.

Keywords Science communication · Science popularisation · Indian science · Science and society

Introduction

Inquisitiveness of human beings to understand the secrets behind the wonders occurring in their surroundings paved way for the emergence of 'science', which brought in miraculous impact on the humanity. Ancient hunter-gatherer who struggled to escape from the furies of nature has evolved into the modern human who can manipulate the environment according to his or her imaginations with the

---

Views expressed in the chapter are the author's own and not necessarily shared by the editors.

V.V. Binoy (✉)
School of Natural Sciences and Engineering, National Institute of Advanced Studies, Bengaluru 560 012, India
e-mail: vvbinoy@gmail.com

P. Bagla and V.V. Binoy (eds.), *Bridging the Communication Gap in Science and Technology*, DOI 10.1007/978-981-10-1025-5_1

golden touch of science. Currently, there is hardly any facet of life left uninfluenced by science. It is uncontested that the dependency of the society on science and its outcome technology is increasing day by day, and these two elements also play a key role in determining the growth and position of a nation in the global playing field (Agre and Leshner 2010).

Access to scientific information can enhance the quality of decisions made both at individual and at population level and could shape the dynamics of a society (von Winterfeldt 2013). Thus, only a populace well informed of the advances in science and technology and its impact on human beings and the environment could fully comprehend the purpose behind a policy, and 'scientise' and democratise the policy making by their active involvement (Collins and Evans 2002). Meanwhile, the presence of pre- and misconceptions about scientific concepts or technologies in a society is as undesirable as the lack of information, since both factors can ignite conflict among different stakeholders (Brossard et al. 2005). Hence, most of the nations give priority not only to make various sections of their population aware of the developments occurring in the field of science and technology, but also to empower them with a culture of scientific temper in order to catalyse the sustainable development.

The relationship between science and society has been a topic of in-depth analysis because society considers science as a process that could provide solutions to enhance the quality of life outside of the pure scientific results (Weigold 2001; Fischhoff 2013). However, bridging the public, who are heterogeneous in their interests and values, with scientists, who focus on specific problems and communicate their findings in a language filled with 'subject-specific jargon and terminologies' that are unhandy even to their own counterparts from other disciplines, is not a trivial task (Jackson 2014). Due to the limited interaction with citizens, in many contexts, scientists do not get the real picture of the public perception of ideas on which they are working . Such a chasm existing between the producers of knowledge and laypersons fails the latter in many contexts in comprehending the benefits of new technologies and policies. Furthermore, availability of an affordable and trustworthy source of information in eliminating the anxieties of the people about the discoveries of science and novel technologies is essential to avoid the misdirection or exploitation of the uninformed sections of the society by agencies with vested interests as well as to avert unnecessary conflict. Unfortunately,the media, the common bridge between multiple stakeholders of a society, is feeble in communicating science and technologies to laypersons due to the lack of specialist journalists who could translate complex concepts of science into a language appreciated by the layman, without losing the crux (Amend et al. 2014). Hence, a platform that could facilitate the crosstalk between scientists and various stakeholders of a community is essential to reach scientific information unbiased up to the lower-most strata of social pyramid in order to assist public and policy makers to arrive at better decisions.

In India, the largest democracy of the world, where the Constitution proclaims "adopting scientific temper, humanism and spirit of enquiry" as citizens' duty, the gap between science and public is still conspicuous . In this nation with a population of 1.27 billion people, being a culmination point of culture and language diversity, bringing science and scientific temper to multiple segments of the society is a mammoth task (Mazåzonetto 2005; Subodh 2013). Existence of rigid power relations and belief systems, compartmentalisation prevalent in various sectors and the visible digital divide between urban and rural populations further worsen the scenario. However, in this regard, various programmes are being carried out successfully by government agencies (e.g., National Council for Science and Technology Communication (NCSTC) and *Vigyan Prasar*), various non-governmental organisations (NGOs) and interested individuals to familiarise the general public with science and its methodologies. Interestingly, the plurality, the hallmark of India, can be seen reflecting in the science-popularisation strategies popularised by different agencies, because communities differing in values and belief systems demand diverse methodologies for effective communication. Hence, a systematic collation of concepts popular in the field of science communication, strategies followed by successful science-popularisation initiatives and the experiences of experts into a single volume would not only be a useful reference material for academicians and students from diverse branches of science, social science and humanities but also be a priceless guide for designing effective science communication strategies and implementing them successfully in any multicultural multi-stakeholder society.

This book is an attempt to compile the experiences of leading science communicators from India, coming from diverse backgrounds, such as policy makers, executives, public servants, academicians, researchers, science journalists, media persons and activists whose interventions have significantly contributed to make the bridge between science and society meaningful.

Content of This Book

In the second chapter of this book (Responsibilities of Science, Responsive to Society: A New Dialogue) Jairam Ramesh, former cabinet Minister of India, describes four major contexts where, as he puts it, “different branches of science collided with the Indian society” during his ministerial tenure. He discloses the strategies utilised to formulate policies by his ministry on the sensitive issues of Bt Brinjal, atomic power plant at Jaitapur, mining in Niyamgiri Hills and the report of Intergovernmental Panel on Climate Change (IPCC) suggesting that glaciers of Himalayas would disappear by 2035 due to climate change, and proposes five basic principles for taking forward the dialogue between science and society.

Pallava Bagla, a senior and popular science communicator, points out many firsts and some missed opportunities in the efforts of Indian Space Research Organisation (ISRO) for communication and public outreach during *Mangalyaan*—India’s maiden mission to Mars—in chapter “India’s Maiden Mission to Mars: Many Firsts and Some Missed Opportunities in ISRO’s Efforts at Public Outreach and Communications”. In his words, the “hugely symbolic satellite mission has failed to capture the imagination of the country. Unfortunately, the public outreach and communication around the mission did not live up to the expectation. Mission successful, outreach much to be desired, is the outcome so far”. In the next chapter (“Challenges in Communicating about Defence Research: Insights in Defence Research and Development Organisation's Media Strategy") Ravi Kumar Gupta illustrates how a well-planned communication strategy transformed the image of Defence Research and Development Organisation (DRDO) from that of 'a dead, nonperforming organisation no better than a white elephant' to a technology leader. In his view, modernising the systems for correspondence with media and public helped DRDO in being recognised as a noteworthy contributor to the national economy and to evolve as a destination of talented youth. On the other hand, even if the policies take genuine care of the interests of the general public, lack of two-way communication between the government and stakeholders could lead to unnecessary agitations and litigations. The chapter by Meenakshisundaram ("Communicating Issues Related to Land and Natural Resources') looks at the challenges in disseminating information on land-related issues through three case studies—diversion of public lands in Karnataka for research and development institutions, acquisition of private lands for industrial growth by the government in West Bengal and the allocation of private and public lands to extract minerals for mining projects in Odisha.

Reporting on 'risks' to the public is considered as an art in the field of communication. If the warning messages are not adequate to convey the nature of a threat and its likely impacts, the recipients would neither pay proper attention to it nor take precautionary measures to face the disaster, ending up in mass destruction of property and loss of lives. At the same time, warning messages should not lead to panic among the masses, because that would negatively affect the mitigation activities. In this chapter ("The Art and Science of Communicating Risks of Natural Hazards"), Ajit Tyagi explains how the messages of hydrological hazards (tropical cyclone, heavy rainfall, floods and drought)—one of the two major kinds of natural hazards—are framed and broadcasted to protect the people from the catastrophe. Although a lot of progress has been made to understand earthquakes, a low-probability high-risk geological hazard, accurate prediction of this phenomenon to safeguard the lives of people have remained a distant dream. R.K. Chadha, in chapter "The Challenges of Earthquake Risk Communication to Public", presents earthquake hazard scenario in India and describes the challenges in effectively communicating earthquake risks to the public.

According to K.S. Parthasarathy (chapter "The Queer Case of Communicating Risks Associated with Use of Mobile Phones and Neighborhood Mobile Towers: Are People Contracting More Brain Cancers?"), ignorance of people about the scientific facts can be exploited by agencies with vested interests for commercial benefits by creating confusion and a fear of risk. Exposure to radiation, no matter whether it is from mobile phones or an atomic reactor, is often associated with the danger of causing cancer. His chapter illustrates how few commercially interested individuals with academic credentials spread baseless yet scary misinformation—through Internet and television interviews—that exposure to radiation from mobile phones could induce brain cancer, while the government agencies struggled to clear the air by providing accurate scientific data. In chapter "Atomic Energy: Reaching Out the People for Perception Management" S.K. Malhotra throws light on the need for understanding and considering opinions of diverse stakeholders to eliminate the fear of the risk associated with radiation from atomic plants. This chapter also describes the initiatives taken by the Department of Atomic Energy to generate a situation where 'rather than pushing nuclear energy to the society, a pull for the same from the society can be created'.

In this context an important question arises: What is the status of Public Understanding of Science (PUS) in India? Gauhar Raza and Surjit Singh share the results of their 25 years of research on PUS based on the surveys during every *Kumbh* and *Ardh Kumbh Mela*, the events where a multitude of people from various sections of the Indian society gathers at *Sangam*—confluence of two physical rivers, Ganges and Yamuna, and the 'invisible' Saraswati River—at Allahabad. Their chapter ("Current Status of Public Understanding of Science: Results of Kumbh Mela Survey Studies") discusses shifts in cultural distances that has come about in the past 12 years and the measurement of the efficacy of various news media such as newspapers, radio and television channels in reaching meaningful information on developments in the field of science and technology to the common man. In India, effectiveness of print medium in communicating science has not diminished although modern means of communication are swiftly gaining popularity even in rural regions of the nation. Taking the case of the monthly popular science magazine in English *Science Reporter* published by the Council of Scientific and Industrial Research for the past 50 years, Hasan Jawaid Khan, editor of the magazine, in the chapter "Communicating Science for a Better Tomorrow", highlights the institutional efforts in the country to popularise science through the print medium, challenges faced by the science magazines and what needs to be done to make press a more potent instrument for raising the scientific awareness of the citizens of India. A vital problem faced by the media while attempting to report science in a multilingual community like India, given the wide communication gap existing between citizens who speak vernacular languages and scientists who read and write in English using numerous technical jargon, is the topic of the subsequent chapters by A.S.K.V.S. Sharma ("Challenges of Communicating Science in Regional Languages: Experiments in Kannada") and Nimish Kapoor ("Challenges Faced by Science Journalists and Communicators Working in Vernacular Languages and Insights Pertaining to Science Communication Courses"). Sharma explains the challenges of communicating science effectively in regional languages in the light of various experiments being conducted in the regional language, Kannada, whereas Kapoor suggests various strategies to bridge the linguistic divide and tackle the challenge of translating scientific jargon into vernacular languages for reaching science to common people.

Since 1990s India has witnessed a revolution in the sector of information technology (IT), which marked its reflection also in the media depended on for communicating science and technology. Television, Internet and social media have become an integral part of life of people and the dependency on newspapers and radio, the vital sources of information in the last century, has dwindled down. The chapter by T.V. Venkateswaran ("Enhancing Science Content on Indian TV: Status, Issues and Way Forward") discusses the need for science and technology to mark its presence on television, one of the most popular and influential media, as a matter of routine so as to be a part of the culture, and to bring scientific temper in the public sphere. According to T.V. Padma (chapter "SCIDEV.NET Advent of Online Science Journalism in India") the Internet era has led to the emergence of a nonlinear model of flow of information with multiple sources, converse from the traditional models of science communication 'in which science journalists were the intermediaries who relayed information faithfully from privileged sources to diverse publics'. The theme of her chapter is 'SciDev.Net' an open-access, not-for-profit international science website focusing on development-related science with an aim of empowering individuals from both developed and developing countries to participate in the dialogue between science and technology and society. In the next chapter ("Smart Websites: Insights from the Department of Biotechnology Initiative"), Archita Bhatta, citing the new website of Department of Biotechnology (DBT) of Government of India as an example, explains the need for shaping government websites as attractive, informative and user friendly by incorporating modern tools of IT. In her opinion, an informative website could function as an effective interface between public and government and is essential for building confidence in the government departments as also to dispel prevailing myths about science and technology. Sandhya Sekar and H.S. Sudhira invite the attention toward social media, another Internet-based communication tool and highlight the need for utilising the ever-growing popularity of this medium for broadcasting developments that are taking place in the field of science and technology. Their chapter ("Using Social Media for Research and Reaching Out") provides an overview of various social media tools and portrays how each of these platforms could be utilised as an effective tool for disseminating science to the public.

After independence India has witnessed numerous initiatives by both central and state governments, NGOs and interested individuals to make science and scientific thinking popular. One unique experiment was *Vigyan Rail*—a train carrying scientific exhibits on various themes, which crisscrossed the length and breadth of the country for eight months during 2003 to 2004. This landmark event in the history of science communication of the country was conceived and implemented by *Vigyan Prasar*, and executed in collaboration with the Ministry of Railways, with active participation of eighteen departments and ministries of Government of India. Vinay Kamble, the then director of *Vigyan Prasar* shares the story of conceptualisation, implementation, impact and the lessons learned from *Vigyan Rail* project in chapter "Vigyan Rail: Science Exhibition on Wheels".

It is without question that activities meant to generate enthusiasm in science, and scientific temper should primarily target youngsters and students, the hope and future of science and the nation. Students showing interest in science and research should be provided not only the support and training, but also the venues for communicating their findings to their peers and the established academicians. The chapter ("Organising Children's Science Congress: Challenges and Opportunities") by Anuj Sinha elucidates the aim and structure of the National Children's Science Congress, a multilevel selection process conducted annually to find potential young researchers below the age of 18 years from every state of India and to publicise their research. A network of educational institutions—schools, colleges and universities—bridged with the researchers' community can provide a common platform for interaction between members of various segments of academics along with encouraging joint knowledge creation. This network could function as a low-cost and effective instrument for bridging various actors of a multi-stakeholder society for the dissemination of scientific knowledge and gathering data from the grassroot levels of the community. In chapter "Bridging Educational Institutions for a Citizen Science Project: A Case Study from Malappuram District, Kerala, India", Binoy, Radhakrishna and Kurup share lessons learned while developing a 'student network' by bridging schools and colleges with professional researchers for conducting a citizen–science programme to monitor the local mammalian diversity.

The last chapter of this book (Communicating Science to the Common: Perspectives of the Science Communicators from India) is a compilation of the views and experiences of established science communicators in India. Balasubramanian, a scientist and science communicator discusses the points to remember while communicating novel ideas of science to the public and policy makers. According to Milind Watve, a researcher, science teacher and an enthusiast of science communication, scientists need to interact with students since ignorance is a highly valuable contributor to science; when nonspecialists or naive audience ask questions about any field, it is very likely that the questions are very fundamental and induce rethinking of the conceptual foundations. The need for establishing a national strategy for the communication of science and utilising diverse strategies, such as science fests, museums and online education tools, for popularising science in a technology-driven society is elaborated by IT leader of the nation, S. Ramadorai. Kasturirangan, an icon of Indian space science, summarises the methodologies followed by the Indian Space Research Organization (ISRO) to generate enthusiasm among people about space sciences—planetary exploration in general, and India's first moon mission *Chandrayaan-1* in particular. Kudankulam Nuclear Power Project (KKNPP) is one of the unique events in the history of science and technology in India, which describes the clash between public and science caused mainly due to inadequate communication. R.S. Sundar, Site Director, KKNPP elucidates the successful communication strategies followed by the authorities of KKNPP to make people aware of the safety measures of the nuclear power plant as well to build trust in the scientific community.

"It is a responsibility of the media, both print and audiovisual, as an institution to bring science to the people, but in the troll of private capital very few takers are available for science, which lacks the potential to drag huge profits from advertisements". However, in the view of Prabir Purkayastha, a popular science journalist and a successful science communicator, "community should tap the potential of Internet to correspond the news from the field of science, which Indian media does not cover generally". He also shares his endeavours to reach science to public through an Internet-based platform called 'newsclick.in'. T.V. Jayan, a senior science journalist resonates with Purkayastha: "When it comes to the coverage of science and technology Indian media is far from satisfactory and most of the science articles Indian newspapers carry are about research done in the Western world". He also attracts attention to the reluctance of Indian scientists to discuss their work with media and the structure of science communication courses offered by the few institutes and universities in India.

Making a society aware of the potential of science and technology and empowering citizens to obtain sustainability in their ecosystem and economy by utilising scientific information is a dream of every science communicator. Rajendra Singh, popularly known as the 'waterman of India', has shown through the activities of his organisation *Tarun Bharat Sangh* that implementation of traditional methods of water harvesting and community-driven approach to popularise them can green villages, rejuvenate local rivers and promote financial security to the villagers by doubling the crop. According to Singh, mobilisation of civil society for scientific management of natural resources and conservation is a dependable method to save the environment and bring prosperity to farmers in Indian villages.

In India it is said, "schoolchildren have to mug up science and spit it out in the exams", as there are no laboratories available in most of the Indian village schools and often the elite schools in cities do not freely allow their students to use the laboratory because chemicals and apparatus are expensive. In the last part of the chapter, Arvind Gupta, one of the living legends of science communication in India, describes some low-cost teaching aids and toys made from local materials that would help children to imbibe complex scientific concepts effortlessly, stimulate their creativity and make science in classrooms interesting.

Two themes arise from this book. First, that communicating science to multiple stakeholders of a society and bringing scientific temper among them are not trivial tasks. Effective communication of the scientific information, ensuring its perception without losing the essence and seriousness and utilisation of such information in the decision-making process by members of every strata of a society requires a multidisciplinary approach integrating knowledge from science, social sciences, education and communication studies. Second, both scientists and society have to initiate intense two-way communication with mutual respect and readiness to hear out each other's disagreements and to reach consensus in the contexts of conflicts. Only then will a science aiming for the well-being of the society, and a society happily providing resources for flourishing of science, sustain.

References

Agre, P., & Leshner, A. I. (2010). Bridging science and society. *Science, 327*, 921.

Amend, E., Capurro, G., & Secko, D. M. (2014). The use of science journalism models to clarify the impacts of alternative forms of production. *Journalism Practice, 8*, 789–808.

Brossard, D., Lewenstein, B., & Bonney, R. (2005). Scientific knowledge and attitude change: The impact of a citizen science project. *International Journal of Science Education, 27*, 1099–1121.

Collins, H. M., & Evans, R. (2002). The third wave of science studies: Studies of expertise and experience. *Social Studies of Science, 32*, 235–296.

Fischhoff, B. (2013). The sciences of science communication. *Proceedings of the National Academy of Science, 110*, 14033–14039.

Jackson, A. (2014). http://blogs.nature.com/soapboxscience/2014/02/03/marcus-du-sautoy-communicating-science-within-the-sciences-and-to-the-public/. Accessed 20 June 2015.
Mazåzonetto, M. (2005). Science communication in India: current situation, history and future developments. *Journal of Communication, 4*(1), 1–6.
Subodh, M. (2013). A perspective on scientific temper in India. *Journal of Scientific Temper, 1*(1), 46–62.
von Winterfeldt, D. (2013). Bridging the gap between science and decision-making. *Proceedings of the National Academy of Science, 110*, 14055–14061.
Weigold, M. F. (2001). Communicating science: A review of the literature. *Science Communication, 23*, 164.

Part I
The Big Picture: Communicating Science to Win the Hearts and Minds

Responsibilities of Science, Responsive to Society: A New Dialogue

Jairam Ramesh

Abstract In India, a mosaic of languages, cultures and political systems, where stakeholders differ drastically in their beliefs, value systems and expectation from the government, introducing a policy to implement or discard a novel technology is a Herculean task. In this chapter, comprising an edited version of the author's lecture to a scientific community followed by questions and answers, the author shares the challenges he had to face while taking decisions as a cabinet minister in four different contexts where science collided with Indian society: Bt Brinjal (India's first genetically modified food crop), Jaitapur nuclear power plant, climate change in Himalayan glaciers and mining in Niyamgiri Hill. Based on his personal experiences the author proposes five principles that would form the basis of a new dialogue on making 'people more science-friendly and scientists more people-friendly', and elaborates the role of these principles in understanding and explaining various dimensions of science–society interface.

Keywords Bt brinjal · Jaitapur nuclear power plant · Climate change · Himalayan glaciers · Niyamgiri hill · Public–scientist interaction

I want to tell you four stories on the science–society interface, which I had to grapple with, particularly when I was Minister for Environment and Forests. These stories are of Bt Brinjal, which is a story of biological science colliding with society; of *Jaitapur* which is a story of nuclear science colliding with society; of climate change in Himalayan glaciers, which is an instance of atmospheric science colliding with society and, lastly, about mining—Vedanta, POSCO, and the associated *Go's, No-Go's*—where ecological science collides with society. I shall conclude the lecture covering these four case studies by deriving five principles—lessons, which

Views expressed in the chapter are the author's own and not necessarily shared by the editors.

J. Ramesh (✉)
Former Union Minister of Rural Development and Environment and Forests, Government of India, New Delhi, India
e-mail: jairam54@gmail.com

P. Bagla and V.V. Binoy (eds.), *Bridging the Communication Gap in Science and Technology*, DOI 10.1007/978-981-10-1025-5_2

I hope would form the basis of a new dialogue. How do we take this dialogue forward? How do we make science more responsive to society—and conversely—the society more responsible towards science? This is a subject that has long fascinated me, and I just discovered in my archives, a special issue of *Dædalus* published in 1974, titled, 'Science and its public, the changing relationship'. The issues of science–public relationship have been discussed by eminent scientists including Nobel Laureates like Steven Weinberg largely, of course, in the Western context. But the theme is pretty much the same: How do you take science to the people? How do you make people more science-friendly? And how do you make scientists more people-friendly?

So, over four decades later, we are discussing pretty much the same theme, albeit in the Indian context, and particularly relating to the decisions I took as a Minister. Let me start with the first example, and I am going to be frank. I am not going to be saying anything different than what I have said earlier—and I have never shied away from controversy even when I was a Minister—but I may add few comments based on the developments since the time those decisions have been taken.

Bt Brinjal was the first issue. I took over the Ministry on 29 May 2009. The very next day reporters from ten TV channels stormed my office and asked, "What are you going to do about Bt Brinjal?" I said, "What is Bt Brinjal? What is this Bt Brinjal all about?" Eventually, I called my officers and asked them, "Is there something on Bt Brinjal going on around here?" Indeed, we had an immediate case for the commercialisation of Bt Brinjal. "Bt Brinjal," said my officers, "is very essential for the food security of our country." I joked with them whether those who did not like *Baingan Bharta* (a mashed Brinjal dish) would get to like Bt Brinjal *Bharta*!

In the days to come, I was inundated with questions on what the government was going to do with, or what was going to be the scheme of commercialisation of, India's first genetically modified food crop. I was aware of the debates on Bt Cotton. I had written about Bt Cotton, I knew about the political debates, the economic debates and the technical debates, and I was also very sensitive to the fact that over 94% of Indian farmers had adopted Bt Cotton. They cannot be all irrational or wrong, and, as an analysis subsequently showed, about 20% increase in the yield that India had 2000s was attributable to the introduction of the Bt Cotton varieties. India has gone from being the third largest cotton producer to become the second largest cotton producer in the world. Undoubtedly, barring the State of Maharashtra, where cotton is largely rain fed, wherever you had irrigated cotton, Bt Cotton has been a success story.

But Bt Brinjal presented a completely different set of issues. For one, the civil society mobilisation was far more systematic, serious and extensive in the case of Bt Brinjal than it was in the case of Bt Cotton. I think Bt Cotton was seen to be something between the seed companies, the research laboratories and the farmers, whereas Bt Brinjal brought in the consumer. Thus, there were many consumer organisations, a number of advocacy organizations, NGOs, environmental activists and farmers' organisations presenting a whole new set of issues. And then of course, you had the scientists, who said that without Bt Brinjal, Indian agriculture was doomed, and food security, as I mentioned, was going to be dependent on Bt Brinjal.

So what did I do? My response after all the briefing I received was something, which was not liked by my colleagues in the government; I started a process of public consultations. I had seven public consultations in seven parts of the country: Chandigarh, Bhubaneswar, Ahmedabad, Kolkata, Nagpur, Bengaluru and Hyderabad, in which about 8000 people participated. These were open consultations organised by Kartikeya Sarabhai's Centre for Environment Education. Nobody could accuse Kartikeya of being anti-science, and it was his organisation that invited people, scientists, farmers, consumers and the general public. The whole idea was just to listen to the people articulating different points of view.

And these public consultations, of course, became very acrimonious. Because nobody was talking *to* each other, everybody was talking *at* each other; the scientists were arrogant in their certitude and the members of the civil society were equally obnoxious in their cocksureness because civil society felt that they had a monopoly on social conscience and scientists believed that they had a monopoly on knowledge. Therefore, actually these public consultations did not become a dialogue. They instead became occasions on which people aired their views. Scientists said what they had to say, activists said what they had to say, and as it so often happens, activists were far more articulate and louder than scientists. Therefore, by and large in each of these consultations, the message that I received was 'hasten slowly'. By all means, 'do not close the door for Bt Brinjal, but do not rush into it either', like we had rushed into Bt Cotton. These public consultations were one way of bringing science and public together.

The second route I adopted was to write to 70 top scientists of the world. I called Pallava Bagla of NDTV, New Delhi and obtained Bruce Alberts' e-mail address. I wrote, "Professor Alberts, here is my problem, here is the Bt issue, please tell me 60 top scientists across the world who are competent to judge what India should be doing in this area," and I got responses with some scientists saying we should go ahead and many others saying "Go slowly". Some scientists, of course, said, "Don't do it at all." So this was the second route of trying to bring science and society together, getting a much larger viewpoint from the scientific community. The third route I adopted was to contact the six scientific academies of India and, perhaps for the first time in their history, all the six academies came together to the table—and were civil and cordial with each other, which was a very rare occurrence—to share with me their views on what India should be doing on Bt Brinjal. Dr. M.S. Swaminathan, considered one of the key players of India's first green revolution, cautioned me saying that India was a centre for genetic diversity for brinjal, and that I must go slow on this issue.

Now, I had many views before me including those strongly in favour, including those of my very good friend Prof. Pental who was then the Vice Chancellor of Delhi University. On my own part, I had very strong views that we should commercialise Bt Brinjal because that would send a signal that we were serious about biotechnology in agriculture. On the other hand, I had equally strong views from scientists, from activists something that you may not immediately appreciate, but very important in the Indian context and from every state government in India opposing the introduction of Bt Brinjal; *every* state government in India! You know, 60% of brinjal cultivation is in three states, West Bengal, Bihar and Odisha, and they were opposed to Bt Brinjal. The states such as Maharashtra and Madhya Pradesh were also opposed to Bt Brinjal. What came to me as a complete surprise was that Gujarat, then under the chief-ministership of the present Prime Minister Mr. Narendra Modi, was also opposed to Bt Brinjal! Every state gave me in writing, "Do not commercialise Bt Brinjal." This was the dossier that I had before deciding the issue.

What was I to do? What I did was to put out a 20-page speaking order and I put all these views in the public domain (MoEF 2009, 2010). I put whatever I received —letters and views, including what Dr. Swaminathan had written to me—into the public domain, and said, "This is what these people have said." My speaking order said, "Let's have a Moratorium on Bt Brinjal," not an indefinite moratorium but a very definite, time-framed moratorium, until three conditions were fulfilled: (1) all states or a majority of states should be on board, (2) there should be an independent regulatory mechanism—a regulatory institution of independent professionals, that will certify that all the protocols and tests have been taken care of and (3) there must be a reasonable degree of scientific consensus that all the tests that are required on toxicity, safety, efficacy, and so on, have actually been conducted. I was quite surprised that the scientists themselves were divided on this issue. I had visited the National Centre for Biological Sciences, Bengaluru, and the Centre for Cellular and Molecular Biology, Hyderabad. The scientists I met there gave me sceptical, though not hostile, views. Therefore, my third condition was to let the scientific community, as a community, develop a broad consensus. And if these three conditions are fulfilled, the government of the day should remove the moratorium. I went out of my way to assure the public, "This is not a moratorium on biotechnology, this is not a moratorium on genetic engineering, this is not a ban on genetic engineering, but it is a limited issue of moratorium on commercialisation of Bt Brinjal for a variety of reasons."

I would not go into the detailed reasons, which were well argued in this 20-page note. One of the important reasons brought out in one of the presentations was the fact that a particular company controlled the bulk of the Bt Brinjal seeds, and that was an important issue, a sort of a red flag to all of us. So, to cut a long story short, a moratorium, a conditional moratorium, was imposed on Bt Brinjal. The moratorium imposed on 9 February 2010, continues in 2016. I left my office in 2011 and the expectation was that the day I left, the moratorium would be lifted. But it has not happened because, I think, there was a persuasive case for 'hastening slowly' laid out in my speaking order. There was a persuasive case for making commercialisation conditional. Those conditions, unfortunately, have not been fulfilled. Also, the states continue to oppose Bt Brinjal. The independent regulator is not in place, and the scientific consensus continues to elude us. Thus, the story on Bt Brinjal is one, I would say, of an initiative that got taken up with great enthusiasm and, no doubt, with great potential, but without asking the fundamental question that a political animal like me would ask: Is Bt Brinjal essential for Indian food security? That question, unfortunately, never got asked. Bt Cotton always had very persuasive strategic arguments but those arguments fell apart in the case of Bt Brinjal, which then became an exemplar of science being a solution in search of a problem.

Now, let me move on to Jaitapur, where we had nuclear science colliding with society over the proposed nuclear power station. I was a hero to environmentalists till Jaitapur issue came up; then I became a demon to the same environmentalists. This was because as a minister I took the view that from a climate change perspective India cannot afford to ignore, or soft pedal, nuclear power which currently constitutes less than 5% of our overall energy supply. It must go up to 10% to 20% in a reasonable time frame of two decades. That this nation of 1.2 billion, going to be 1.7 billion by 2040, will meet its energy demand through its 'renewables' is a romantic notion. Even today, all the renewables account for something like 8% of the capacity and less than 2% of India's energy supply; and that is going to be the case for a foreseeable future. Therefore, for the next 20 to 25 years, India has to be a major player in nuclear energy, and I am not even getting into the 'thorium argument', which only gives an extra weight to what we should be doing in the nuclear sector. Jaitapur came and instinctively I wanted that Jaitapur should get the green signal (NPCIL 2010; DAE 2011). But everything went wrong on Jaitapur. Fukushima happened and after Fukushima, the entire discourse on Jaitapur changed. Local people came to me and said they did not know where Fukushima was—they would say, "Fuku-Fuku-Fuku-Fukushima"!—and they protested against Jaitapur. Added to this was the fact that Jaitapur was a frontier cutting-edge technology, which had not been used anywhere else except in Finland and perhaps in China. We were to be one among the first countries where a nuclear power plant of 1600 MW capacity was to be installed (my sympathies with Prabir Purkayastha; see chapter "Communicating Science to the Common People: Perspectives of the Science Communicators from India" for further information) on terms, which were considered to be of such great strategic importance that only very few knew what those terms were. We could not be completely open in Jaitapur for legitimate reasons. Part of the problem was that the transparency, which I could bring into Bt Brinjal, I could not bring into Jaitapur, because of the nature of the problem. However, the issues faced were identical. These issues were of confidence in the nuclear establishment, of safety in the wake of the world seeing what happened in Japan and of introduction of a new technology—a completely new technology. Another issue that came up, and still remains, was of indigenisation. India had gone for 40 years on the *Can do* route although the joke was that the *CANDU* had become the *Can Don't*! But the Can do route was the one we had adopted, and now we were going to go into a completely different nuclear track. Why not perfect the 700 MW *CANDU* reactor and proliferate the Indian design reactor rather than depend on somebody from outside? All these issues were raised when Jaitapur became a public issue. So what were we to do? How were we to get out of it? The easiest way would have been to say, "We are not going to give an environmental approval to Jaitapur." But that went against my education, my training, my instinct, my head and my heart—it went against everything.

Thus, I came up with the classic Indian solution: attach 36 conditions to the clearance. So this was a 'Yes', but a 'Yes-But'. If you see the Jaitapur's speaking order—all this is in the public domain—it was an environmental clearance for

Jaitapur subject to '1 to 36 being fulfilled'. Of course, Jaitapur now has not progressed to the extent that we expected it would progress. However, I think the imposition of the conditions was one way of demonstrating from the government's side—from the ministerial side—that we were sensitive to public concerns of safety, of outreach and of compensation. The issue that came up in Jaitapur included compensation to be paid to farmers whose land was being acquired and to the fishermen who were going to be resettled. Therefore, these conditions epitomised our commitment to embedding the concerns of people in the clearance process. But we cleared the project subject to the fulfillment of these conditions, and we also created a mechanism to ensure the monitoring of these conditions; not just you take that piece of paper and then you go and ignore what the conditions are, but actually make periodic reports and periodic audits of the extent to which these conditions were going to be fulfilled. Thus, while Bt Brinjal was a case of a 'No', a conditional 'No', Jaitapur was a case of a conditional 'Yes'.

Now let me turn to the third issue, climate change. When I came to the Ministry, one of the first things I read was that the Himalayan glaciers would be disappearing by the year 2035 (IPCC 2007). Dr. R.K. Pachauri, the Chairman of Intergovernmental Panel on Climate Change (IPCC) was credited with this statement. I found this something remarkable and wondered if Indians at large knew this was going to happen. I called a couple of my scientist friends and asked, "Have you seen this report of the IPCC?" They said they did, but also said they had grave doubts about it. I asked, "What are your doubts—scientific or non-scientific?" When they said they had scientific doubts, I persisted, "Why haven't you spoken up? Why haven't you communicated what you believe are your doubts?" Scientists having doubts is very rare! When they do have a doubt, the media would definitely run to them. They said, "No, Sir, because we couldn't, … IPCC, a global body … Government of India …" I said, "Look. I want India's top glacier experts to come. Let us have a round table, and let us discuss the issue of Himalayan glaciers."

Thus, I got the scientists for a round table. Dr. Kasturirangan (Former Chairman, Indian Space Research Organisation, ISRO) came, and I also got other people from ISRO. I got a 75-year-old gentleman, V.K. Raina, a geologist from Geological Survey of India. He has, in the course of 55 years, visited something like 3000 glaciers. Not a Ph.D., but India's authority on glaciers, and called 'India's Mr Glacier!' I said, "Mr. Raina, please give me what your views are on the glaciers. And I give you two months. You prepare a report for me and I want to be educated and I want to educate my colleagues." He came out with the report, and it created a furor internationally! I got hate mail from every corner. The only person who supported me was the senior science journalist Pallava Bagla (2014). What did the report say? First, the report said that there are about 10,000 glaciers on the Indian side of the Himalayas. Second, most glaciers are retreating but some glaciers, like the *Gangotri* glacier, are retreating at a decelerating rate! Very interesting! *The Gangotri glacier is retreating at a decelerating rate!* Third, there are some glaciers, which are actually advancing! The famous *Siachen* glacier, for example, is actually advancing! And last, which was the most important part of the report, the clincher, that Himalayan glaciers are fundamentally different from Arctic glaciers, because their snout, that is, their lowest point, is at least 3000 metres above sea level! And that therefore the behaviour of the Himalayan glaciers a priori cannot mirror the behaviour of the Arctic glaciers, in response to perturbations. We put out this report and I stuck my neck out: "Look", I said, "Indian science, you know!" I believe science is science, I don't believe in Western science, Indian science, Vedic science and all these, but this was one instance where I said that Indian science, the science done by Indian scientists, has established that Himalayan glaciers will NOT vanish by the year 2035! The situation is, however, very complex. It is a dangerous situation because the health of the glaciers is very poor. The glacier melt can adversely affect water supply, but let us not get into alarmist evangelism: here, I wish to cite an example of how alarmist evangelism can be dangerous for India.

In 1990, the United States came out with a report that methane emissions from Indian agriculture were 38 million tons per year. Alarm bells rang all over India. We can manage CO_2 emissions because CO_2, as you know, is man-made. But anthropomorphic methane is derived from cattle: we cannot do much about 250 million cattle! You can manage a billion people but you cannot manage 270–280 million cattle! However, this methane and wet paddy—paddy being staple in India—set out alarm bells. I must complement the late Dr. A.P. Mitra (National Physical Laboratory, NPL, New Delhi) who took up the challenge and said, "This measurement is wrong." He started, I think through NPL, a programme for monitoring methane emissions in Indian agriculture and by 1994 came out with a report which said that methane emissions from Indian agriculture was 4, not 38, million tons per year! Today, the internationally accepted figure is 4 to 6 million tons! Had Dr. Mitra not taken that giant leap of faith, we would have been under enormous pressure internationally to manage methane emissions; because, from a climate change point of view, carbon dioxide may be more preponderant but methane is more potent. So, Dr. Mitra and this glacier are a similar story. We depended on somebody else's science. Science which had had not just a scientific agenda: This notion that scientists are politics free, value free—it used to be said of George Nathaniel Curzon, he was a superior person—now this notion that they are superior persons bereft of human emotions like politics is a myth! Much of the science in today's context is political, to drive political agendas, both internationally and domestically. Himalayan glaciers was one such example where clearly the scientific community was behaving like evangelists and not like scientists till a report came out of India. Even my Prime Minister asked me, "Are you sure about what you are saying?" I said, "Sir, I have consulted all the experts, this gentleman swears by what he says, and I am prepared to back him 100% to the hilt." The day we released the report, Dr. Pachauri put out his press release calling it Voodoo science! NGOs burned effigies in Copenhagen. But to cut a long story short, subsequent events proved Mr. Raina right. IPCC for the first time issued a retraction saying they were wrong, their peer review process was wrong and slipshod and that the Indian Government's position was the more nuanced and right position.

Thus, in the case of climate change, when science collided with the public, it was not public that was wrong. It was science that was at fault. It was science that had created this great atmosphere that glaciers were endangered. And when somebody challenged that science, the truth came out differently. The point I want to make is that this scientific method is the method of questioning. The scientific method is a spirit of enquiry. It is a spirit of doubt. If we are to accept everything coming out of scientific establishments, as gospel, I think that goes against the very grail of science. Dissent is part of the scientific DNA, which will be tried to quell at every step. And in the case of the glaciers, if we had not allowed that small window for dissent, we would not be in this situation today where the world community recognises the complexity of the Himalayan glaciers' issue and does not get into sensationalism and fear mongering, that the Himalayan glaciers are going to disappear by the year 2035 (Bagla 2014).

Let me now turn to the final issue of mining, which is perhaps the most complicated issue; these are not just scientific issues; these are human issues and livelihood issues, as Dr. Meenakshisundaram has pointed out (see chapter "Communicating Issues Related to Land and Natural Resources"). In the case of the mining issues, an issue of faith was also involved. In the case of Vedanta, bauxite was going to be mined from a mountain that is considered to be religious. I would like to ask of the urban Indians, "Suppose diamonds were discovered underneath Ram Janma Bhoomi—the sacred shrine believed to be the birth place of Lord Rama—by some miracle; would India tolerate mining at Ram Janma Bhoomi?" No! If the Niyamgiri Hill is considered sacred for the Dongria Kondh tribal people of India who are we to determine what is right for them? That was the issue in Vedanta. Only 5 % of the total bauxite supply for that 6 million ton alumina refinery was to come from this mountain. They could have found this bauxite anywhere else. But it was the certitude. This is what we want; this is what we will get. This is *zabardasti*, to use a Hindi word for brute force. In the case of Vedanta, it was a mountain that was considered sacred. It was not a scientific issue, but an issue of faith, an issue of culture and an issue of lifestyle. To my mind, a public issue is as important as a scientific one.

In the case of POSCO, which Dr. Meenakshisundaram shared with us, in which again I had the misfortune of being entangled, it was a straightforward case of land acquisition. People grew betel leaves in the region and these leaves were supplied all over India. And they did not want to part with their land—it was as simple as that. Seven villages' land was to be acquired, five villages agreed, two villages did not agree and that is how the POSCO issue came up. It was a straightforward case of livelihoods being lost. Added to this was our poor track record in providing alternative livelihoods as well as resettlement and rehabilitation (R&R). It is learnt that R&R for Tarapur as well as for Biladila iron ore project that started in the 1960s is still incomplete. R&R for Bhakra Nangal is still underway. Now, in the face of this reality, what is the confidence level that anybody is going to evoke in the public when you say, "I am going to acquire your land and give you compensation and R&R"? Nobody is going to believe it! And that is what has happened in POSCO.

In the case of coal mining, the issue is a little more complex. It is related to ecological science. All the coal that India needs to extract in future is in thick, dense forest areas. Therefore, the more the coal you are going to extract, the more the natural forests you are going to destroy. You can always have artificial plantations, but you are going to be destroying hundreds, if not thousands, of hectares of high-density, good-quality forests. It is noteworthy that although 21% of India is under forest cover, of this 21%, only 2.5% constitute good-quality forests; 40% of Indian forest cover is *Naam ke waste* (just make-believe) forest cover, namely degraded open forests. What we really call forests is only 2.5% and that is where most of the coal mines exist. Now if you are going to extract coal at the pace being envisaged, then the country will pay a certain ecological price; maybe the country is ready to pay that price, but the consequences of paying that price is what the public is responding and reacting to. The result is, lands are being acquired but compensation is not being paid and R&R is not being completed. Then there are mines being worked, finished and not being restored to their original conditions; and that is not just some isolated case in India. These are the realities as far as mining is concerned. And in mining, I must say, the issues go much beyond science. They go into issues of legality, of land acquisition, resettlement, rehabilitation and so on and so forth (Saxena et al. 2010).

Now, from these four examples, Bt Brinjal, Jaitapur, climate change and mining, what are the general principles that one draws? These are nice, interesting stories, and for those of the readers who are interested, all the material on each of these is available in the public domain. I have not said anything, which is not in the public domain. To draw up the principles, the first thing, I think we must recognise, is that growth is essential—there is no questioning that India must grow at 7.5-8 % per year. Between 2004 and 2014 the average GDP (gross domestic product) growth was 7.7%. For those who know the compound interest tables by heart know that if something is growing at 7% it is doubling every 10 years. This means, basically, a GDP growth of 7.7% over a 10-year period has meant that the GDP has doubled in the last decade, even though there has been a slow-down in the past 2 years. We must get back to an 8% growth trajectory on which there can be no doubt whatsoever. But, in the process of getting on to this 8% growth trajectory, issues that I have highlighted will become essential to the choices that we make. This 8% growth will not be without contradictions, complexity and conflict. I think it is the job of all of us to clarify how these contradictions, complexities, and conflicts are going to be managed. There are going to be, undoubtedly, choices that have to be made: investment choices and technology choices. Those choices will involve tradeoffs, and what those tradeoffs are will have to be made explicit and, as Dr. Meenakshisundaram puts it, made completely transparent so that decisions can be taken as a part of the democratic process. Now, what is this democratic process? This democratic process is not the government sitting with a huge mandate in Delhi or in a state capital but the locally elected bodies. Ultimately, we must recognise that if these locally elected bodies, *Gram Sabhaas* or *Palli Sabhaas*, as they are called in some states, are not a part of the democratic process of decision making, these conflicts, complexities and contradictions will never get resolved.

This resolution cannot be a top-down process. We are not China; I know there is a great fascination for China these days. We are not Japan either. We are India where chaos and anarchy is a part of the system. If we think that we are going to mastermind a top-down approach to managing these complexities of conflicts, we are sadly mistaken. This cannot be done in a centralised command and control mode. It has to be done by involving local communities and local-elected representatives. I venture to suggest that scientists in India more than any other country in the world are a pampered lot. From Jawaharlal Nehru's days they have enjoyed unprecedented prestige, unfettered access and the Prime Minister himself and nobody else has always been the *karta dharta* (executive authority) of the scientific establishment. Therefore, the scientists have never felt the need to engage the democratic processes at the local level. With great difficulty they are engaging at the state level. Because if anything goes wrong, you can always run to the Prime Minister, whoever the Prime Minster is and the Prime Minister will set it right—whether it is biotechnology, nuclear, space, defence or agriculture. We need to understand that in this process of growth, to meet these complexities, contradictions and conflicts, and engagement with the democratic process at the grass roots is very much important. That is where the scientific community really needs to do far greater work irrespective of the area that we are talking about.

Now I derive five basic principles before I end my lecture. I derive these five basic principles for taking this dialogue forward. We recognise the value of science. We all know that science has much to contribute to the growth process. We all know that major investments have been made in science. But we are also aware that we are operating in an environment of scepticism in which there is questioning of science. There is an environment of distrust in many cases. So what is the way forward? I would suggest the following five principles for adoption, both by the scientific community—as a community—and by civil society which is loosely organised, but recognised as the spokesperson of the people's point of view, whatever that people's point of view may actually mean. But before we embrace these five principles, both sides have to accept that neither side has a monopoly on truth, patriotism or social conscience. Let us assume that truth, patriotism and social conscience are something common to all of us. Hence, the first principle: science needs to be less arrogant and society needs to be less suspicious. The second principle: science needs to listen, and society needs to hear. As I would like to tell my good friend, Rajendra Singh of Tarun Bharat Sangh of Rajasthan, *aap ko sunna hai!* (You must listen!) *Aap sunte nahin hain!* (You never listen!). *Aur jab maine aapko Kanpur ke IIT mein professor se milwaaya tha, aap ne mujhe kaha tha, "Yeh IIT waaley kya jaanenge?"* (When I introduced you to the professor in IIT, Kanpur, you said, "What do these IITeans know?") *Tees saal se Ganga cleaning ki baat ho rahi hai—tees saal se! Aur jab meiney saaton IITs ko pehli baar kaha, ki aap ek-jut hokar Ganga ke leiye ek action plan tayyaar keejiye, tab mere mitra, Rajandra Singhji ne pehli baar uthkar kaha "Yeh scientists kya kahenge? Hum se poochiye! Hum jaante hain Ganga ko kaisey saaf karna hai!" Toh aap ko sunna hai !* (The talks for cleaning of the river Ganga are on for the past 30 years—*for the past 30 years*! And when, for the first time, I asked all the seven IITs to unite and come up with an action plan to clean up the Ganga, my friend Mr. Rajendra Singh rose for the first time to say, "What would these scientists say? Ask us! We know how to clean up the Ganga!" Well, you [the civil society] have to listen.)

The third principle: science needs to be open to scepticism and society needs to be open to questioning. Today, we cannot question Sunita Narain (of the Centre for Science and Environment, New Delhi). She is a Goddess when it comes to environmental issues. You can't question Dr. Kalam (Former President of India) either! He is a God when it comes to scientific issues! Maybe I tried the wrong examples of the scientists, but with due respect to Dr. Kalam, science does need to be open to skepticism and society does need to be open to questioning. The fourth principle, which is what we are discussing here, is, science needs to be proactive in communication, not just active in communication! I presently explain what I mean by that. And the society needs to be less of a sensationalist and conspiratorial that Monsanto, for example, is out to enslave India! The Vandana Shiva (an Indian scholar, environmental activist and anti-globalization author) of the world. At the same time, Pental *Sahab* , you have to be proactive in your communication. Not when there is a crisis on Bt Brinjal; you start writing articles daily (when the crisis appears)! Where was the biotech community before Bt Brinjal? It was only when this Bt Brinjal crisis arrived that the scientific community woke up. Communication cannot be crisis driven; it has to be constant! Communication has to be done not necessarily with the view to convince, but it has to be done with a view to educate, which necessarily takes time. Finally, the fifth principle: science needs to develop trust. Nobody believes the Department of Atomic Energy on the future of the nuclear power generation: we say we are 3 % on nuclear energy and we will go to 6 %; they say 4500 MW will go to 10,000 MW, and we have heard this for the last 30 years. Society has to nurture confidence in the science we do and that will come when there is trust. I find the scientific community notoriously laggard in sharing information, I repeat, notoriously laggard! The more information you share, the more open you are, and the more trust you will build; also the more criticism you will get, no doubt. But I think science has to get out of the cocoon, if the scientific community has to develop the trust among the people.

So these are the five principles, and I recapitulate. First, science needs to be less arrogant, and society needs to be less suspicious. Second, science needs to listen, and society needs to hear. Third, science needs to be open to skepticism, and society needs to be open to questioning. Fourth, science needs to be proactive in communication, and society needs to be less sensationalist and conspiratorial. Five, science needs to develop trust, and society needs to nurture confidence. I believe that if we follow some of these basic principles, and if we get people together on an ongoing basis, we will be able to show a dialogue forward. The great weakness in the Indian system is the lack of institutional forums, where people from different points of view can actually sit down and debate with each other. Most of the time, it is the converted preaching to the converted. We value biodiversity, but when it comes to intellectual discussions, we make sure that there is no *bio-diversity*. There is a mono-culture. I venture to suggest to the scientific community that if we break out of this syndrome and provide institutional forums, these debates will become more and more significant in the years to come. There will be different points of view—some informed, some uninformed—but I think we do not provide the space. We have to provide the space for different points of view. We have to provide forums for reducing the differences between the two different points of view. But I think the real challenge is, those in public life, when called upon to make decisions must have the intellectual honesty and the integrity to stand by these decisions; to put out these decisions in the public domain, to be held accountable for them, to be criticized for them or applauded for them, as the case maybe, and have the courage to say, “I have been wrong in some cases.” That, “I should have done things differently.” I have not yet reached that stage. We need many such interactions and of these interactions will come sensitization to both sides—not just to communicators, but to the activists also—and sensitization to the scientific community as well about the challenges they face in the larger public arena. Debates on nuclear power, renewables, environment, genetically modified crops, nanotechnology, stem cell research and so on, you name it, we don’t know! I am just taking the examples of the technologies we are familiar with. These debates should continue. So I would like to thank Pallava—although he has not shared the proceeds of the award that he got on the Himalayan glaciers—and to all the people who are here. In the case of nuclear, I am a hero to you but because I am a hero to you, all the environmentalists think that I have sold out to you! So they are not easy choices to make—you are damned if you do and you are damned if you don’t. So thank you very much.

Responses of Science Communication Experts

V. S. Ramamurthy (*Director, NIAS Bengaluru*): Thank you. And I think this is a closest view of the decision-making process in the government we could have had. Life was much simpler 50 years ago, one Bhabha and one Nehru would decide everything and everything else would be in place. But today, we are evolving into a system where both the sides have to interact on a continuing basis.

Shiv Vishwanathan (*Vice Dean, O. P. Jindal Global University*): Okay, Jairam … brilliant! You are the perfect crowd pleaser! As a storyteller you are brilliant! But your conclusions are absolutely trait! And I think that is the disappointment. If you look at it, each of your stories had better morals than you could conclude. Sometimes, maybe the storyteller isn’t the best philosopher of his own story. Now, let me take this thing. You seem to operate for the model of rationality, which has to be questioned. Because what was really questioned in science was the rationality and the nature of knowledge itself. You have never raised that question. You seem to think whether this is Bt Brinjal or nuclear power they are all equivalent problems. They are not! Two, you seem to ignore the ironies of the system. To a certain extent, these can be zero-sum games. I mean, it is okay making fun of Sunita Narain or Vandana Shiva. But I think what they are trying to say is that your ethical position is much more complex than you make out to be. And I think there, you have to be much more complex. I think politicians, unlike scientists and activists, tend to make the world simpler than it actually is. There you have to take a certain responsibility. Because you are brilliant in the analysis of activism and science. But you black boxed yourself and I think that is something you will have to answer.

Jairam Ramesh: No, I don't disagree with you, Shiv, but I didn't have the luxury of rejecting the assumptions of, you know, the system, of which I am a part. I had to find a way forward for Jaitapur. Nicolas Sarkozy (Former President of France) was coming. He was signing this agreement. I am telling you facts. I had to find a way! I could not say no to Jaitapur. We needed nuclear power, climate change, you know, we had a huge investment in nuclear energy, I mean, I am just giving one example. So, I do not have the luxury, I am part of the system, okay? I am part of the problem. I do not have the luxury of being anarchist. I cannot say, I am not willing to say no to biotechnology. But I am willing to say "Put a moratorium on Bt Brinjal." I am not prepared to say, "After Fukushima, India must abandon nuclear power." But I am willing to say, "Let us do nuclear power subject to 36 conditions." I agree with you, that it is not a completely satisfactory position to take. Maybe now that I am out of power, I might actually do the things that you are suggesting but even so, I think, Shiv, if you look at the growth imperative for India, you will be hard put to take positions that are extreme in nature.

The Indian reality, and I haven't really spoken about this, but let me just take two minutes on this. You know, why should green growth be important for India? This is a question that has often bothered me for a long time, and frankly, when I look back on it, there are four reasons why green growth is important. India must be an exception to the 'grow now, pay later' model. All countries in the world follow the model of 'grow now, pay later'. America has followed this, China has followed this, Brazil has followed this and India is also following this. 'Grow now, pay later.' That's our 8 % growth for 20 years and don't worry about chemical contamination, pollution, asthma and cancer, 20 years from now. Why should India be different? And these are very important questions. All of us know, some youngsters are here, I am sure all of you are from the ecological community, we must ask this hard question, 'Why green growth?' Why should India be different other than the fact that we are the land of the Mahatma, who is the patron saint of all environmentalists? What is the first reason? The first reason is demographic. Most countries in the world are facing population declines. India is the only country in the world that is going to add 400 million people … in the next 30 years! That is our demographic *karma* (fate). Four hundred million people constitute one-third of our population, virtually! We are going to add in the next 30 years. Now, if your population is declining, you can maintain the current standards of consumption but if you are going to add 400 million people, you have got to worry about not just your consumption but also about what prospects you are going to leave behind for future generations. Are you jeopardizing their growth prospects? I mean India is the only country of her kind in the world! We are adding 10 million people every year to our labour force! 10 million! No country in the world … the United States and United Kingdom would also have population decline, had there been no immigration from India! It is the immigrants from India who are keeping their birth rates up: of course, India, Bangladesh and Pakistan.

So the demographic reason is number one. Second, there is a climate change reason. There is no country in the world, which is more sensitive, more vulnerable to climate change than India. But none of us recognize it. We all think that climate change is some Western plot to keep us poor! Our dependence on the monsoon. Our dependence on the glaciers. Our dependence … you know, 300 million people living in coastal areas are vulnerable to increase in mean sea levels. And the fact that all our natural resources are in our forests! Third, today, environment has become a public health issue. It is no longer an environmental upper-class, upper-middle-class elitist pastime. It is a public health concern. Look at the incidence of asthma in children in Bengaluru—documented studies. Look at the chemical contamination of our rivers, Ganga and Yamuna—the disposal of toxic waste. Municipal sewage has led to serious deleterious public health concerns. And fourth, you recognize this more than anybody of us, in India environmentalism is not a lifestyle issue—it is a livelihood issue; 250 million tribals depend on forests for their livelihood. It is a livelihood issue. Hence, when people are agitating, we know, we may be calling it environmental in nature but what are they agitating for? Water rights, forest rights and land rights; they are bread and butter issues. Even so, green growth for us is not some luxury, *yaar* (friend), it has to be something fundamental to us.

And I think a part of our failure is that today in this current environment of growth and technology triumphalism that you see in the media and in the political discourse, I am afraid, the prospects for sustainable growth are now going to take a backseat. People are going to say, "Hmm … you know, let us grow and we will think about sustainability later and so on and so forth …". So the challenge for the scientific community and the challenge for the civil society in larger public, including media, is to reinforce this message of why green growth is important. Rajendra Singh would give you one example: *Teen saal pehle inhoney ek andolan shuru kiya tha ki Alakananda par jo baandh ban rahe thei, unko aap khatam karo*. (Three years ago, he launched an agitation to stop construction of dams that were being built across the river Alakananda.) Government had spent rupees 4000 crore already building these dams and he started an agitation saying that there is no water in the river anymore if you are going to build dams indiscriminately like you have been doing. And you know what we did? We stopped those dams! We stopped three projects. Everybody was furious but we actually stopped the construction of those dams because we said, "It is more important to have water in our rivers rather than generate some 1000 MW of power which we can find somehow." So these types of choices are going to be important, but the type of choice that you are asking me to make, which is, reject the green revolution, the poster that Pental showed us, "Reject the green revolution, reject private transport," I mean these are, I think, fundamentalism that nobody will buy in today's context.

Deepak Pental (Former Vice Chancellor, University of Delhi): First of all I don't think of you a villain for rejecting Bt Brinjal, I wanted to make that clear. But I would say that some mistakes have been made and I would just like to point them out in a minute. When you realized that there is so much opposition, and this Bt Brinjal is really not that important for India's food security, then what was the great fun in bringing the whole science of biotechnology and distribute by going to urban centres, opening it to these die-hard people who don't want to understand science, whatever gains we can make out of genetic engineering later on. And secondly, when you said that you wanted an independent regulatory body, who had objected to it? The government of such bright people let it lapse! And I fully agree with you that we should question, we should be skeptical, this is the way sciences has moved on. I fully agree that technology deserves some questions what is the relevant technology and so on, but here is a technology which could be a game changer for India's 47 % population; then you talked about demographics, dividends or disasters whichever way one wants to see it, the fact is today 47 % of India's population is on agriculture. And in 2023, we will be 42 %. The actual number will be a little more than what it is today. So who is going to improve the farm incomes? Where from the increase in farm incomes will come? Only through low-input, high-output agriculture, and genetic engineering is one of the techniques. Why did we do some grandstanding within the ministries and why we didn't develop a cogent framework to say, "Look, this is no good, we will move to the next one …"; and I want to inform you that for one and a half years the GEAC (Genetic Engineering Appraisal Committee) has not been meeting! Now, that shows, you know apathy towards things; it is not hasten slowly. Something should have been cleared, and if nothing was worthwhile in India, everything should have been rejected. Now, that would have been like an acting government. We don't know where we are navigating, and we are doing it for the purpose of pleasing only the skeptics and who were comfortable with hasten slowly; and Dr. M. S. Swaminathan belongs to that category of scientists who believe in hasten slowly. This maybe great wisdom, but I don't think this is the way to provide leadership from the political class.

Jairam Ramesh: No, you see, I am sensitive to your point of view because had we hastened slowly, the first green revolution wouldn't have taken place. The first green revolution took place because we didn't hasten slowly. We just went ahead and imported 18,000 t of seed from Mexico and from the Philippines and we got the first green revolution going. But I think the difference between the first green revolution and the Bt Brinjal case, as you yourself brought out is, the first green revolution was entirely in the public domain. The seeds were available from CIMMYT (International Maize and Wheat Improvement Center) and IRRI (International Rice Research Institute). The seeds came to IARI (Indian Agricultural Research Institute), and it was entirely in the public domain. But now we are not talking about public domain science any longer. We are talking of private science. We are talking of highly oligopolistic science you know if not monopolistic. And that is what has created a lot of fears. I agree with you that if by this decision, India is going to abandon biotechnology in agriculture, then certainly Mark Anthony is proved right: "the good is often ended in their bones, the evil that men do lives after them," you know? The evil that ministers do lives after them, I agree. If that is going to happen, but I hope that is not going to happen. Because time in and time out again, I kept saying, "It is not biotech, it is not Bt general, it is Bt Brinjal case," and you use the Bt Brinjal case now. Why didn't we come up with the regulator? I don't know. It shouldn't take long for the parliament to pass (a bill to that effect); why was not this consensus created using state governments? I don't know! And Dr. Pental you would be surprised, state after state irrespective of political party—

Congress state, BJP state, Left Front state, BJD state, Karunanidhi followed by Jayalalitha—they never agree on anything but on this they agreed! I think it was a complete communication disaster, complete communication disaster from Day One. I will tell you something, with all due respects to all the scientists present here, you know when scientists retire, they get a different point of view than when they are in active service. Scientists become more society-friendly after retirement than they were in service. But this is not true only of scientists. It is true of civil servants also. Civil servants also become very friendly towards public and civil society activists after they retire! Actually civil servants, one year before their retirement, to prepare for their transition, they become more enlightened, you know? And they reach full enlightenment after their retirement, you know? So scientists are no exception to that. But I agree with you, it was a complete communication disaster!

K. S. Parthasarathy (*Former Secretary, Atomic Energy Regulatory Board*): As a minister at that time, were you not worried, why is it that all the state governments are against it?

Jairam Ramesh: Yes! I was completely flabbergasted! Let me tell you, Mr. Parthasarathy, if you read the Bt dossier decision on the website, every state government had an agricultural scientist giving a point of view to that state government. You know, Lord Cairns once said, a very famous quotation, "People in public life who think that they are free from the stranglehold of idea are usually prisoners of some defunct, economist." This quotation came to my mind. Because every state government had some retired Magsaysay or some award guy giving them the advice! It is amazing that not one state government said, "Go ahead". Even Punjab, Haryana!—it was shocking to me.

Umesh Srinivasan (*PhD student, National Center for Biological Sciences, Bengaluru*): I was just wondering, very often the media is the interface between the civil society and the scientific community. And if you read a scientific paper for example, it always nuances with hedges and caveats and you know context-specific things, say that this might be applicable to particular situations, it might not be applicable elsewhere and so on. But the nature of media, especially TV media, where you know, the nuances and the contexts are not brought out, I mean, what are the principles that you would draw? You have drawn principles for civil society, you have drawn principles for the scientific community. What are the principles you would draw for the media in terms of communicating science to civil society and vice versa?

Jairam Ramesh: Very good point. You don't depend on interlocutors. You must have direct contact between the scientific community and the constituency that you are engaged with. Do not depend on the media because the media is driven increasingly by a different cycle than what the scientists want and what, maybe, the people want. So I think it is important for a scientific community. Let me give you an example of what I am saying here, and I will take the example of a professor from Indian Institute of Science (IISc), a retired professor—Madhav Gadgil—and how he handled the Western Ghats report. He didn't depend on the media. He did direct interaction, to use an ugly phrase but a true phrase, with different stakeholders. And he communicated directly. Of course, Madhav's great advantage, which very few scientists have, is the ability to communicate in non-English languages, you know? He communicates beautifully in Kannada, Marathi and Hindi, and you know he is not a typical scientist. He wears *khadi* (handspun cloth). Scientists are all people who come in suits, boots and all those things. He looks like a person who you can trust. Many people think the media is the interface. I would not trust that interface in today's day and age. I would strongly plead for direct interactions. If it is Jaitapur, between Nuclear Power Corporation of India Limited (NPCIL) and locally elected bodies, locally elected institutions, local organizations, local civil society organizations, and very often I find that the agreement that can be reached is far more probable when the media is absent. Because when the media is present, grandstanding takes place from both sides, you know? The scientists want to prove something and the civil society wants to prove something else. But this is not to say that we have to make an investment in upgrading the level of skills of the media. Pallava Bagla talked and took us through how (numbers of) scientists communicating have come down but science communication has also come down. Both of them have come down. Over the years, you do not have a critical mass of people who are communicating complex scientific issues in languages that people can understand. The number of people can be counted. And therefore I think, science communication is an investment, is an externality, is an investment that we must make. I don't know how many institutions run specialized programmes for science communicators, I am sorry that most scientific organizations that I am aware of think science communication as drafting a press release and releasing it; that is not science communication. Science communication is actually engaging the media, educating the media, explaining to the media what it is all about and also be prepared for some hostility and some skepticism from the media. Let us take an extreme example, Dr. Baldev Raj (Former Director of Indira Gandhi Centre for Atomic Research, Kalpakkam) takes a media team to Kalpakkam, to show what a great job we have done on our fast breeder reactor and we have done a fantastic job. Not many people know about it, but somehow he takes a media team. I can bet you if he takes ten people, there will be four people who will not write about the fast breeder. They will write about the increase in cancer incidents of the people living around Kalpakkam. Immediately he will get angry and he will say "*Media ko jaane do!*" (Don't care about the media!) This has happened. I am giving you a concrete example. This has actually happened. People have written, they have not written positive things about Kalpakkam and they have said about how people living around Kalpakkam have actually become more cancer prone, how people living in Tarapur have become more cancer prone and so on. So the scientific community has to be prepared for some backlash when there is openness. When you let sunlight in, not everything is going to be hunky-dory. There will be some criticism. There will be some degree of questioning which our scientific community must be prepared to accept; I am not sure that our scientific community is actually prepared to accept that degree of openness.

Ashish Navneet (*PhD student working on policies and politics of genetically modified crops in India*): My simple question to you is that within the government, Ministry of Agriculture and Ministry of Environments and Forests, aren't they failing to generate consensus among themselves? Why they are coming with different opinions? Why they can't form common consensus?

Jairam Ramesh: I disagree with the premise of your question. The job of the Ministry of Agriculture is to promote genetic engineering, rather low-input, high-output technology, of which genetic engineering is one. The job of the Ministry of Environments and Forests is to do its diligence on the application of this technology. The job of the government is to create the consensus. If the Ministry of Environment is going to create a consensus, then it becomes a *thappa* (rubber stamp) ministry, isn't it? So, I take a different view. My view is that every person has to fulfill his/her *dharma* (obligations). The mandate of the Ministry of the Environment and Forests is to protect India's natural resources, protect livelihoods, ensure sustainability and not to promote genetic engineering. But the mandate of the Ministry of Agriculture is to increase agricultural output for which genetic engineering is one technology. Now the job of the government is a collective, is to create this consensus. Consensus is different from unanimity. Please mark my words. We can never have unanimity in these views. But we can have a consensus. Consensus means, *Aap ko sun liya, meri baat bhi ho gayi hai, yeh karna hai!* (I have heard you, I have said what I had to say, now we do this), that is consensus. Unanimity in my view is 'my way or the highway'! That is not possible in government. So what is lacking in India is forums which create these types of consensus that are durable. Sometimes they get done through strong-arm tactics, sometimes they get done through debate and discussion, sometimes they get done from outside the government, but please do not fall into this trap of thinking that all ministries must be monochromatic, that all of them must sing the same tune, that all of them are saying, "genetic engineering, *wah, wah!* (kudos!) Bt Brinjal, *wah! wah!*" That is not the way I feel democracies function. Democracies function best when each one articulates their mandate's view and then as a collective, we are able to come out with a consensus.

Kavita Kuruganti (*Alliance for Sustainable and Holistic Agriculture: ASHA*): I had a couple of things to say. One is that your whole presentation is somehow devoid of a political economy framework which once again is something that says science and technology for growth is what we should be discussing about—not science and technology for sustainability, social and environmental justice, plurality and you-name-it different kinds of justices that are missing out there today. Maybe because you are speaking with a group of public sector scientists, from the names that I gathered while standing here. Maybe it seems like a correct sort of a thing to discuss. But in a policy setting, which favors only growth, in a policy setting where every space is left to the private sector, how do you expect various other kinds of investments to take place? Coming from the field of agriculture, I would like to know who would invest in agroecology. It will not add to the GDP (gross domestic product) because people will not buy things in *dabbas* (packaged goods), pesticides will not be bought, fertilizers will not be bought, maybe they will use nature's processes, it won't add to the GDP. Maybe people will not fall sick. And maybe hospitals won't run. It will not add to the GDP. What do we do in terms of some of the real solutions that don't get adequate resources from governments that don't think about actual plurality? May we see a situation where particular technologies are favored than particular other technologies? How do we explain this?

Jairam Ramesh: Kavita, I am a prisoner of the system. I told you right in the beginning. I am not questioning the assumption of growth. I am not questioning the paradigm of growth. I am not questioning the model of economic growth. All I am pleading for is that recognizing the growth imperative, how do you maximize ecological benefits? How do you maximize public confidence in that growth process? And how do you minimize the pains of the growth process—whether it is land acquisition, whether it is deforestation, or whether it is pollution. So I am not questioning the premise of this energy-intensive, resource-intensive model. I am not, because I am not a romantic. And I am not an anarchist. I am a realist. You know, you have to find jobs for 10 million youths who are entering the labour force every year. You have to generate 60,000 to 70,000 MW of power every year to meet energy demands. You need water. You know, I don't have to spell it out for you. So you can go the Schumacher route and question the entire model of growth. I know many of you might want to do it. But you know, I think if Gandhi lived today, Gandhi might have had a different viewpoint! Hind Swaraj was a product of 1904, it was a product of his times. I think you know the circumstances and the challenges in which we are living force us on a certain growth path. Electricity is essential, transportation is essential, jobs are essential, low-input high-output agriculture is essential. You have to feed people! And thinking that India is going to produce, 270 million tons of food grains with cow dung and cow urine, I think is romantic to the extremes!

Shiv Vishwanathan: Then don't talk about green growth! It is utterly hypocritical. Your argument is coming apart, at one level you are using the same causalities you accuse your opponents of. The way out is, you put back what you said. Because the minute a question is asked, you talk beautifully of green growth and then you say "I am a realist" as if you are competing with Montek Ahluwalia.

Jairam Ramesh: You know, I have often said Shiv has to be heard, he has to be read, but not believed!.

Prabir Purkayastha (*NewsClick and Delhi Science Forum, New Delhi*): I think between green growth and green capitalism, we must also be clear. I suspect that what you call green growth is green capitalism and in that sense, it is capitalism with a human face or the neoliberal economy of the human face. If that becomes an issue, then we really have problems. I do appreciate the fact that the scientific community in this country is not used to speaking to the people. And the whole tradition of the people's education in science, public communication of science—how to bring it back—is a big challenge for us. Otherwise, we will have a one-sided debate in which the scientists are going to lose, and lose badly.

Jairam Ramesh: Let me start with your last comment that scientists are losing the plot. They are losing the plot on nuclear, they are losing the plot on genetic engineering, they are losing the game on a whole variety of technical issues because apart from the fact you know, the civil society is well organized. Some of them get funding and the real issue is what you have put your finger on. Unwillingness and the inability of the scientific community to come out, engage, not just communicate —engage! I find that to be the single greatest weakness.

Prabir Purkayastha: Jairam, this whole issue of foreign funding of NGOs is becoming a big issue. Now the question is Monsanto is here, Areva is here, we have an Enron over here and a lot of 'public education' within quotes. So don't blame Green Peace if you bring in all the others! Don't bring the opposition to Walmart if you bring in Walmart!

Jairam Ramesh: Let me tell you an interesting story about Green Peace that has not come out. A file came to me in February 2010 on Dhamra Port in Odisha saying that Dhamra Port which was being built by Tatas was actually forest land. They did not get forest clearance and this would lead to a massive, destruction of Olive Ridley Turtles. So I did my due diligence. Got all the files, got all the paperwork talked to all the officers who went to Dhamra, did a field visit. I came back and I recorded a two-page note, "Why Dhamra should get the clearance." Okay? 1, 2, 3, 4, 5 … Green Peace got hold of that through RTI (Right to Information), no problem, I have no hassle, they asked for it and I gave it. And before I know anything there was an advertisement in all the newspapers of India including *The Times of India, The Hindu, The Indian Express and Hindustan Times* reproducing my note and saying that I am an agent of the Tatas, that I have taken money from the Tatas to clear the Dhamra Port. Now, the accusation against me is that I have reduced GDP growth by 2.5 % points. Nobody has accused me of taking money from the Tatas. Greenpeace was the first organization to accuse me of clearing a project! Normally the accusation against me is over not clearing project. This was an occasion in which they went to town saying I actually cleared a project! Now, having said this, on Bt Brinjal, they made my life miserable. Even though my final product turned out to their satisfaction, during the process, they would sit outside my house, they would bombard my e-mail address, fill up my inbox with memorandums from all over the world, made my life miserable. So I have had a particularly rocky relationship with Greenpeace. But let me say this for the record publicly, I disagree with putting any form of restrictions on Greenpeace. I disagree with what they have said, I don't find their methods also agreeable at times, but you cannot choke off dissent like this, *yaar!* (friend!) These are people … you know, they are raising issues! And the government must have the courage and the integrity to address those issues! They used one of my reports on the Mahan Coal Field to raise issues. And Essar has filed a defamation case against them. I would defend Green Peace on that. So I don't think that we should reduce the great danger. I feel, in the next couple of years the room for dissent will shrink. The tolerance for different points of view will shrink. The growth mania will be accompanied by envirophobia, you know? And I think that is what we have to ward against and I will speak up for Green Peace publicly. I have already spoken up publicly. We should allow protest, we should allow organizations to have different points of view, and if it means delays in projects, so be it. But this type of fundamentalism that we will ride roughshod over all forms of protests and agitation, I think, will not do justice to the growth process and it certainly will not make the growth sustainable, socially.

Pallava Bagla (*NDTV, New Delhi*): Mr. Narendra Modi (the current Prime Minister of India) seems to be following a lot of what you have said; Mr. Modi seems to be following very closely your footsteps. Are you enamoured and in love with him now?

Jairam Ramesh: In public life, you cannot take intellectual property rights on ideas or on words, there are many words I may have used which others have used. But yes, when I first said that Indian women have more mobile phones than toilets, women's organizations jumped on me. When I said that in India, toilets are more important than temples, VHP (Vishva Hindu Parishad) and RSS (Rashtriya Swayamsevak Sangh) came and urinated outside my house. But when Narendra Modi said, "*Sauchaalay, Devaalayon se bhi zyaada mahatv rakhtein hain!*" (Toilets are more important than temples!), big deal, *yaar!* (friend!) I mean he has recognized the value, all power to him. And I hope something comes out of it. On environment, I am not sure. He says that environment should be integrated with growth, but in actual practice, if that integration means making choices, which means no dam on the Ganga, or which means no coal mine near tiger reserve, then I am not sure that Mr. Modi will follow what I would recommend. So I am not sure. And third, the difference between him and me is I am far more comfortable with Green Peace than he will ever be. I am far more comfortable with the Kavita Kuruganti's, I don't know for how many years she has been arguing with me. We disagree on most things. But you know, we have a most agreeable relationship here. I don't think Mr. Modi is capable of that.

Pallava Bagla: The Intelligence Bureau report, which we are talking about today, came out at the time when your government was in power and Mr. Manmohan Singh was the Prime Minister, and he himself said in an interview, very sharply against whatever is being said. So it seems to be going across governments. So, I don't think Mr. Modi should be blamed for that. Equally, if the blame has to go, it has to go to both UPA (United Progressive Alliance) and the NDA (National Democratic Alliance) for not wanting dissent.

Jairam Ramesh: Intelligence Bureau is a contradiction in terms, you know? It is an oxymoron. It is like military intelligence—they don't go together! The amount of intelligence that is shown in this report indulges not in casual empiricism, but spurious empiricism. There is empiricism of a casual variety, but there is empiricism of a spurious, and a bogus variety when it says that India's GDP has been reduced by 2–3 % points. First of all, I don't know whether it is 2–3 % or 2–3 % points. There is a fundamental difference between the two, which will go beyond the IB capability to comprehend. But you know, they say that India's GDP has come down 2–3 % because of environmentalism. How do you get this estimate? Who has done this estimate? Where is the data? Why is it not in the public domain? Why is it not peer reviewed? It is another Himalayan glacier type of a bogus statement, you know? So if the IB (Intelligence Bureau) has done some path-breaking econometric analysis, let it be out in the public domain. Let it be debated. But you cannot, on the basis of innuendo and insinuation say that Indian environmental activism has reduced India's GDP by 2–3 % points. It is bogus, *yaar!* (friend!) It is bizarre and bogus!

Parthasarathy: I just don't want to join issue with you, I was in the first environmental Ministry of Environment Committee for nuclear projects. When one of the power projects was considered, they had put 40 conditions not 36! And some of them were, you know very loose. Again, as you said, now the way it was put, you know okay, let us have more conditions added. And I was responsible for putting one condition, a very innocuous condition that "In addition to the MoEF (Ministry of Environment, Forest and Climate Change) conditions, AERB (Atomic Energy Regulatory Board) conditions also should be fulfilled!" So, I am sorry you need not take credit for putting 36 conditions in the case of Jaitapur because 25–27 years ago, probably in 1986 or 1987, when the first environment MOEF Committee worked on it they had already put 40 conditions. Now, I didn't look at these conditions, your conditions and their conditions. So you are not escaping by putting conditions that was what I was telling you.

Jairam Ramesh: Yeah, conditions are what Shiv would call a "cop out", you know? Yeah, it is a cop out, in a way. You have given me an interesting thought I will get back to you. I will compare those 40 conditions with these 36 conditions. These 36 conditions, we tried to make them as site specific as possible and I suspect, I might be wrong Mr. Parthasarathy, but the 2010 conditions, would have more to do with ordinary people's concerns about resettlement, rehabilitation, acquisition of lands, rise in temperatures and loss of livelihood, those types of issues. So, I suspect that there is some difference between the two.

Srikumar Banerjee (Former*Chairman, Atomic Energy Commission of India*): I don't want to go into the Jaitapur issue because a lot has been discussed. I just want to make one statement that you made regarding the expectation that how much would be the nuclear contribution. I think, a kind of a wrong perception prevails on this because Indian Government has never committed to install reactors to generate 10,000 MW by 2000—it is from a power point presentation by a chairman earlier—that is all! Because it has never been a sanctioned project, it is not a government policy. So if the government investment is not there, there is no way you can get that done. I think it is a wrong notion. Actually whatever has been committed by the government in the form of a sanctioned project, I think there is not a single one where one can say that there has been a failure.

Jairam Ramesh: Well, you know Dr. Banerjee, my first job in the government was in the advisory board on energy. And one of the first things I sat through in that was a presentation by Dr. Raja Ramanna and Dr. M. R. Srinivasan, who made a perspective plan for nuclear energy: 10,000 MW by the year 2000. That was not government policy. I agree with you. It was a perspective prepared. Then it became 10,000 MW by 2010, then it became 10,000 MW by 2020.

Srikumar Banerjee: In many minds this is there. See, technically I am not talking about sustainability. If today we talk of sustainability involving thorium, immediately the question will be that thorium we have been listening from Homi Bhabha's time. So why thorium is not in action today? Now this has a technical bearing. It has nothing to do with Bhabha has told or not. It essentially depends on whether you have installed several fast reactors or not. Like you mentioned about the cancer cases around, I have given exact data surveying over 150,000 people. So that is why there are perceptions. It is like where you get the maximum deformities? Around the places of worship; so you wouldn't associate these deformities of people with the shrine. It is the same kind of logic.

Jairam Ramesh: It is a very interesting analogy.

Prabir Purkayastha: A simple point I would like to make that you know, one of the reasons for nuclear programme got into problems was uranium mining. Funds were cut and the two people who were responsible for cutting that were the then Finance Minister Manmohan Singh, and Montek Singh Ahluwalia, who was the Finance Secretary at that time.

Srikumar Banerjee: The issue is that we really pay a very heavy price for uranium in our country because we are having an ore which is as low as 0.03 to 0.04 % in uranium content. This is not considered to be an economically exploitable material. But Homi Bhabha decided way back to open a mine in 1966. You see the wisdom of that. Only because of that India has independence in the overall nuclear activity. Otherwise, we would have perennially been dependent on import of uranium. So there are issues of this kind; so it requires one-to-one discussion.

Jairam Ramesh: Sure.

References

Bagla, P. (2014). http://blogs.scientificamerican.com/guest-blog/ipcc-finally-acknowledges-its-e2809chimalayan-blundere2809d/. Accessed June 22, 2015.

DAE. (2011). http://dae.nic.in/writereaddata/2238_lsus090311.pdf. Accessed June 22, 2015.

IPCC. (2007). https://www.ipcc.ch/publications_and_data/ar4/wg2/en/ch10s10-6-2.html. Accessed June 22, 2015.

MoEF. (2009). http://www.moef.nic.in/sites/default/files/Report%20on%20Bt%20brinjal_2.pdf.

MoEF. (2010). http://www.moef.nic.in/downloads/publicinformation/minister_REPORT.pdf. Accessed June 22, 2015.

NPCIL. (2010). http://www.npcil.nic.in/pdf/press_29nov2010.pdf. Accessed June 22, 2015.

Saxena, N. C., Prasuraman, S., Kant, P., & Baviskar, A. (2010). Report of the Four Member Committee for investigation into the Proposal Submitted by the Odisha Mining Company for Bauxite Mining in Niyamgiri [submitted to Ministry of Environment and Forests].

Part II
The Indian Landscape of Communicating Science and Technology

India's Maiden Mission to Mars: Many Firsts and Some Missed Opportunities in ISRO's Efforts at Public Outreach and Communications

Pallava Bagla

Abstract India's maiden interplanetary mission to Mars has been a technological marvel not just because it was the satellite fabricated in the shortest time in India—in less than 15 months—but also a result of frugal Indian technology made at a cost of INR 4500 million. Yet, at the same time, the hugely symbolic satellite mission has failed to capture the imagination of the country, at least till June 2014. Unfortunately, the public outreach and communication around the mission did not live up to the expectation. The Indian space agency did make a beginning by making a foray into using social media but a burning desire to connect with the people seemed to be lacking. Mission successful, outreach much to be desired, is the outcome so far.

Keywords *Mangalyaan* · Mars orbiter mission (MOM) · Science communication · Participatory research

India's mission to the Red Planet began well. The journey from the Red Fort (in Delhi, where the Mars mission was first announced) to the Red Planet started just as promised, and the dreams of a billion plus people went riding on a small 1350 kilogram satellite, which headed for Mars: a space mission which aims for Mars, yet has national pride written all over it, and a deep desire by India to become the first Asian nation to orbit Mars (Bagla 2014).

This chapter is based on the many writings by the author, which are copyrighted with him, and is mostly updated till June 2014.

P. Bagla (✉)
New Delhi Television (NDTV), 72 Samachar Apartments, Mayur Vihar-I,
Delhi 110091, India
e-mail: pallava.bagla@gmail.com

and

Shri T V Raman Pai Chair Visiting Professor of Science Communication,
National Institute of Advanced Studies, Bengaluru, India

P. Bagla and V.V. Binoy (eds.), *Bridging the Communication Gap in Science and Technology*, DOI 10.1007/978-981-10-1025-5_3

It is incomprehensible why the Mars Orbiter Mission (MOM, Figs. 1 and 2) satellite does not carry India's flag anywhere on the spacecraft. Dr. K. Radhakrishnan, the then Chairman, Indian Space Research Organisation (ISRO) confirmed this when asked at a press meet in Sriharikota saying, "Yes, it does not carry India's flag but everybody knows it is an Indian mission." It was an opportunity lost to have sent India's tricolor all the way to the Martian orbit. In contrast, *Chandrayaan-1* not only carried India's flag to the moon but also sent it hurtling down to the lunar surface as one of its first major achievements. The country rejoiced in unison when it happened. It should be made mandatory that India's name, emblem and flag be placed on every satellite the country launches.

According to ISRO, as of 9 April 2014, MOM had travelled 337 million kilometres in its long elliptical path to Mars, and was about 39 million kilometres away from earth. It was travelling with a velocity of 28 kilometres per second or about 100,800 kilometres per hour and is currently under the influence of the Sun. On 9 April 2014 the spacecraft "crossed the half-way mark of its journey to the Red Planet along the designated heliocentric trajectory', said ISRO. Since then the mission has successfully entered the Martian orbit and orbits Mars conducting observation.

Fig. 1 A simple schematic of the Mars Orbiter Satellite or *Mangalyaan*. *Credit* ISRO

Fig. 2 Pallava Bagla inside the clean room of the ISRO at Bengaluru with the *Mangalyaan* satellite in the background. *Credit* Pallava Bagla

Ironically, this iconic mission has by and large failed to capture the imagination of the Indian people. It is possibly a tale of missed opportunities that ISRO failed to capitalise—opportunities in the shape of goodwill and public interest the mission could have generated had the space agency reached out to try to ignite the minds of the young generation. It almost seems like ISRO did not have a carefully crafted communication strategy in place to ensure that the Mars mission could become a powerful vehicle to enthuse India's swelling student community.

It is important to analyse as to what has gone amiss, for the mission is simply not on people's radars any longer. Except for the date on which it was launched into space on 5 November 2013, and one or two other milestones of the journey that found limited mention, the mission seems to have also fallen off national and regional news pages and broadcasts as well. The latter, it may be argued, is not a bad thing since the lay media tends to focus on what goes wrong, so the fact that the mission is not currently making news is because it is proceeding normally.

Incidentally I should be the last person to complain since I was given the unique opportunity to spend a lot of time with the mission team and observing the actual MOM spacecraft in the clean room of ISRO, but then as they say, "one swallow doesn't make a summer". I remain grateful for the opportunity that was given to New Delhi Television, but then my desire is to see that such a unique mission (see 'More on *Mangalyaan* and Its Context', at the end of the chapter) gets the maximum mileage it can, since it provides an opportunity for us to celebrate our local technological capabilities.

Any communication strategy needs to be targeted carefully, if there is need to engage with the news media, to issue press releases and for press conferences to work well. If the target is the young generation who are connected to the Internet, then social media is a viable alternative. If schools and colleges have to be targeted, then there is no alternative but to engage them with live interactions such as model-making workshops. On the same lines it may not be a bad idea for the space agency to hold town-hall meetings especially in cities where it has a substantial presence so that the local community becomes an active stakeholder and not a mere bystander. When locals have a buy in, it can do huge wonders, especially to the morale of the technicians who work in secluded laboratories, yet have a social linkage to the community which would feel happy to have such role models in their midst.

It is surprising that one hears of Indian students repeatedly winning competitions held by National Aeronautics and Space Administration (NASA), USA, be it to make satellites or suitable human habitats on other planets; often students incur huge personal costs to participate in these international events. Unfortunately, our own space agency does not conduct enough events to attract such enthusiasm. There could be millions of youngsters who, since they may not be able to afford to go to NASA, could at least reach Ahmedabad and Bengaluru.

A limited number of students can always be invited to watch live rocket launches, if necessary, after participating in and winning competitions. Students could have been invited to draw and paint what a future Indian colony on the Red Planet would look like. Even messages from students could have been carried on a simple pen drive weighing some 20 grams tucked in a corner on the satellite; imagine the friends ISRO would have won if 100,000 Indian students were to receive one-page certificates saying that his/her message was carried to the Red Planet by India's MOM. These are only some ideas and much of these can still be implemented at a minimal cost in upcoming missions.

Students and community groups can also be invited to view live rocket launches at Sriharikota, obviously from a safe distance and albeit at a small ticketed fee. The thrill of watching a rocket launch remains etched in the mind for the whole life. They will become friends of ISRO for life, becoming unpaid brand ambassadors for India's space ambitions.

To be fair to ISRO, the Mars mission was the first time the organisation opened its account on the use of the so very powerful social media by starting a dedicated page on Facebook on 22 October 2013, called 'ISRO's Mars Orbiter Mission'. As on 1 June 2014 when the online assessment was made, this page had 313,498 'likes', a measure of interest shown by users of Facebook who regularly use this globally-popular social network to communicate. In comparison, NASA's Mars Curiosity page had some 670,000 likes. Comparing agency to agency, NASA had accumulated some 3.2 million likes on its Facebook account and the addition in 2014 of an official Facebook page by ISRO, called ISROOFFICIAL, had some 84,962 likes.

A good beginning on the social media platform, but the last serious update was as far away as on 9 April 2014 at 9:50 am, where it said, "India's Mars Orbiter Spacecraft crossed the half-way mark of its journey to the Red Planet." Think about it—a network where a young user might login several times a day, at the very least needs constant nurturing of its community with exciting updates and ISRO seems to lag behind on that front. The learning curve on how to use and deploy social media is steep, but it must be said that compared to many other Indian government agencies where a conservative approach is second skin, ISRO has certainly broken new ground to explore this cyberspace. Maybe many more will learn and follow suit.

Moving further on cyber-media, on 19 December 2013, ISRO started its official Twitter account with Twitter handle @ISROOFFICIAL with a tag line 'bigger and better' and as of mid-2016, 151 tweets had been sent out by this account, with some 9708 following it. In comparison, NASA's official Twitter handle @NASA had 6.63 million followers and had put out 31,000 tweets. Another useful comparison is viewership statistics of ISRO's official film on the Mars mission. Posted on YouTube it had been viewed by some 43,944 when last checked, while the film "Seven Minutes of Terror" on NASA's Mars Curiosity rover had been viewed by more than 1.1 million people. A good beginning has been made by ISRO, but a dedicated team of experts catering to the needs of social media needs to be put in place. As of 12 September, 2016 @ISRO-OFFICIAL had 834,000 followers with 711 tweets and @MARSORBITTER, ISRO's Twitter handle for MOM had 502,000 followers and had made just 34 tweets with just three tweets in 2015 and one in 2016.

Despite ISRO's maiden foray into social media, there are probably many reasons why MOM may have failed to connect with the people of India. Its official announcement came in very late; the members of larger Indian scientific community were mostly left as bystanders, having little or no role or contribution in the mission; while it falls in the class of a scientific mission, no single lead scientist was identified by ISRO and there is no international partnership in the scientific payloads, hence it remains an 'Indian mission'. In its wisdom or probably due to a lack of resources ISRO failed to conduct road shows in Indian universities and educational institutions to galvanise the momentum and human curiosity the launch certainly had the potential to generate.

The first official announcement that India was heading to Mars came only in 2012 when Prime Minister Manmohan Singh in his Independence Day speech delivered from the ramparts of the 17th century Red Fort gave a clarion call while also christening it, and said, "*Mangalyaan* will be a big step in science [and] technology". Since then, in record-breaking 15 months, ISRO built the satellite from scratch, the least fabrication time the organisation has taken for such a task.

There was a passing mention of the MOM in the 2012–2013 budget documents, but no official information was released by ISRO till well after the Prime Minister made the official announcement. To be precise, it was only in August/September 2013, just two months before the actual launch, that the official announcement was made. Till as late as August 2013, ISRO's official website did not have any details of the Mars mission and a search on its website using the word 'Mars' only yielded results related to "S.K. Shivaku'MAR'", the director of ISRO's satellite centre! Interactive websites are such a powerful way of communicating with one and all who may have access to the Internet and have an inclination to learn about India's space adventures.

This was very much unlike India's maiden mission to the moon, *Chandrayaan-1*, which was first announced on 11 May 1999 and then formally communicated to the Indian public by Prime Minister Atal Bihari Vajpayee in his Independence Day speech of 2003, even though the mission finally lifted off only in 2008 (Bagla and Menon 2008). During this time, ISRO took the Indian scientific community into confidence, even wrote some introductory articles in *Current Science*, and in a huge gesture even accommodated international scientific instruments from countries like USA, UK, European Space Agency and Bulgaria, which made *Chandrayaan-1* 'truly an international mission where India was the captain' (Fig. 3).

In contrast, the MOM remains a totally Indian mission. In fact, all the five payloads have been designed and fabricated by internal teams of the Indian space agency or its affiliate laboratories and none from any university. There was almost no public engagement with the larger Indian scientific community to bring them on board, even though a very small group of scientists from outside ISRO were involved in some internal reviews of the mission. There was no 'announcement of opportunity' seeking participation of international teams. While this may have been because of time constraints, as is often explained, it has clearly been a missed opportunity.

Both geopolitical and planetary configurations have played on the minds of Indian policy makers who took the Martian plunge. S.K. Shivakumar, director of the ISRO Satellite Centre in Bengaluru admitted that the making of the Mars satellite was indeed 'fast tracked'. Therefore, some suggest, this is the start of a 21st century Asian space race where India and China—the two regional rivals—are locked in a modern-day interplanetary marathon. Japan, the third aspirant to reach

Fig. 3 Igniting India, celebrating *Chandrayaan-1*, the moment when India's maiden mission to the moon landed India's tricolor on the moon surface on 14 November 2008. A moon globe with India's flag firmly on it is being rejoiced with. *Left to right*: Dr. G. Madhavan Nair, Chairman of ISRO; Dr. A.P.J. Abdul Kalam, President of India; Dr. T.K. Alex, Director of the ISRO Satellite Centre; Dr. K. Radhakrishnan, Director of the Vikram Sarabhai Space Centre. *Credit* Pallava Bagla

Mars, is also jogging alongside, with its 1998 maiden effort using a satellite called Nozomi having failed.

China may have beaten India in almost every aspect in space, conducting ahead of New Delhi its manned mission in 2003 and even its mission to the moon, which came before India's. Yet, Mars could be the space event where India could take a lead. In a fortuitous turn of events for India, the maiden Chinese orbiter to Mars being piggybacked on Russian satellite Phobos-Grunt ended in disaster in November 2011, after it failed to be boosted into space. This gave India a small window of opportunity to march ahead of China. Indian policy makers did not fail to capitalise on this opportunity, and so was born India's maiden mission to Mars.

A satisfied Radhakrishnan had said after the successful launch, "we are not racing with anybody and the Indian Mars mission has its own relevance", but had admitted that there was an element of 'national pride' involved with the mission.

It is this very 'national pride', a passionate emotion, that seems to have been underutilised through compelling and consistent communication. There was a tremendous opportunity for ISRO to have reached out to the public to showcase the organisation's capability and to attract better talent for itself, but in the 19-month period during which the Mars mission is on its long journey to Mars, only a handful of concerted outreach activities have been conducted by ISRO. Few talks by

Dr. M. Annadurai, the program director for MOM, stand out as an exception, and to his credit, many of his talks are in Tamil. Many more live interactive events where use of 'water rockets' and 'can satellites' to be made by engineering students could have been attempted.

In fact, it may come as a surprise that the entire programme of public outreach for ISRO is handled by all of three technical persons with a budget of INR 15–16 million a year of the INR 56,150 million budget—in effect the budget for outreach amounts to merely 0.000002% of the entire budget of ISRO. In the 2013 fiscal year, NASA's budget on educational activities was pegged at USD 136 million (INR 8190 million), and this year after the budget cuts as part of 'sequestration' it has dropped to USD 100 million (INR 6023 million). If India desires to be a space power to reckon with, investments in communication have to go up manyfold.

Space is of abiding interest and space exploration is known to galvanise nations into a unity of purpose like no other activity; the Apollo missions by USA are a good example. There is an immense and urgent need to strengthen the Publications and Public Relations Unit of ISRO. Communications, outreach and education are very specific activities, and there is need to man them with trained people. While the spokesperson's job is very specific and it can neither be replaced nor undermined, a cadre of specialised young people is needed who can conduct such outreach activities that would increase the public understanding and appreciation of science all the year round.

Using the ongoing Mars mission as the vehicle ISRO could have, if it wanted, reached out to countless engineering colleges and schools, but unfortunately it failed to capitalise on the gains; general outreach events were conducted only in two to three cities. Planetariums of the country and science museums could have been tapped to hold special exhibitions and talks on Mars, but the vigour lacked, it seems.

Interestingly, it is reported that some 20,000 Indians signed up for a ;one-way ticket' to the Red Planet as part of the bizarre plan floated by an overseas group called *Mars One* that claims it will set up a colony on Mars by 2024; of these 1058 were actually shortlisted by the Dutch group as prospective candidates or hopefuls many of these could well have become 'friends of MOM'!

The mission itself began with a bang and on the day of the launch there was almost saturation coverage on most Indian television channels, but it soon tapered off. It was a balmy Tuesday afternoon, as usual hot and sultry at India's hi-tech rocket port at Sriharikota on the coast of Bay of Bengal when at 2.38 pm, as scheduled, India's Polar Satellite Launch Vehicle (PSLV) (Fig. 5) lifted off, blazing a trail that will remain etched in the minds of the select few who watched India make history. The ground shook at the jam-packed viewing gallery and the sound was loud as India's maiden voyage to the Red Planet began with a perfect liftoff; 45 minutes later and almost three-quarters of the globe away the *Mangalyaan*, as the satellite is nicknamed, had a 'very precise' injection in its earth orbit (Fig. 4).

ISRO prefers to call it the Mars Orbiter Mission, and there has been a needless controversy on why some called it *Mangalyaan*. Incidentally, the then Prime Minister Manmohan Singh called it *Mangalyaan* in his speech read in Urdu from

Fig. 4 MOM looks at mother earth. This lovely image of India was captured by *Mangalyaan* on 19 November 2013. *Credit* ISRO

Fig. 5 The Polar Satellite Launch Vehicle of ISRO in its 25th flight hoisted the *Mangalyaan* into orbit on 5 November 2013, in a flawless mission. *Credit* ISRO

the ramparts of the Red Fort on 15 August 2012. But, despite the latter name being much more evocative in a country where the ancient language of Sanskrit belongs, the niggling over MOM versus *Mangalyaan* has led to yet another missed opportunity for meaningful communication. Maybe ISRO should hold a competition among the public to come up with names for its future missions. It is high time India's rockets, the PSLV and the Geosynchronous Satellite Launch Vehicles (GSLV), got simple Indian names like calling the PSLV the 'Raman Rocket' named after India's Nobel laureate in Physics, Sir. C.V. Raman, and calling the GSLV the 'Vikram Rocket' named after the father of India's space program, Dr. Vikram Sarabhai.

There is an urgent need to address the lack of capacity among Indian scientists to communicate with lay public, and communication—which is actually a science and an art—is still undervalued as a skill, particularly by scientists. Skill building is very necessary to create a cadre of people who can help spread the message of science to all shades of lay public. All in all, *Mangalyaan* remains a mesmerising Indian Martian adventure that unfortunately failed to ignite minds of lay people.

More on *Mangalyaan* and Its Context

Mangalyaan is an unmanned satellite that has been conceived, designed and fabricated by Indian scientists, and was launched using an Indian rocket from Indian soil. The cost of the entire maiden Mars mission is about INR 4500 million, and about 500 scientists had toiled to fabricate it from scratch—a low-cost mission no doubt.

The then ISRO chairman, K. Radhakrishnan, had explained this Indian Mars mission as "really a technology demonstrator", essentially showcasing to the world that India can undertake "interplanetary leaps". Till date, only Japan, China, Russia, USA and the European Space Agency have even attempted space travel to Mars and of these only the latter three have succeeded. Since 1960, some 51 missions have been launched, about 27 of these have ended in disaster, the most recent being the Chinese failure in 2011. When India made it to Mars, it was the third nation in the world to have done it all on its own after USA and Russia; the space consortium or the European Space Agency has also reached Mars. It is for this reason that MOM should have been made a big deal and used as an effective vehicle to mesmerise the minds of Indians.

The tiny Indian satellite made at a cost of INR 1500 million is carrying with it five Indian-made instruments weighing about 15 kilograms that will sample the thin Martian atmosphere about which we know so little. The global scientific community is very excited about India's effort to send the first dedicated methane gas sensor to Mars. The presence of methane gas, also called 'marsh gas', on earth is one of the clinching signs of the presence of carbon-based life forms. Therefore, in a way without even landing on Mars India hopes to provide an answer to that big question: Are we alone in this universe? The global scientific community, it seems, is more excited about India's search for methane on Mars since no other earlier mission to Mars has ever attempted getting a global picture of the presence of methane in the Martian environment (Bagla and Menon 2014).

India's venture to the Red Planet had an unconventional start. Since the PSLV could not provide enough thrust to send *Mangalyaan* on a direct path to Mars, the satellite revolved around earth for several weeks, and using the satellite's on-board rocket motor and 850 kilograms of fuel stored on the satellite enough velocity was imparted so that on 30 November 2013 it began its over 780 million kilometre journey becoming the first Indian spacecraft to escape from earth's influence.

India's workhorse rocket in its 25th launch sent the satellite into space. The PSLV rocket stands about 44 metre tall or as high as a 15-storey building and weighs nearly 320 tons, about the same weight as a fully loaded 747 Jumbo Jet. With this launch the PSLV has had 24 consecutively successful launches.

When it neared Mars after its 11 month long journey on 24 September 2014 *Mangalyaan* was slowed down in a tricky manoeuvre so that it could be captured by the gravity of Mars. Once that was done, it followed a highly elliptical orbit and has been studying Mars.

Therefore, is this a giant leap or a foolhardy step by a nation that still cannot provide electricity to 400 million of its population and where 600 million people still defaecate in the open? It all depends on which side of the divide you belong. But nobody doubts that the *Mangalyaan* is truly the cheapest interplanetary mission to be ever undertaken by any country, and now it may well pave the way for low-cost access to Mars.

Radhakrishnan, speaking to this writer had said: "This Mars mission is a historical necessity; after having helped find water on the moon, looking for signatures of life on Mars is a natural progression. India is now demonstrating its capability to undertake interplanetary travel with end-to-end technological prowess in space."

The satellite is the size of a 'Tata Nano' car, and it is hard to believe that it can navigate all that distance. India's historic journey to the Red Planet began well and when it reached the Martian orbit it created history by India becoming the first nation to reach Martian orbit in the maiden attempt itself.

ISRO embraced the social media, both Twitter and Facebook, in a big way after September 2014 when Prime Minister Narendra Modi visited the control centre. Since Modi himself is very fond of using social media many government departments, including ISRO, followed suit. Yet there is a lot of reluctance in walking the talk in reaching out by ISRO on the social media.

Further Reading

Bagla, P. (2014). http://www.thehindu.com/opinion/op-ed/a-faster-cheaper-mars-orbiter/article6442854.ece. Accessed 16 December 2015.

Bagla, P., & Menon, S. (2008). *Destination moon: India's quest for the moon, mars, and beyond.* India: HarperCollins.

Bagla, P. & Menon, S. (2014). *Reaching for the stars: India's journey to MARS and beyond.* Bloomsbury Publishing India Private Limited.

https://www.facebook.com/isromom

https://twitter.com/isroofficial

www.isro.gov.in

www.nasa.gov

www.ndtv.com/mars

Challenges in Communicating about Defence Research: Insight into Defence Research and Development Organisation's Media Strategy

Ravi Kumar Gupta

Abstract Communicating nuances of defence science and technology to common man is not easy; communicating these and the challenges associated with the development of defence technologies to media was indeed a delicate and uphill task. In spite of numerous achievements and accomplishments that had been realised in the face of numerous stonewalling challenges, Defence Research and Development Organisation (DRDO) had to pass through a difficult phase marked by adverse publicity drawing flak from all quarters for a considerable period of time, damaging its reputation and denying its rightful place in the minds of masses and stakeholders. Insufficient and ineffective communication mainly due to absence of a single window agency with authorisation to interact with media was identified as the major cause for a widening perception gap resulting in the rapidly worsening situation. How creation of the Directorate of Public Interface, as the single window for interacting with media as well as public in general, and implementation of other well-planned communication strategies aimed at bridging perception gaps, particularly with media, helped in transforming DRDO's image to evolve into a major contributor to the nation building and a potential driver for national economy, attracting the country's best talent, is discussed along with a brief background of DRDO's evolution.

Keywords Defence research · DRDO · Public interaction · MBT arjun · Electronic warfare · Media intervention · Guided missiles

An organisation that has placed India among top few nations of the world possessing demonstrated capability to design, develop and produce long range strategic missiles, nuclear powered submarines, two-layer ballistic missile defence systems, four-plus generation combat aircraft, contemporary main battle tanks, wide range of

---

Views expressed in the chapter are the author's own and not necessarily shared by the editors.

R.K. Gupta (✉)
Public Interface, DRDO Headquarters, Room 117, DRDO Bhawan Rajaji Marg, New Delhi 110011, India
e-mail: dpidrdo@gmail.com; dpidrdo@hdr.drdo.in

P. Bagla and V.V. Binoy (eds.), *Bridging the Communication Gap in Science and Technology*, DOI 10.1007/978-981-10-1025-5_4

sophisticated radars and electronic warfare systems and sensors and equipment for defence against nuclear, biological and chemical (NBC) warfare among others, till about 2007–2008, was painted as a dead, nonperforming organisation no better than a white elephant with nearly 30,000 employees and a budget around 80,000 million. A look at this organisation's achievements and contributions made till then towards nation building against numerous challenges, including an isolation imposed through denial regimes by global powers, would prove that such an image was incorrect and far from reality. Today, the image of Defence Research and Development Organisation, now popularly known as 'DRDO' has transformed into that of a technology leader capable of developing state-of-the-art weapon systems, military platforms and equipment with unique and significant contribution to nation building and a potential driver for national economy, attracting country's best talent. What magic would have changed the image of DRDO in a span of few years? *The catalyst for transformation was strategic communication taking its accomplishments in research and development (R&D) to masses*. But before the strategies to achieve this change are discussed it would be prudent to understand DRDO as an organisation, its genesis, the challenges faced by it in a dynamically changing scenario and its strengths.

DRDO: Genesis and Journey to Success

Defence Science Organisation (DSO), an organisation created primarily with an advisory role was formed in September 1948 under the leadership of Prof. Daulat Singh Kothari, a highly accomplished theoretical physicist. Post-Independence, the nation, *reborn* and struggling to absorb the shock of partition, preceded by over 200 years of colonial rule had missed the industrial revolution. Its own time-tested traditional systems of industry, trade, education and skill development had been systematically destroyed. The country not only lacked the knowledge base, skilled human resource and R&D infrastructure required for developing indigenous defence technologies or to design military hardware, but also the industrial environment needed to transform technologies into products. Although some scientific, technical and industrial institutions were set up in colonial period, largely due to compulsions of the two World Wars, their role and function were limited to fulfilment of commercial and strategic interests of the rulers. The first Prime Minister of India, faced with the challenging task of building a defence R&D and production base in the country, sought the advice of Prof. Patrick Maynard Stuart Blackett, a British Nobel Laureate with proven credentials in defence R&D. He had proposed, in his classified report, creation of the DSO with a purely advisory role and advocated against indigenous R&D in weapon systems—at least for the time being. Ten years later, DRDO was formed with just ten laboratories, with amalgamation of Technical Development Establishments (TDEs) functioning under the Directorate of Technical Development and Production (DTDP) (created by British rulers mainly for maintenance and support of defence equipment) with DSO (Shenoy 2006).

The next 20 years were more a period of shaping up the organisation, learning and generating capabilities. DRDO laboratories during this period became more focused, grew in size and numbers, and began to evolve into R&D establishments generating technologies, products and processes. Also in focus were generation of capabilities and setting up infrastructure for design and development of military hardware as well as for their testing and evaluation. Good numbers of products were developed, and many of them inducted into service by the end of 1970s. Some examples are 75/24 Pack Howitzer and its ammunition, lightweight nickel–cadmium batteries, sono-buoys, and anticorrosion, antifouling and underwater paints as well as cathodic protection systems. Wire-guided antitank missile was also developed and nearly 700 flight tests were carried out, including about 100 as part of user trials. The production value of DRDO-developed items by this time had already crossed INR 1050 million—not a meager sum, keeping in view the then prevailing value of the rupee.

The decade of 1980s became a turning point in DRDO's history with the launch of mega-projects and programmes including the Integrated Guided Missile Development Programme (IGMDP) and the programme-definition phase for taking up development of light combat aircraft (LCA), main battle tank (MBT) Arjun, unmanned aerial vehicle Nishant and aircraft simulator. A host of relatively smaller projects covering broad requirements of Indian armed forces were also taken up. The Integrated Test Range (ITR) and Aeronautical Development Agency (ADA) were established in 1986 and 1988, respectively. The period also gave DRDO the benefit of leadership of legendary leaders such as Dr. V.S. Arunachalam and Dr. A.P.J. Abdul Kalam. A number of products spanning almost all areas of technology were led to production, taking the production value of DRDO-developed systems to over INR 20,000 million. The decade also witnessed the launch of the first Agni and Prithvi ballistic missiles marking India's entry into the exclusive club of countries having such capabilities.

The last decade of the 20th century marked development of new-generation systems. Full-scale engineering development of LCA was sanctioned during mid-1993. The development of electronic warfare programmes Samyukta and Sangraha for land and naval platforms, Pinaka multi-barrel rocket-launching system and Kaveri gas turbine engine was taken up. Other development projects included NBC systems, AB steel for naval applications, radars, stealth and camouflage technologies and next-generation sonar and torpedo systems. This was a period of intense technology activity in DRDO when assessment of the existing capability and a move towards enhancement of human resource through recruitment and training was given sharp focus. In 1998, as the fallout of Pokhran tests, USA imposed technology sanctions—perhaps, knowing India's potential they feared future competition in the global arms market. Worst to be hit was the LCA, programme. However, converting challenges into opportunities, DRDO launched indigenous efforts leading to creation of much stronger knowledge and skill base. Three years later, in January 2001, the first flight of LCA took place amid much fanfare as DRDO demonstrated to the world that India was unstoppable. Although the production value of DRDO-developed systems crossed INR 88,570 million, much more significant was the creation of capabilities and generation of 'we can do' attitude.

The first decade of 21st century and the years that followed have marked the beginning of a new phase of induction and 'productionisation' stages of DRDO's flagship programme. Strategic missiles Agni 1, Agni 2 and Agni 3 were inducted; Agni 4 is under induction, while intercontinental ballistic missile (ICBM) Agni 5, after two highly successful flights, is on its way. Indigenous nuclear-powered submarine 'Arihant' will soon be undergoing sea trials along with its indigenous underwater-launched strategic missile BO5. Major successes have been achieved in ballistic missile defence programme with the development of two-layer defence system and demonstrated destruction of targets in both exo- and endo-atmospheric regions. Tejas, the Indian LCA, achieved initial operational clearance. Two regiments of MBT Arjun got inducted into service, albeit after a long struggle. An exhaustive range of radars (Rajendra, 3D CAR, Rohini, Revati, weapon locating radar, battle field surveillance radar, Aslesha and Bharani), sonars (HUMSA, USHUS and MIHIR), electronic warfare suites and night-vision devices, armament (e.g., INSAS rifle, Pinaka rocket launcher and multi-mode hand grenade) and ammunition for small arms, guns and rockets (e.g., BMCS, FSAPDS and flares) as well as equally exhaustive range of soldier support systems and NBC defence have been developed and led to production.

The production value of DRDO-developed systems that have been inducted or approved for induction has crossed INR 1,600,000 million (excluding strategic systems that constitute major share of DRDO's expenditure). The average total expenditure of the Department of Defence R&D (including pay and allowances of around 27,000-strong workforce) during 2001–2002 to 2013–2014 has been about INR 67,500 million, just 5.2% of total expenditure on defence (too little as compared to most nations, e.g., 12% in USA and 20% in China during 2013–2014).

The rapidly increasing production value has not been the only indicator of DRDO's success or its only major contribution to nation building. National capabilities and production capacities including modern R&D infrastructure have been created, with enormous employment generation within the country. The organisation has emerged as a technology leader. Its visibility and image reached new heights with the launch of Indian ICBM Agni 5. The countries and multinational companies which, at one point in time, considered India as untouchable as far as technology sharing or sale of most hi-tech equipment in the defence sector were concerned, are now keen to collaborate.

Distorted Image and Visibility in Public Domain

Ironically, the first decade of the millennium also witnessed an increasingly aggressive criticism of DRDO that had reached its peak by 2007. By and large, there was little visibility or recognition. Most people either did not know about DRDO, or had a negative image generated through adverse publicity. Rather than its achievements, DRDO was being talked about more for delays, cost overruns and getting blamed for increasing arms imports ignoring the real factors. Questions were being raised on its performance in almost all forums. The negativism was spreading like wildfire, propagated by numerous media stories presenting a gloomy picture of DRDO and a biased picture of India's self-reliance in defence technologies. Impact of any occasional media report talking about few of DRDO's achievement was more or less masked by the gloomy image generated by negative reporting.

Surprisingly, the adverse publicity was growing even though the performance of DRDO, against stonewalling challenges and struggles in its journey of the first five decades or so, could only be called excellent. Propagation of negative image was becoming increasingly detrimental to organisation. Such negative stories painting one-sided picture of issues indeed have potential to damage morale of the workforce. They also make induction of talent increasingly difficult and lead to high attrition rates. Wide gap in perception about DRDO in minds of various stakeholders, especially the press, was apparent. The need for taking corrective steps was clearly evident. With firm belief in organisation's strength and being proud of DRDO's significant and extremely valuable contributions to nation, peers in DRDO were determined to go to the root of the problem and take decisive steps to change the situation. Initiative was taken to fill the gaps in perception through communication of accomplishments made in defence R&D, leading to creation of the Directorate of Public Interface (DPI) in January 2008. It was also considered prudent to make a quantitative assessment of DRDO's contributions and capabilities. An independent, systematic and rigorous economic evaluation of the returns to the nation from its research outlays over the past 50 years was initiated through the National Council for Applied Economic Research (NCAER).

The NCAER Study

The findings of independent, in-depth study in 2007 by NCAER clearly indicated that investments in DRDO had yielded immense positive returns, not just in terms of generation of capacity, technology and knowledge, but also in terms of economic growth, industrial growth, quality of research and social benefits.

Social Returns on Investment in DRDO : Generally, a benefit-to-cost ratio (BCR) of greater than one is regarded as good for a project. Over the entire period of about 50 years, the social benefits had far exceeded the social cost. Thus, irrespective of time and cost overruns, the investment in DRDO had yielded minimum benefit (all benefits had not been accounted while cost had been accounted in full) of 1.18 times the cost for the period 1962–2007 (with 10% discount rate). The figure was much higher for some major programmes— for example, 1.60 for IGMDP and 1.42 for MBT Arjun, both at 10% discount rates, till 2007. In terms of employment generation alone, the minimum employment supported by the DRDO research outcome was of the order of 221 man-years per INR 10 million of investment in DRDO. All these figures would stand much taller today in view of greater acceptance of DRDO products by armed forces, leading to higher production and resultant higher employment.

Capacity Augmentation and Learning by Doing: A definite and quantum learning by doing over the years, the study indicated, had yielded a BCR of 1.36 (10 per cent discount rate) covering the 25-year period from 1982 to 2007 and had also brought India into thick of the key market segments of global defence industry.

Contribution to Economic Growth: The study suggested that there was clear evidence of positive contribution to economic growth and clear unidirectional causal relationship following from DRDO expenditure to gross national product.

Contribution to Industry: Contribution of DRDO in developing business of industries was found significant both in public and private sectors. It was observed that the *Navaratna* industry (a status bestowed on public sector companies in India that have comparative advantages) like Bharat Electronics Limited was deriving more than 20 % of its turnover from products developed by DRDO. A large proportion of respondents from private sector gave DRDO high to medium ranking to attributes such as improvement in quality standards, development of new products, improvement in access to new markets, management practices, procedures and technological knowledge.

Contribution to Quality of Research: About 67% of intellectual partners gave high rating to the intensity of research at DRDO, while only 35% accorded high rating to other institutions in India in this regard.

The forgoing discussion clearly indicates that all through its journey from inception, DRDO had been contributing in a very significant and unique manner towards the cause for which it was created as well as to the much larger cause of nation building.

DRDO's Image: The Gaps in Perception

As is evident from the findings of the NCAER study as well as number of products that had entered production (or were developed and had huge potential, but for one reason or the other were awaiting users' acceptance), DRDO had already emerged as a technology leader. However, the qualitative assessment revealed a wide gap in perception of DRDO among various stakeholders, particularly the media. It was also observed that the primary reason for these perception gaps was the absence of correct information about DRDO in public domain, mainly due to a communication gap.

Obviously, a change in perception of the stakeholders through a well-crafted communication strategy was the need of the hour. However, the task of communicating scientific accomplishments, their significance to a given cause and the challenges associated in achieving them to masses is in itself a challenging task; more so for a government organisation. DRDO being a government organisation, discussing official matters, technical or otherwise, with media by its employees is prohibited. The Manual of Office Procedure (MOP) stipulates that only ministers, secretaries and other officers specifically authorised in this behalf are allowed to interact with press. Any other government employee, states the MOP, "will direct them to the Press Information Bureau or will seek the permission of the Secretary of the department before meeting the press". Thus, unauthorised interaction, howsoever well intentioned, is likely to attract provisions of the Central Civil Services Conduct Rules, 1964 and the Official Secrets Act, 1923 (DARPG 2010).

In a democracy, people want to know about defence preparedness as well as the technology and military strength of the country as it affects their own security and pride. Further, the taxpayer wants to know how their hard-earned money is being utilised. Media plays a key role in democracy and is a powerful and effective medium for communicating to masses as well as specific stakeholders. Prior to 2008 DRDO had no authorised channel for interacting with media. If correct information about a large organisation such as DRDO were to be communicated, there had to be a single window with authorization to interact with public, especially the media. The DPI was created at the corporate headquarters of DRDO in Delhi, in January 2008 to fulfil this requirement. The main objectives were to bridge the gap in perception with the stakeholders, generate true image of the organisation and act as single window agency for interacting with media and the public in general. The directorate was headed by a carefully selected experienced Defence R&D Service scientist of the level of Joint Secretary who was also required to act as the DRDO spokesperson. Having a scientist as the head was important to have a clear understanding of technical issues, but he was also required to have good communication skills, capability to make the common man comprehend even the complex technical issues and be able to reach out to the public as a *scientist*, shedding the cover of authority. Suranjan Pal with over 28 years of experience as an experimental scientist in the area of laser technology became the first scientist to head this newly created key position. As its extended arms, DPI also has regional public relation centres at Bengaluru, Hyderabad and Pune, besides a public relations officer in each laboratory.

Communication Strategy

With the establishment of DPI as a single window agency, a detailed strategy for reaching out to the masses as well as specific stakeholders was chalked out. Media, both electronic and print, during the present times has rapid and deep reach covering people from all walks of life, including armed forces, bureaucracy, polity, students, academia, industry, general public and, most importantly, DRDO's own human resource. In addition to the intense drive for communication through media as discussed in the following paragraphs, *direct communication* through series of other measures was also pursued. These included DRDO exhibitions wherein live exhibits, mock-up models, posters, videos and so on were displayed. An important feature of such exhibitions was day-long interactions of the scientists working on different projects with the visitors— such visitors included a large number of students and teachers besides the common public. Technical festivals were organised in IITs, NITs and leading engineering colleges, and lectures by senior DRDO scientists including representatives from DPI were held at such leading educational institutions. Permanent DRDO galleries were set up displaying a variety of DRDO products and technologies (Gopalaswami and Reddy 2013).

Special attention was given to media. However, information on anything related to defence is generally considered as sensitive. In order to avoid inadvertently releasing any sensitive information, often there is a tendency to withhold all information, leading to communication gaps. Thus, before making an attempt to enhance the level of communication and transmit information on defence technologies to the masses, it was essential to understand the need of striking an appropriate balance between the extent to which public scrutiny could be allowed and fulfilling the needs of information security. A careful scrutiny brought out that a major portion of information pertaining to DRDO, its major projects, its achievements and accomplishments, the unique challenges it had been facing and the significant contribution of DRDO's efforts towards nation building could be shared in the public domain.

Communicating through Media: Challenges and Opportunities

> The media's the most powerful entity on earth. They have the power to make the innocent guilty and to make the guilty innocent, and that's power. Because they control the minds of the masses.
>
> —Malcolm X

If challenges and opportunities in communicating through media were to be summed up in one sentence, in light of Malcolm's quote, it would be: "Stronger a medicine, deadlier is its damaging potential, if dealt without due care." With no formal training in journalism, the following basic principles were adopted by DPI:

- Media needs to be recognised as an important *pillar of democracy* and accordingly given *due importance and respect.*
- A long and sustainable relationship with media has to be based on *truth*; credibility is extremely important.
- Media functions in a competitive environment with 'breaking news' syndrome. The news media is looking for news.
- Media houses are run on commercial lines.
- Timely flow of information about events, decisions, processes and opinions is important.
- Each journalist is a unique person.

- Criticism should be respected and treated, after scrutiny, as a feedback about wider perceptions.
- The adverse/critical reporting in media could be categorised into two main classes: (i) factually incorrect or misrepresentation of fact and (ii) selective criticism based on one side of the facts with disregard to the other equally or more important aspects of the issue. Each requires different response.

Broadly, the following strategies were adopted to take along media in attempt to fill the perceptional gaps:

- *One-to-one rapport*: The primary aim was to establish one-to-one rapport with the journalists generating goodwill and understanding of each other's perspectives. Not only the achievements and activities of DRDO were discussed during such meetings, the circumstances under which DRDO had evolved and was presently functioning were also discussed to the extent considered appropriate. DRDO's strengths, achievements, the partial successes as part of learning process, the mistakes, shortcomings and limitations, all were discussed communicating correct and authentic information. Adherence to such principles helped in establishing credibility. Such meetings were organised on a day-to-day basis.
- *Lab visits*: Needless to emphasise, most activities of DRDO are highly technical in nature. Understanding the nuances of such activities during a discussion is obviously not easy. Carefully planned visits of media personnel to DRDO laboratories, conducted under the guidance of DPI personnel, were organised. Special permissions were given for allowing the media personnel along with their equipment to facilitate these visits. Such visits contributed a great deal in bridging the communication gap since journalists could witness many of the challenges faced by the scientists in developing cutting-edge defence technologies and the extraordinary efforts being made by them. For many of the journalists, it was a surprise since they did not expect such hard work and dedicated efforts from people working in government departments.
- *Press releases*: Media coverage of events appropriate for bringing them to public notice goes a long way in enhancing visibility and image of the organisation and motivating its human resource . Such events include development trials of major systems (such as missiles, aircraft, rockets, combat vehicles and weapons), new milestones in product development and recognition of DRDO scientists in national and international forums. Carefully crafted press releases highlighting the importance of the event as well as the essential details and names of key personnel were issued regularly. Care was taken to:

  - ensure that the content is in the form of an interesting story in simple-to-understand language avoiding technical jargon (when context demanded, explaining the technical terms and their significance) and not dull statements;
  - include matter that enhanced the news value of the press release;
  - attach print quality photographs, and if possible, short video clippings;
  - ensure that press release was issued promptly, preferably before 4 pm;

- alert key journalists over telephone or through SMS;
- be available over telephone, practically 24×7, to journalists seeking clarifications or further detailed information.

Coverage in newspapers Is generally a function of news value of the content (in the eyes of the editor) and availability of space in the publication. In most cases, these efforts ensured fair coverage. In any case they acted as a means to spread information about the success among the journalists. Adequate coverage of DRDO activities and events was ensured through:

- *Exclusive interviews*: One-to-one interviews by journalists from print and electronic media with DRDO leaders (e.g., Scientific Advisor to *Raksha Mantri* (Defence Minister), Chief Controllers R&D, individual DRDO laboratory directors, project directors and senior scientists) were organised on a regular basis.
- *Press conferences*: On select occasions, such as maiden launches of Agni 3, Agni 5, and Prithvi Air Defence (exo-atmospheric interceptor against ballistic missiles) as well as a part of the 'meet the press' events, journalists were invited for press conferences to cover specific technology areas. Apart from standard, known processes adopted for organising press conferences, care was also taken to provide the journalists with ample background information in the form of technical briefs, photographs, video clips, brochures and so on, stored on a DVD.
- *DRDO episodes on science channels*: Detailed half-hour television programmes, each focusing on specific major products, technologies or infrastructure facilities were prepared in collaboration with science channels. Thus, a six-episode series and a 20-episode series were produced and telecast on Discovery and Khoj channels, respectively. Subsequently, these episodes were also played during DRDO exhibitions, seminars and similar events and shared with the media.
- *Damage control*: There were occasions when prompt and specific actions to contain adverse publicity or rumors were needed, as follows:

  - *In response to events such as accidents (and incidents)*: DPI promptly issued crisp, clear and factual statements, before the news spread, which helped in avoiding speculative (mis)reporting, taking away the 'first to report' and 'breaking news' advantage.
  - *In response to adverse reporting with selective criticism*: In most such cases, primary strategy was to avoid direct open confrontations or rebuttals. A one-to-one discussion with such authors with exchange of views reflecting other side of the coin was preferred and yielded good results.
  - *In response to adverse reporting with factually incorrect or distorted facts*: Rebuttals were issued and, in most cases, were published. Another approach was to bring out articles that directly and indirectly negated the wrong information and presented facts in right perspectives.

The following two case studies are presented in brief to give an idea of how these strategies were applied and the way they worked to enhance the organisational image.

Case Study 1: MBT Arjun—Image Transformation

Development of MBT Arjun, whose two regiments (Mark-I) are now pride possessions of Indian Army, with its Mark-II version undergoing user evaluation, is a saga of design and redesigns for the changing user requirements, followed by extraordinarily long phases of user trials spread over many years and finally, its induction on 29 May 2009, making India join the club of seven countries in the world possessing capability to design, develop and produce an MBT. Arjun had to face adverse publicity that peaked during 2007 to 2008, perhaps unprecedented for any military platform. The facts of the matter were as follows:

- Arjun was the best tank India ever had in terms of technology, mobility, firepower and protection.
- Arjun was a contemporary tank comparable to MBTs operating in advanced countries.
- Arjun had already undergone exhaustive, prolonged user trials for many years, far more than any of the imported counterparts.
- The challenging task of developing MBT, a highly complex vehicle, had been initiated in an era when *country was yet to develop any automobile or even its components*.
- Above all, it was indigenous, *designed by Indians for the Indians* and India's terrain and climate. Induction in army in reasonable numbers resulted in not just self-reliance and national pride but also tactical advantage, development of Indian industries, economic advantage and much more.

Still in its formative stage, DPI pitched in to bridge the perception gap about DRDO through tactical communication, in a last ditch attempt to take Arjun to the glory it deserved. Besides DRDO's efforts internal to the ministry, a campaign to correct the image and fill perception gap was initiated. The strategy adopted could be summarised in following actions:

- Facts, figures and status of Arjun Tank highlighting its strengths in comparison with the tanks available globally, and the real reasons for its delay, largely beyond DRDO's control, were collected.
- Dialogues were initiated with journalists covering defence, discussing with them threadbare the various aspects of the project.
- Many journalists were invited to visit Combat Vehicles Research and Development Establishment (CVRDE), the nodal DRDO laboratory engaged in development of MBT Arjun. The capabilities of Arjun were demonstrated to journalists during such visits, besides giving them detailed presentations about the system.
- Short films on MBT Arjun were produced and shown at appropriate forums.
- Series of articles on Arjun were prepared and published.

Perception about MBT Arjun began to change. Strong critics of Arjun became its strongest supporters. Indian MBT was inducted; Arjun had won.

Case Study 2: Vibrant Coverage and Global Message

The Prime Minister, long ago during a visit to DRDO's missile complex at Hyderabad, had once emphasised the need to have an ICBM for achieving credible deterrence commensurate with India's threat perception. Although India had declared herself as a 'nuclear weapon capable' state in 1998 and had a variety of platforms to deliver strategic payloads, the dream of demonstrating ICBM capabilities was realised on 19 April 2012 with the maiden launch of Agni 5. A decision was taken to share the landmark achievement with masses, not just the success but also the journey behind it that was full of challenges and struggles, leading to *indigenous* development of a missile that was comparable to the best in world and the most advanced India had developed. In order to make this happen, detailed preparations were initiated several months before the expected launch. These included interviews with scientists from project teams, audio–video coverage of demonstration of key technologies and stock footage of trials and subsystems. Efforts were made to avoid technical jargon, and present the contents in a simplified and attractive manner, enriched with anecdotes. Concepts such as the reentry heat shield and its successful performance were innovatively displayed in the laboratory. Some of these stories were telecast few days before the launch. The maiden launch, its exhaustive preparations and the postlaunch jubilation received extensive coverage in India and abroad generating keen interest among people and significantly enhancing the organisational image.

The strategies were showing results. Slowly but steadily the perception of media about DRDO had begun to change. More and more people were taking interest in DRDO. Its pavilions in exhibitions were drawing large crowds.

Impact on Organisation and Its Performance

Evidently, the efforts discussed in the preceding paragraphs brought about very encouraging results in terms of transformation of DRDO's image to that of technology leader performing to deliver. The impact of changing image is also reflected in every aspect of DRDO's performance.

Best of talents from top engineering institutions have begun to get attracted and flow in through DRDO's recruitment processes. This is amply reflected, among other things, in a sharp increase in the number of fresh IIT graduates opting to join DRDO in spite of attraction of much higher perks and packages in multinational companies, and greener pastures abroad. Other schemes of lateral induction also have been receiving far better response. The attrition rates, high in the past, have dwindled. Indian industries are now responding with increasing enthusiasm and coming forward for participating in development process and taking up production of DRDO's products. The confidence level in DRDO's technologies has clearly shot up. Researchers in academic institutions and other science and technology (S&T) organisations are showing increasing interest for participation in result-oriented research with DRDO. Above all, the motivation level within DRDO has improved considerably (Murugaiah 2011). DRDO scientists are participating in events such as technical and professional conferences and exhibitions with higher enthusiasm and pride, boldly facing queries from the media and public. They are also experiencing greater respect at public platforms. Increasing interest is being shown by more and more international collaborators from world's leading technology giants. Recent display of DRDO's products in South Korea, for exploring their export potential, attracted unprecedented response. Perceptible positive change can also be seen in attitude of armed forces in terms of higher level of confidence and rising level of acceptability of DRDO's products. The positive vibes are also leading to delivery of one product after the other from all technology clusters, in much shorter development cycles. The first flight of Agni 5 took place in just three years after project sanction. Mark II of Arjun rolled out in less than three years. Arjun catapult was ready for trials in much less than a year from the word *go*. The list of such examples is long and many innovative products evolved in a short time are in advanced stages of development. Products developed and those under development are cost effective, contemporary or even better and have great export potential. The future for DRDO is sparkling with greater challenges and opportunities.

Conclusion

Communicating perspectives of defence R&D effectively to people is a challenging task. Communicating through media with innovative strategies based on mutual trust and transparency, and calibrated release of information at appropriate time and occasion can yield fruitful and decisive results without compromising on associated security aspects.

Acknowledgements The author wishes to express his gratitude to Shri Avinash Chander, the former Scientific Advisor to Raksha Mantri, Secretary, Department of Defence R&D and Director General DRDO for his valuable guidance, suggestions and permission to contribute this chapter. Thanks are also due to Mr. Suranjan Pal, Outstanding Scientist and Director, Centre for High Energy System and Science (CHESS), former Director, DPI, for very fruitful discussion on the topic.

References

DARPG (2010, September). Security of official information and documents, manual of office procedures (13th ed., Chap. XIII, p. 64, para 119). Ministry of Personnel, Public Grievances and Pensions, Department of Administrative Reforms and Public Grievances (www.darpg.gov.in).

Gopalaswami, R., & Reddy, G. S. (2013). Strategic perspectives on growth phases and long-term techno-economic performance of India's DRDO. *Journal of Defence Studies, 7*(4), 63–100.
Murugaiah, R. (2011). http://dspace.pondiuni.edu.in/jspui/bitstream/1/1799/1/T05453.pdf. Accessed 22 July 2015.
Shenoy, R. P. (2006). Defence Research and Development Organisation, DRDO Monograph Series.

Communicating Issues Related to Land and Natural Resources

S.S. Meenakshisundaram

Abstract Land rights and governance are central to the livelihood of most people in rural India. They determine the allocation and efficient use of the scarce natural resources. Any attempt by the government to assign lands—either private or public—for development or other purposes generally results in conflicts. While some actions get accepted over time, others escalate into law and order issues. Even when the actions of the government are genuinely in the interests of the people, lack of a two-way communication between the government and the stakeholders does lead to unnecessary confrontation. This chapter looks at the challenges in disseminating information on land-related issues through three case studies—first, the diversion of public lands in Karnataka for research and development institutions; second, the acquisition of private lands for industrial growth by the government in West Bengal; and the third, allocation of private and public lands to extract minerals for mining projects in Odisha. All these cases reveal the appalling lack of communication between the government and the local stakeholders resulting in avoidable agitations and judicial disputes. The courts feel that such disputes have to be settled amicably between the contending parties, in the interest of long-term peace and harmony. If there has to be an amicable settlement, the parties have to discuss the issues in public and negotiate—for which information is the key requirement.

Keywords Land right · Conflict · Amrit Mahal Kaval lands Challakere · Tata Nano project Singur · POSCO · Vedanta · Niyamgiri hills

Views expressed in the chapter are the author's own and not necessarily shared by the editors.

S.S. Meenakshisundaram (✉)
National Institute of Advanced Studies, Indian Institute of Science Campus,
Bengaluru 560 012, India
e-mail: meenakshi54@hotmail.com

P. Bagla and V.V. Binoy (eds.), *Bridging the Communication Gap in Science and Technology*, DOI 10.1007/978-981-10-1025-5_5

Introduction

Land has always been a source of economic, social and symbolic security in India. Associated land rights and governance are central to the livelihood of most people in rural India, particularly the poor—those belonging to the category of small and marginal farmers. They also determine the allocation and efficient use of the scarce natural resources. Besides being a source of revenue, land continues to be a significant tool for implementing welfare and development programmes of the government. The roots of land administration in India lie in the colonial era and to this day, they act as pillars of land governance. Under this system, rights recognition and enforcement of individual land, commons and forests; land-use planning, land management and taxation; management of public lands, including expropriation and assigning land for investments; and dispute resolution and conflict management have been the responsibility of the government. Any attempt by the government to assign lands—either private or public—for development or other purposes does result in conflicts with the existing equilibrium. While some actions get accepted over time, others escalate into law and order problems. Even when the actions of the government are genuinely in the interests of the people, lack of a two-way communication between the government and the stakeholders leads to unnecessary confrontation.

Secondly, cases of increased disputes while preparing land records, sale/transfer of large tracts of lands to private entities or encroachment create social instability. Information regarding rights such as tenure rights, transfers and land acquisition has socioeconomic implications. In a developing country like India, legislations relating to property rights, transfer of property for industrialisation and so on are rapidly changing. Lack of awareness of these changes can cause social tension and friction with the government. Land-transfer disputes have more often than not caught the national attention when it involves natural resources and customary rights of indigenous groups. Although institutions have been established to resolve conflict, governments could persevere to reduce prevalent information asymmetry in these matters thus easing potential disputes. Informational asymmetries are less problematic when there is detailed interaction between parties involved in the transaction. There is therefore a need to create avenues for the dissemination of information on issues relating to land and natural resources management which in turn will lead to better transparency in all land-related matters, thus paving way for a more efficient and accountable system of land governance.

It is in this context that we look at the challenges in disseminating information on land-related issues in this chapter through a set of case studies and the lessons they offer. The rest of this chapter consists of three sections. The first section covers three case studies—first relating to the diversion of public lands in Karnataka for government institutions to facilitate growth and development, the second relates to the acquisition of private lands for industrial growth by the government in West Bengal and the third deals with allocation of private and public lands to extract minerals for mining projects in Odisha. The next section draws out few action points from these case studies to facilitate better land governance. In the concluding section, we attempt to identify the challenges in communication between the governments and the local citizens affected by the proposed land-use changes and discuss methods by which there can be better public involvement in matters relating to land and other natural resources.

Diversion of Public Lands in Challakere

Our first case relates to the allocation of Amrit Mahal Kaval lands in Challakere Taluk of Chitradurga district in Karnataka to different government institutions.

The Wadiyar dynasty, the erstwhile rulers of Mysore State reserved large grasslands available in different parts of the state, for grazing of the herd of cattle comprising the local varieties and those brought from Vijayanagara Empire. Such lands were called 'Amrit Mahal Kaval' (Kaval meaning pasture land). Amrit Mahal Kavals are located in places like hillocks, slopes of hills and catchment area of various tanks or man-made water bodies and in areas of limited rainfall so that ecology and biodiversity are maintained. Such reserved grasslands were 240 in number, and their total area was 413,529 acres. After the dynasty of Wadiyars, Nawab Hyder Ali and his son, Tipu Sultan utilised vast Kavals for grazing and training the selected breed of cattle from the herds for using them during wars for pulling canons.

After the 4th Anglo–Mysore War (1798–1799) between the Kingdom of Mysore (of Tipu Sultan) and British East India Company, these Kavals came under the jurisdiction of British rule. Subsequent Maharajahs of Mysore reestablished Amrit Mahal breed of cattle and utilised the Kavals for grazing. However, after independence, substantial extent of these Kaval lands were utilised in the entire state for distributing them to landless agriculturists and other developmental works. As of now, the Department of Animal Husbandry and Veterinary Services claims possession of 65,925.36 acres (27,468.9 ha) of Amrit Mahal Kaval lands in 62 locations in 6 districts including Challakere Taluk in Chitradurga district. These Kavals have preserved the natural vegetation without disturbance for centuries, and the practice of letting the cattle graze in these lands have been continuing. Over the years, the strength of Amrit Mahal breed of cattle decreased conditions in the Kaval areas of Challakere became nonproductive due to repeated failure of monsoon and subsequent drought, and the strength of Amrit Mahal Kaval breed of cattle dwindled further.

During 2009 to 2010, around 10,000 acres of this precious, biodiversity-rich land protected for centuries as Amrit Mahal Kaval were recommended for allotment to Defence Research and Development Organisation (DRDO), Indian Institute of Science (IISc), Indian Space Research Organisation (ISRO), Bhabha Atomic Research Centre (BARC), Karnataka State Small Industries Ltd. Corporation (KSSIDC) and also for an integrated solar park development. These decisions also got the approval of the Karnataka government. When the local people came to know of these proposals, they formed an association and started agitating against these allotments, stating that the entire exercise was mala fide and undertaken without involving any of the locally elected government bodies, local members of legislative assembly (MLAs) and relevant departments dealing with forests, lakes, agriculture, animal husbandry, horticulture, commons, planning and so on (Report of the fact-finding committee on Amrit Mahal Kaval 2013).

The agitators pointed out that such massive and unprecedented allocations of ecologically and culturally sensitive landscapes, which also support a variety of livelihoods of local communities, have been made in violation of the Environment (Protection) Act, 1986, the Wildlife (Protection) Act, 1972, the Forest (Conservation) Act, 1980, the Biological Diversity Act, 2002, the Scheduled Tribes and Other Traditional Forest Dwellers (Recognition of Forest Rights) Act, 2006, various laws and policies relating to land use and land allocations, the directions of the National Green Tribunal and the High Court of Karnataka, various judgements of the Supreme Court of India, and fundamental principles of environmental jurisprudence including the Public Trust Doctrine, the Precautionary Principle, the Polluter Pays Principle, the Principle of Intergenerational Equity, the Principle of Sustainable Development, the Principle of Prior and Informed Consent and so on. The issue has now been taken to the Green Tribunal which has stayed all further proceedings in the matter, pending adjudication. The contending parties, namely the villagers' association and the Chitradurga district administration have been filing voluminous documents before the tribunal in support of their contentions. These documents reveal the following:

(a) From time immemorial, the Kaval lands at Challakere have not been inhabited by human beings, and there are no records to indicate human settlement in these Kaval lands. The villages are located around the Kaval lands, and, therefore, the diversion of the land Amrit Mahal Kaval, Challakere to the projects will not affect the present human settlement in these villages. However, once the projects are completed and their operations commence, there may be migration of people from other areas to this place for employment/business opportunities.
(b) From the time of existence of villages close to and around the Kaval lands, the villagers were collecting firewood, wood for agricultural tools, sand and mud for construction work, fruits and other edible greens, medicinal plants, palm leaves for preparing baskets, mats and brooms and limestone from these Kaval lands. Since Challakere is a predominant area for sheep rearing, the shepherds and the villagers who maintain sheep were using these Kaval lands for grazing the sheep. The cattle/buffalo/goat owners of the villages were also using these Kaval lands for grazing of their animals. During drought season, sheep/goats from villages located far away were also brought to this place for grazing. Shepherds of the villages were observed to collect the wool from the sheep, reel and weave them into blankets. These blankets were sold in the market at Challakere, thus making their livelihood.
(c) While the people who are now utilising Kaval products and land for grazing activities are going to be deprived of these livelihood facilities, the

establishment of education/research/commercial organisations will increase the employment opportunities for both skilled and unskilled labour in the surrounding villages, which can improve their standard of living.

(d) Establishment of residential quarters for the personnel of all these organisations and the housing facilities by Karnataka Housing Board would lead to the establishment of good schools and colleges for the children of employees. As a result, children of the villages may also get the opportunity of exposure to better education facilities. Human settlement would also lead to the establishment of medical facilities by the recipients of the land. This would provide access for local villagers to get timely medical assistance.

(e) With increase in population, the demand for vegetables, groceries, milk, meat and other local products would increase. This would directly increase the economic opportunities for the villagers surrounding the Kaval lands. This will be in addition to the job opportunities provided by the organisations. Development activities of organisations such as BARC, DRDO, ISRO and IISc would also lead to minimising the migratory pattern of the local population.

(f) The land allotted for each of the organisations is sufficient to initiate developmental activities. IISc is planning to use only 40% of its land for development. The land allotted to DRDO is a composition of plain lands and valleys. The soil is suitable for developing good runway for testing of easy takeoff and landing of unmanned air vehicles. ISRO will be contributing nearly 31 acres of land for laying a 3-kilometre-long 40-metre- wide road to meet the societal needs of existing villagers. The land allotted to BARC is not cultivable, and the area is sufficient for its activities.

(g) The water sources located in the Kaval land will get blocked due to the construction of boundary walls and villagers will have no access to it even for their animals.

(h) Establishment of BARC and enriching of radioactive materials would lead to possible health hazards to the local people.

The arguments for and against the proposed diversion may be equally strong and valid, but it is not within the ambit of this chapter to pronounce a judgement on this issue. The limited point relevant to this chapter is the appalling lack of communication between the government and the local stakeholders. Admittedly, no attempt was made to convince the local people and carry them along while making these proposals. A frank discussion in the *gram sabhas* (village councils) of the affected villages would have certainly resulted in the identification and allotment of minimum required lands to any or all of these government institutions which have all been set up for the benefit of the people and at the cost of the taxpayer's money.

Tata Nano Project in West Bengal

The next is the well-known Singur case of West Bengal where the then Left Front Government had acquired substantial extent of private lands for establishing a 'Nano' car project by the Tata group.

Singur is located in Hooghly district, around 45 kilometres from Kolkata. In May 2006, the West Bengal government decided to acquire 997 acres of agricultural land for Tata Motors' small car factory in Singur. According to available official data, Singur block belongs to a highly fertile agricultural area with 83% of the land irrigated and a crop density of about 220%. The crops produced are mainly paddy and potato, but jute and a variety of vegetables were also produced in some fields. Almost 6000 families, including agricultural workers and marginal peasants, faced eviction threats and subsequent loss of land and livelihood because of this land acquisition.

The case of the landholders is that the land-acquisition process was without any prior informed consent; the state government had overruled the decision of *gram sansad* (a village level decision-making body in West Bengal) to reject the project; no adequate compensation and/or rehabilitation packages were offered to project-affected people; and no compensation was offered to the landless agricultural workers, unrecorded *bargadars* (sharecroppers) and other rural households who were dependent on acquired land and related agricultural activities for their livelihoods. That a large number of landowners in Singur were absentee owners, and the land was actually tilled by the sharecroppers/agricultural labourers, made the situation quite complex, as the little compensation offered by the government was received by absentee landowners in most cases.

The case of the state government is based on the need to identify alternate avenues for livelihood to the rural poor who could no longer be supported by agriculture. The Nano project by Tata Motors would open up employment opportunities and also encourage other entrepreneurs to establish similar industries in different parts of the state. The Left Front government was eager to erase the impression that the political environment in the state was unfriendly to the industry.

As livelihoods of real tillers were lost, the people of Singur refused to give up their struggle against this land acquisition. The Singur movement got the support of larger civil society organisations and also human rights organisations (Halim 2013). A public interest litigation (PIL) was filed in the Calcutta High Court. After a long battle, Tata company decided to leave Singur because of the *failure of the state to give protection to the company workers*. However, Tata group has retained Singur land to date and when the successor government of West Bengal demanded that land back from Tata group, the Tata company asked for an exorbitant compensation for the same land, which it received almost for free from the previous government (Chandra 2008).

As the land was still with the Tata group under a lease agreement, the state government passed a bill in the State Assembly in 2011 to take possession of the entire area, and the Singur Land Rehabilitation and Development Act, 2011 came into effect on 20 June 2011 vesting the land with the state government free from all encumbrances. However, the legal battle did not end here. Tatas filed a case in the Calcutta High Court challenging the new Act, and the High Court declared the Singur Land Rehabilitation and Development Act, 2011, unconstitutional. The state government challenged the decision of the High Court before the Supreme Court, and in 2016 the Supreme Court ruled that the land be returned to the state government.

The Odisha Mining Case

We now move on to the allotment of private and public lands for mining projects in Odisha. Our National Steel Policy (NSP 2005) aims at accelerating growth in production and consumption of steel to achieve global competitiveness. POSCO, which is one of the world's largest steel manufacturers with presence in over 16 countries, entered India in 2005 with a project proposal to build a 12 million-tonnes-per-year steel plant in Odisha, with a captive port and iron ore mines. Their proposal was widely celebrated as the single largest infusion of foreign direct investment (FDI) after Indian economy was liberalised in 1991. POSCO intends to set up a special economic zone (SEZ) in Odisha to manufacture superior steel and export 6.3 million tonnes of its production per annum. This would help in achieving the target for exports set by NSP 2005—annually 26 million tonnes by 2019. Estimated at USD 12 billion, the project was acclaimed by Odisha government to 'bring prosperity and wellbeing to its people' by embarking on major industrialisation based on exploitation of its natural resources. National steel policy has identified inadequate infrastructure and high cost of debt as one of the biggest weaknesses of the steel sector. Addressing this, POSCO intends to invest INR 77.9 billion in the domestic tariff area (DTA) and INR 453.3 billion in the SEZ with planned investments in roads, water, power, townships and so on. This project, however, faced strong resistance from a vigorous people's movement on the ground, comprising villagers apprehensive of losing their lands and livelihoods. Consequently, five years after the project was launched, it was yet to acquire a single acre of land, and has been embroiled in legal, logistical and procedural difficulties.

The economic impact of the project is multifold. The success of an industry has multiplier effect on associated sectors as well as the economy. Studies indicate that rather than mere extraction of existing iron ore, it would be more beneficial to the economy if the firm established steel production as well. While these are the key arguments put forward by the state, the local citizens, nongovernmental organisations (NGOs) and academics raise several other arguments against the project which include:

1. Violation of the Forest Rights Act by demarcating 3096 acres of forest land for use by the firm.
2. Increased use of natural resources such as water (from Mahanadi and Hansua rivers) and land will affect the livelihoods of the local farmers and fishermen.

3. Customary rights of the indigenous people living in this area would be affected significantly; a large proportion of the tribal people living there depend on minor forest produce for sale and consumption. Several village councils had therefore rejected the proposals put forth by the firm.
4. The economic impact assessment conducted by the government failed to include the loss of natural habitats to endangered animals like the olive ridley sea turtles (*Lepidochelys olivacea*); Coastal Regulation Zone Notification of 1991, which protects fish-breeding grounds, has in fact been violated since the area designated for POSCO's captive port is classified as an 'ecologically sensitive area'.

There have been several impediments to successfully engaging the villages and the people involved to ensure a smooth transition of natural resources from the public to a private industry to establish an SEZ, and the issue remains unresolved.

In a similar vein, the Vedanta bauxite mining project was rejected by the Ministry of Environment and Forests after 12 village councils in Niyamgiri, Odisha overruled it. Vedanta is one of the largest mining and nonferrous materials companies in India. In March 2003 Vedanta Alumina Limited applied for environmental clearance for their alumina refinery project to the Ministry of Environment and Forests to be set up in eastern India. The clearance was granted on September 2004, clearly stating that the project did not intend to clear forests. However, the alumina refining project depended heavily on mining in the Niyamgiri hills, occupied by the Dongria Kondh tribes. It would require 3 million tonnes of bauxite per annum, which it intended to source from the Niyamgiri hills. The project would require about 1440 hectare of land in the hills for both the mines and the refineries. Kalahandi district where the hills are located and where the mining was to take place is one of the most economically backward regions in Odisha; it however has rich reserves of Bauxite (Saxena et al. 2010).

The project was opposed by the locals on the grounds that the mining and other related activities by Vedanta would disrupt the Dongria Kondh's way of life as they had cultural and spiritual attachment to the Niyamgiri hills; the work on the project has begun without requisite clearances and that the tribals were forcibly removed from their habitations; and the Niyamgiri hills are historically recognised as a rich sanctuary for diverse flora and fauna (Venkatesan et al. 2007). The refineries and mines would pose a threat to this environment through excessive dumping, pollution and deforestation and the water regime in Odisha that is protected by the Niyamgiri hills will be critically damaged by the project (Goswami and Mohanty 2014).

In April 2013, Supreme Court of India upheld the ban on mining in the Niyamgiri hill range and ruled that the rights of the Dongria Kondh communities must be taken into account in deciding whether the mining project may go ahead. In August 2013, all the 12 *gram sabhas* of the tribal villages voted against Vedanta's project in the Niyamgiri hills. In January 2014, the Ministry for Environment and Forests decided to withdraw the environmental clearance for this project, which brought a closure to this mining project.

What Do the Case Studies Reveal?

Although the three cases discussed in the previous sections relate to different types of land-related disputes, they do have certain commonalities. To start with, the state governments play a crucial role in all these cases and, irrespective of the political party in power in that state, their response to the disputes is almost similar. The stakeholders who constitute the epicentre of the conflicts in all the cases feel that they were not consulted at all and the decisions have been taken behind their back. When they do not get any redressal from the executive, they knock at the doors of the judiciary. They do get some temporary relief from the courts, but the final decisions take a very long time. Some courts also feel that such disputes have to be settled amicably between the contending parties, in the interest of long-term peace and harmony rather than opting for a legal solution. If there has to be an amicable settlement, the parties have to discuss the issues in public and negotiate—for which, information is the key requirement since lack of relevant information, or sometimes misinformation, is generally the underlying cause for most of these conflicts.

The usual obstacles to the smooth transaction of any land include lack of well-defined property rights, high cost of access to ownership records, environmental legislations and customary rights of indigenous groups. If land transactions extend beyond community members, it becomes more difficult to ascertain who holds legitimate rights. Usually the nature of land-related disputes falls into any or some of the following categories:

- Title/ownership of land
- Possession of land
- Disputes relating to tenancy
- Lease and mortgage
- Usage of land, whether public or private and so on
- Usufruct and other rights relating to land.

Increased access to relevant information, legislation and so on would adequately address the aforementioned problems. For a decision maker faced with uncertainty regarding the quantum and location of land required for a particular use, increased access to information coupled with stakeholders' views is more beneficial than making such decisions in isolation. Transparency is a critical component in a well-functioning land administration framework in view of the scarcity of clear and credible information on land availability and transactions, and the poor dissemination of public information on land rights and policies. This argument is highlighted both in the Challakere case involving public lands and the Singur case where Land Acquisition Act was evoked to set in motion large-scale industrialisation discounting the value of highly fertile agricultural lands without the consent of the land users. Similar is the case relating to the environmental law violations in the dealings of POSCO in Odisha as well as the several illegal mining cases in Bellary district of Karnataka. In the backdrop of successful establishment of the Tata Nano plant in the state of Gujarat subsequently, it will be advantageous to reflect on the strengths and good practices of land acquisition and transfer process. All said and done, when in place, transparency can encourage civic engagement and stakeholders' accountability by rendering the public decision-making arena more accessible. This in turn strengthens confidence in government and public agencies, and has a positive economic impact.

How to Ensure Transparency and Adequate Flow of Information?

Because transparency is the hallmark of an efficient revenue administration, we need to institutionalise it instead of leaving it to the whims and fancies of the individuals. To start with, land records must be prepared with the full knowledge and participation of the stakeholders. To ensure greater public participation during preparation of land records, awareness campaigns among the general public through local print and electronic media have to be organised. Complete digitisation has to be undertaken and the availability of land records and information regarding acquisition and transfer of agricultural lands as well as those that are demarcated as forests should be put in the public domain and also exhibited at the office of the *gram panchayat* (the government body at the village level). Once the inventory of lands is available in the public domain, there has to be periodic updating which has to be undertaken through the *gram sabhas* in an open meeting. This will ensure transparency.

Historically, the revenue department has always been headed by the collector (or the deputy commissioner as her or she is called in some states) who has been saddled with several other regulatory responsibilities including protocol duties. Time has perhaps come to have a separate officer looking only at land-related issues at the district level. This will help not only in better supervision and monitoring, but also facilitate quick information flow under the Right to Information (RTI) Act to the needy.

Thirdly, the scope of the district information office (which is already in existence in every district) can be enhanced from organising government-related functions and keeping the media informed, to information dissemination to the people particularly when private lands are proposed for acquisition or public lands are proposed for allotment.

As a matter of policy, the provisions of the Land Acquisition Act, 2014 in so far as they relate to consultations with the stakeholders should be implemented in letter and spirit. Before acquiring private lands or taking away any land assigned for a public purpose, the affected parties have to be heard and their opinions recorded.

One final recommendation has to be the empowerment and involvement of *gram sabhas* in all land-related issues, as these local institutions can play a crucial role in providing the necessary link between the governments and the people.

Conclusion

Our case studies reveal that adequate information regarding activities relating to land acquisition, transfer or natural resource exploitation has not so far been sufficiently recorded or monitored. Increased attention to such activities in the initial stages of development would have ensured a smooth progression of events and prevented illegal activities. While increased diligence on the part of the Ministry of Environment and Forests as well as the state departments is a perquisite to positive policy changes, it is only through collective action on the part of the government and the people can the economy achieve the target of reducing information asymmetry in conjunction with the sustainable use of scarce natural resources. Even if proper policies and legislation are in place, bad governance and an ineffective/corrupt state could render such policies futile. The only way to prevent such a situation is to empower the people through dissemination of information and knowledge.

References

Chandra, N. K. (2008). Tata Motors in Singur: A step towards industrialization or pauperization. *Economic and Political Weekly, 13*, 36–51.

Goswami, U. A., & Mohanty, M. (2014, January 11). Environment ministry rejects Vedanta's mining proposal in Niyamgiri. *Economic Times*.

Halim, U. (2013). *From Singur to Sompeta: People's resistance against land grabbing gaining power in India*. Pesticide Action Network Asia and the Pacific: Speak Out.

Report of the fact-finding committee on Amrit Mahal Kaval (2013, July). Challakere, Chitradurga District, Karnataka, Submitted to National Green Tribunal-South Zone, Chennai.

Saxena, N. C., Prasuraman, S., Kant, P., & Baviskar, A. (2010). *Report of the four member committee for investigation into the proposal submitted by the Orissa Mining Company for Bauxite Mining in Niyamgiri* [submitted to Ministry of Environment & Forests].

Venkatesan, R., Wadhwa, W., Saluja, M. R., Basu, S., & Rastogi, R. (2007). *Report on social cost benefit analysis of the POSCO steel project in Odisha*. New Delhi, India: National Council of Applied Economic Research (NCAER).

The Art and Science of Communicating Risks of Natural Hazards

Ajit Tyagi

Abstract Every year natural hazards cause significant loss of life and adversely impact development gains by destroying crops, properties and infrastructure. Of all the reported disasters, about 90% are directly or indirectly related to weather and climate. Over the past 50 years vulnerability to disasters has increased as more and more people and assets have been located in areas of high risk. Climate change is expected to increase the frequency and intensity of the most severe weather-related hazards in the decades to come. Identification and communication of risks associated with natural hazards to stakeholders are therefore vital in minimising and mitigating adverse impacts of natural hazards. Success or failure of managing risk associated with natural hazards is dependent on reliable early warning, effective communication and appropriate response mechanism on ground. Early warning of natural hazards lies at the core of the disaster management system. Warning messages are to be simple, containing location- and time-specific useful actionable information to enable proper responses that will help safeguard lives and livelihoods. In near future improved seasonal and climate predictions will help in identifying climatic risks and assist in strategic planning and long-term investments.

Keywords Risk communication · Vulnerability · Natural hazards · Natural disasters · Natural disaster management · Uttarakhand disaster · Cyclonic storm · Phialin

---

Views expressed in the chapter are the author's own and not necessarily shared by the editors.

A. Tyagi (✉)
Ministry of Earth Sciences, Lodi Road, New Delhi 110 003, India
e-mail: ajit.tyagi@gmail.com

P. Bagla and V.V. Binoy (eds.), *Bridging the Communication Gap in Science and Technology*, DOI 10.1007/978-981-10-1025-5_6

Introduction

Natural hazards are severe and extreme weather and climate events that occur naturally in all parts of the world. Natural hazards in vulnerable areas become natural disasters causing loss of lives and destruction of property. Natural hazards can broadly be classified into two categories: hydro-meteorological and geological hazards. Tropical cyclone, heavy rainfall, floods, drought and thunderstorm are major hydro-meteorological hazards. Earthquake and volcanic eruptions are geological hazards. Natural disasters occur all over the world; however, their impact in developing countries is greater due to the geographical location in zones highly susceptible to natural hazards (natural vulnerability) and also due to the different types of economic, social, political and cultural vulnerabilities that exist. Every year natural hazards/disasters cause significant loss of life and property adversely affecting developmental gains especially among poor people. During 1980–2010, hydro-meteorological hazards accounted for 90% of total number of disasters, 70% of causalities and 75% of total economic losses. From 1970 to 2009, 7870 hydro-meteorological-related disasters were reported globally, causing the loss of 1.9 million lives and economic damages of USD 1.9 trillion (Fig. 1). Since 1965, while economic losses associated with hydro-meteorological hazards have increased, there has been a significant decrease in loss of life. This has been

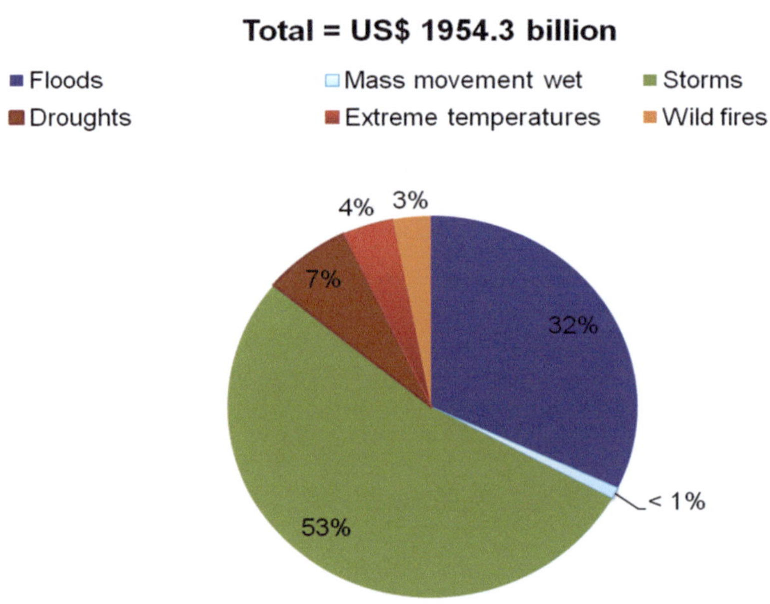

Fig. 1 Economic damages from hydro-meteorological disasters (1970–2009). *Source* D. Guha-Sapir, R. Below, Ph. Hoyois—EM-DAT: The CRED/OFDA International Disaster Database—www.emdat.be—Université Catholique de Louvain—Brussels—Belgium

attributed to real-time monitoring and early warning systems with effective risk communication (RC) and emergency preparedness at national and local levels.

RC is a relatively new field. During the mid-1980s RC came to be recognised as a necessary component in risk management and community decision making. It started in the areas of environmental and occupational health, as the developed nations faced mounting concern over toxic wastes, nuclear power plants, hazardous materials and genetically modified (GM) crops from environmental groups. Unlike conflicting viewpoints on risk management in areas of nuclear energy, GM crops and big hydroelectric power projects, acceptance of RC in area of natural hazard has been noncontroversial and smooth. However, application of RC in the context of natural hazards has not been uniform. Startling difference between the management of heavy rainfall event during 15–17 June 2013 in Uttarakhand and very severe cyclonic storm Phialin (12 October 2012) highlights the difference in the adaptation of RC and associated preparedness and emergency response by two states. Case study of cyclonic storms affecting Bangladesh and India brings out importance and success of RC in saving precious lives.

Risk Communication

Although RC per se is a recent development, communication as such has been part of civilisation right from the beginning starting with sign language to spoken and written languages. Communication, put in the simplest form, is the process of sending and receiving messages or transferring information from one part (sender) to another (receiver). Communication has been used in diversity of disciplines, ranging from religious, social and political discourses to contemporary journalism and marketing. Traditional communication model [described in detail in Lindell and Perry (2004)] has been the basis of the bulk of the RC literature.

RC seeks to inform people about the potential future harm and the associated dangers so that they might take action to mitigate the risk (Seeger 2006). RC is a complex, multidisciplinary, multidimensional, evolving process of increasing importance in protecting the life and property of public. Risk communicators (disaster managers, government officials and community leaders) use RC to give citizens necessary and appropriate information and to involve them in making decisions that affect their safety.

Initially RC was linear and unidirectional with focus on providing official information to increase recipient awareness. It involved an intentional transfer of information designed to respond to public concerns or public needs related to real or perceived hazards. Primary assumption was that the main issue was poor recognition of a risk by public and that official provision of information would resolve the issue (Kasperson and Stallen 1991).

The technocratic unidirectional approach of RC became the subject of social science critique from the late 1980s onwards. Consequently, RC moved to the following two aspects of communication:

- Persuasive communication to change risk-related behaviour of people.
- An emphasis on two-way communication and exchange, in which stakeholders should engage with and learn from each other (Renn 2005, 2006).

Primary reason for increasing interest in two-way communication has arguably been the failing of previous communication efforts to deliver the desired changes in risk-related attitudes and behaviours (Kollmuss and Agyeman 2002).

Purposes of Risk Communication

RC is an enabling mechanism (Hoppner et al. 2010) aimed to

- raise awareness;
- encourage protective behaviour;
- inform to build up knowledge on hazards and risk;
- inform to promote acceptance of risks and management measures;
- inform on how to behave during such events;
- warn of and trigger action towards impending and current events;
- reassure the audience (to reduce anxiety or 'manage' outrage);
- improve relationships (build trust, cooperation and networks);
- enable mutual dialogue and understanding;
- involve actors in decision making.

Principles of Risk Communication

There are seven cardinal rules for the practice of RC, as first expressed by the US Environmental Protection Agency (US EPA; Covello and Allen 1988):

1. Accept and involve the public as a legitimate partner.
2. Plan carefully and evaluate your efforts.
3. Listen to the public's specific concerns.
4. Be honest, frank and open.
5. Coordinate and collaborate with other credible sources.
6. Meet the needs of the media.
7. Speak clearly and with compassion.

Key Elements of Risk Communication

Stakeholders

Communication occurs between individuals, groups, public and private institutions in direct or mass communication settings. Gray et al. (1998) have identified following categories of stakeholders in RC:

- government (local and central) and regulatory agencies;
- political and social leaders;
- scientists and experts;
- industry;
- nongovernmental organisations (NGOs) and groups;
- the exposed/concerned/affected public;
- the general public;
- the mass media such as local and national television, radio and newspapers.

Communication

RC takes place in two phases, namely preparatory and warning phases. Communication during preparatory phase makes use of following modes:

- *One-way communication*: Leaflets, brochures, videos, newsletters, exhibitions, television, radio, newspapers and Internet.
- *Feedback*: Site visits, public meetings and opinion polls.
- *Two-way dialogue*: Community advisory committees, deliberative workshops, citizen panels, consensus building and Internet dialogue.

RC during the warning phase is one way, using direct/hotline communication to key stakeholders, using mass media such as television, radio, Internet and short messaging service (SMS) for wider circulation and to the target population by mobile loudspeaker, warning signals and flood sirens.

Message

The content and format of a message is the key to communication. As regards natural hazards warning and alert, the messages should be simple, clear and specific about the nature of threat and its likely impacts. Warning/alert messages have to be geographically specific to ensure that the warnings are targeted to those at risk only. Warning messages should also convey the impact the disaster is likely to cause in the affected area. Indian Meteorological Department (IMD) has included this aspect

Table 1 The expected damage associated with the cyclonic disturbances of different intensities along with action suggested

Intensity category and the maximum wind speeds	Damage expected	Action suggested
Deep depression 50–61 kmph (28–33 knots)	Minor damage to loose and unsecured structures	Fishermen advised not to venture into the open seas
Cyclonic storm 62–87 kmph (34–47 knots)	Damage to thatched huts; breaking of tree branches causing minor damage to power and communication lines	Total suspension of fishing operations
Severe cyclonic storm 88–117 kmph (48–63 knots)	Extensive damage to thatched roofs and huts; minor damage to power and communication lines due to uprooting of large avenue trees; flooding of escape routes	Total suspension of fishing operations; coastal hutment dwellers to be moved to safer places; people in affected areas to remain indoors
Very severe cyclonic storm 118–167 kmph (64–90 knots)	Extensive damage to *kutcha* (temporary or weakly constructed) houses; partial disruption of power and communication lines; minor disruption of rail and road traffic; potential threat from flying debris; flooding of escape routes	Total suspension of fishing operations; mobilise evacuation from coastal areas; judicious regulation of rail and road traffic; people in affected areas to remain indoors
Very severe cyclonic storm 168–221 kmph (91–119 knots)	Extensive damage to *kutcha* houses; some damage to old buildings; large-scale disruption of power and communication lines; disruption of rail and road traffic due to extensive flooding; potential threat from flying debris	Total suspension of fishing operations; extensive evacuation from coastal areas; diversion or suspension of rail and road traffic; people in affected areas to remain indoors
Super-cyclone 222 kmph and more (120 knots and more)	Extensive structural damage to residential and industrial buildings; total disruption of communication and power supply; extensive damage to bridges causing large-scale disruption of rail and road traffic; large-scale flooding and inundation of seawater; air full of flying debris	Total suspension of fishing operations; large-scale evacuation of coastal population; total suspension of rail and road traffic in vulnerable areas; people in affected areas to remain indoors

Source Author's work

in its cyclone warning bulletins (Table 1). For warning the messages have to be clearly recognisable and consistent, and IMD has developed standard format for warnings of different natural hazards. Warnings are updated at regular intervals, and mechanism is in place to de-warn the community when the threat has ended.

Benefits and Barriers of Risk Communication

Benefits

RC benefits include improved decision making, both individually and collectively. The purpose of the exchange and the nature of the information have an impact on the benefits. Depending on the situation, personal and community anxieties about natural hazard risks can be reduced or increased. For example, a goal might be raising concern about tsunami and prompting action.

Other benefits of the RC process include a better-educated public, an appreciation of limited resources and difficult choices, increased coordination between various levels of government and the development of working relationships between diverse interest groups. As citizens become more involved as participants, they become part of and contribute to the solution.

Barriers

Because the RC process is so deeply embedded in broader social issues, barriers and problems are many. A key barrier is the term 'risk' itself—how it is measured, described and perceived. Interested parties perceive risks differently. People do not believe that all risks are of the same type and size. Many RC recipients do not comprehend the concept of probabilities—a 0.05 probability is less comprehensible than the statement, "5 of 100 people have an increased risk for a disease". Other barriers are:

- conflicting risks and messages;
- difficulty of translating scientific information;
- disagreement on what the risk itself is and how to assess it;
- lack of RC expertise;
- organisational cultures unfamiliar or uncomfortable with two-way processes.

Risk

Risk arises from the combination of hazard and vulnerability at a particular location. In statistics, the notion of risk is often modelled as the expected value of an undesirable outcome. This combines the probabilities of various possible events and some assessment of the corresponding harm into a single value. The simplest case is a binary possibility of accident or no accident. The associated formula for calculating risk is then:

$$\text{Risk} = \text{Hazard} \times \text{Vulnerability}$$

For example, if performing activity 'X' has a probability of 0.01 of suffering an accident of A, with a loss of 1000, then total risk is a loss of 10, the product of 0.01 and 1000.

The concept of disaster risk is used to describe the likelihood of harmful consequences arising from the interaction of natural hazards and the community. Thus, two elements are essential in the formulation of disaster risk: the probability of occurrence of hazard and the vulnerability of community to that hazard.

Vulnerability

Vulnerability is the degree to which a community is at risk from the occurrence of extreme physical or natural phenomena. Vulnerability refers to the way the hazard or disaster will affect human life and property and depends on:

- frequency and intensity of hazard,
- proximity to possible hazardous event,
- population density,
- construction styles and building code,
- governance,
- availability and readiness of emergency services,
- infrastructure and support system.

In India, vulnerability maps have been prepared for all major natural hazards (BMTPC 2015). However, these vulnerability maps have limitations as these provide general idea of vulnerability at macro-level and do not factor socioeconomic conditions of the people.

Vulnerability is a characteristic of individuals and groups of people who inhabit a given natural, social and economic space, within which they are differentiated according to their varying position in society into more vulnerable or less vulnerable individuals and groups. It is a complex characteristic produced by a combination of factors derived especially (but not entirely) from class, gender or ethnicity. Vulnerability also depends on:

- *Livelihood resilience*: the degree of resilience of the particular livelihood system of an individual or group, and their capacity for resisting the impact of hazard.
- *Health*: including both the robustness of individuals and the operation of various social measures.
- *Preparedness*: determined by the protection available for a given hazard, something that depends on people acting on their own behalf and social factors.

There are indeed many other kinds of vulnerability, which are as follows:

- Lack of access to resources (materials/economic vulnerability).
- Disintegration of social patterns (social vulnerability).
- Lack of strong national and local institutional structures (organisational vulnerability).
- Lack of access to information and knowledge (educational vulnerability).
- Lack of public awareness (attitudinal and motivational vulnerability).
- Limited access to political power and representation (political vulnerability).
- Certain beliefs and customs (cultural vulnerability).
- Weak buildings and weak individuals (physical vulnerability).

However, all of these can be classified into four main types of vulnerability: social, economic, political and cultural. This classification indicates that each social entity has different types of vulnerability, and it is not only the result of the human actions, decisions and choices, but also the result of the interaction of the natural, economic, social, cultural and political contexts where people live. Vulnerability cannot be treated as a homogeneous and general term; its dynamism is given by each society, and it is both a universal and particular concept.

Vulnerability arises due to the coupling between the natural and human systems. In this sense, vulnerability can be divided into natural vulnerability and human vulnerability. Natural vulnerability depends on the threatening natural hazard (very much related to geographical location); thus, there are different vulnerabilities, such as volcanic vulnerability, flooding vulnerability, land-sliding vulnerability, tsunami vulnerability, hurricane vulnerability and so on. In contrast, human vulnerability is based on the social, economic, political and cultural systems.

Natural Hazards

A natural hazard is a severe natural phenomenon/event resulting from natural processes of the earth. Earth system comprises solid earth, atmosphere and oceans. Natural phenomena occur as an outcome of natural processes/forces within one subsystems or in combination. Processes in the earth subsystem cause earthquakes and volcanic eruptions. Landslides and avalanches are caused by combination of geological and hydro-meteorological factors. Tropical cyclones form under favourable atmospheric and oceanic conditions. Natural phenomena and associated hazards occur across different time and area scales from a span of few minutes to days and weeks (even months and seasons for droughts) and an area of few kilometre going onto covering areas of thousands of kilometre in the case of droughts.

Some of the natural phenomena may be accompanied by many hazards like heavy rainfall, floods, strong winds and storm surge occurring at the time of landfall of tropical cyclone. As stated in a preceding section, natural hazards broadly fall into two categories, hydro-meteorological and geological hazards. In this context, a third category, namely hybrid disasters has also been defined. A hybrid disaster arises out of combined effect of hydro-meteorological and geological factors.

Natural hazards exhibit varied characteristic features which require matching observational systems, forecasting techniques and RC strategies. Characteristic features of natural hazards based on four attributes are as follows:

1. Size (in kilometres)
   - Microscale: 1–10
   - Mesoscale: 10–100
   - Synopticscale: ~ 1000
   - Planetaryscale ~ 10,000
2. Life period
   - Turbulence: seconds
   - Cloud burst/landslides: minutes
   - Thunderstorms: hours
   - Low-pressure systems: days
   - Cyclonic storms: days to weeks
   - Drought: months, seasons
3. Speed of onset
   - Sudden: earthquake, flash flood, tsunami, landslide, avalanche
   - Moderate: cyclonic storm, riverine flood
   - Slow/creeping: drought
4. Frequency
   - Normal
   - Rare.

Hydro-meteorological, geological and hybrid disasters are as follows:.

Hydro-meteorological disasters

- Cyclonic storms
- Floods
- Drought
- Heavy rainfall
- Heavy snowfall/blizzards
- Thunderstorms/hailstorms/tornadoes
- Heat waves/cold waves

Geological hazards

- Earthquakes
- Volcanoes

Hybrid disasters

- Landslides
- Avalanche
- Tsunami.

Early Warning

Early warning system lies at the core of RC and risk management. Warning system empowers individuals, communities and businesses to respond timely and appropriately to hazards in order to reduce the risk of death, injury, property loss and damage. Continuous monitoring of hazard parameters and precursors is essential to generate accurate warning in timely fashion. To have credibility in the forecasts and warnings, early warning system has to have a sound scientific basis for prediction and must be in operation 24 hours a day, 365 days a year.

Since mid-2010s, proliferation of multisensor observing systems, real-time data communication, high-speed processing using supercomputers and advanced numerical models have led to significant improvements in the skill of weather forecasts (IMD 2015; http://202.54.31.51/bias/). Cyclone Warning Division of IMD also functions as World Meteorological Organizations' (WMO) Regional Specialised Meteorological Centre (RSMC) for tropical cyclones for northern part of Indian Ocean. It provides advisory and guidance to all countries in the region (Tyagi et al. 2010).

The effectiveness of a warning depends on:

- the timeliness of warning;
- effective communications and public information systems to inform the public about the imminent danger;
- the credibility of the source of the warning.

Case Study: Risk Communication of Cyclonic Storms

Cyclonic storm is one of the most severe natural hazards both in terms of causing deaths and damage to property and infrastructure. Countries in Asia are generally worst affected (Table 2). In days prior to improvement in observing and forecasting systems and introduction of systematic approach to disaster management and RC, death toll associated with landfalling cyclonic storms used to be phenomenal.

Difference in response by Myanmar and Bangladesh to cyclonic storms of same intensity brings out the importance of RC in managing natural disasters. Even with accurate forecast of cyclonic storm Nargis (3 May 2008) three days in advance and despite advisories issued to Myanmar government by international agencies

Table 2 Ten top deadliest cyclonic storms in the world history, which affected Asian countries and of which seven storms affected India and Bangladesh

Rank	Name/areas of largest loss	Year	Ocean area	Deaths
1	Great Bhola Cyclone, Bangladesh	1970 (12 November)	Bay of Bengal	300,000–500,000
2	Hooghly River Cyclone, India and Bangladesh	1737	Bay of Bengal	300,000
3	Haiphong Typhoon, Vietnam	1881	West Pacific	300,000
4	Coringa, India	1839	Bay of Bengal	300,000
5	Backerganj Cyclone, Bangladesh	1584	Bay of Bengal	200,000
6	Great Backerganj Cyclone, Bangladesh	1876	Bay of Bengal	200,000
7	Chittagong, Bangladesh	1897	Bay of Bengal	175,000
8	Super Typhoon Nina, China	1975 (5 August)	West Pacific	171,000
9	Cyclone 02B, Bangladesh	1991 (5 May)	Bay of Bengal	138,866
10	Cyclone Nargis, Myanmar	2008 (3 May)	Bay of Bengal	138,366

Source Author's work

including WMO and RSMC, New Delhi, because of the absence of RC mechanism in Myanmar, a staggering number (138,000) of deaths occurred. On the other hand, the cyclonic storm of similar intensity named 'Sidr' (15 November 2007) affected Bangladesh and because of effective RC mechanism, deaths were limited to 4200, a far lower number than for Nargis (Table 3).

Nevertheless, for cyclonic storm Nargis, redeeming feature is that there has been sharp decline in the number of deaths caused by cyclonic storms since 2000. It has been achieved, as mentioned earlier, by systematic improvements in both early

Table 3 Details of the storms Sidr and Nargis

Cyclone event	Sidr	Nargis
Year	2007	2008
Storm surge (m)	5–6	~4
Maximum wind speed (kmph)	245	235
Category (Saffir–Simpson Hurricane Wind Scale, SSHWS)	4	4
Number of affected people (approximate, in million)	8–10	2–8
Mortality (approximate)	4200	138,000
Damages (USD, billion)	2.3	4.0

Source Author's work

warning system and RC. RC played an important role in cyclone preparedness in Bangladesh and India in the following ways:

(a) Creating awareness among the people, government employees and people of other professions for reducing risks during disaster.
(b) Imparting training to the government employees, elected representatives and others on disaster management in cooperation with different ministries, local authorities, training institutions and NGOs.
(c) Supplying books, maps and other information to those associated with disaster management.
(d) Collecting and preserving lists with location, condition and ownership of cyclone shelters, embankments and platforms at higher-than-flood level (flood proofing).
(e) Coordinating the activities of NGOs.
(f) Educating all about weather signals.
(g) Arranging public awareness programmes at regular intervals on radio and television on various disaster-related issues.
(h) Organising meetings, seminars and workshops at national, district and village levels and at police stations to increase awareness about disasters.
(i) Liaising with the Ministry of Education for inclusion of disaster topics in the curricula of schools and colleges.
(j) Publicising cyclone signals in the villages through posters, cultural functions, documentary films and so on.

At the field level, the following actions proved effective:

(a) Organising simulated drills continuously in disaster-prone areas and monitoring the status of preparedness drills participated by the people in April and September every year.
(b) Selecting volunteers and completing their training according to laid-down rules and organising grouping of families for evacuation in times of need.
(c) Making physical verification of equipment allotted to the leaders of volunteers and their condition and undertaking their repairs and replacement, if needed.
(d) Examining the wireless system and keeping it running in order to maintain communication among headquarters, police stations and government offices.
(e) Educating and creating awareness among the local people about the cyclone-preparedness programme and understanding of different warning signals.
(f) Earmarking shelter places and safe raised land, to keep them fit for use and to publicise the evacuation plan.
(g) Making the people and volunteers aware of the dangers of resistance to disaster-preparedness activities, in coordination with police stations and panchayat administration.

Improved Observing and Forecasting Systems

There have been steady improvements in observing systems over ocean and coastal areas with real-time availability of satellite imageries and data from buoys. It has been of great help in monitoring genesis, intensification and movement of storms over the ocean. To detect cyclonic storms approaching Indian coast, IMD had installed 10 radars having the range of 400 kilometres covering the entire Indian coast—six along east coast and four along west coast—in 1980s. These have been replaced by state-of-the-art Doppler weather radars for real-time monitoring, and currently, the real-time satellite pictures and radar images of a cyclonic storm are used by forecasters to monitor location and intensity of cyclonic disturbance (for further information, visit http://www.rsmcnewdelhi.imd.gov.in/index.php?lang=en). Another significant development, which contributed to the improvement in forecasting since 2010, was use of state-of-the-art numerical weather prediction models. Steady improvement in skill of cyclone forecast has created confidence about IMD's forecast among administrators, disaster managers and public. This was amply demonstrated in the case of very severe cyclonic storm Phailin (12 October 2013).

Lead time in providing warnings has a crucial role in disaster preparedness especially for making preparations for safe evacuation of people from likely affected areas. IMD provides cyclone warnings to central and state government officials in four stages:

(i) Pre-cyclone watch
(ii) Cyclone alert
(iii) Cyclone warning
(iv) Post-landfall outlook

The first stage warning known as 'pre-cyclone watch warning' issued 72 hours in advance contains early warning about the development of a cyclonic disturbance in the northern part of Indian Ocean, its likely intensification into a tropical cyclone and the coastal belt likely to experience adverse weather. This early warning bulletin is issued by the Director General of Meteorology in person and is addressed to the Cabinet Secretary and other senior officers of the Government of India including the Chief Secretaries of maritime states concerned. Director General of Meteorology remains in continuous touch with Cabinet Secretary, National Disaster Management Authority (NDMA), Ministry of Home Affairs and Chief Secretary of the states all through the life cycle of a cyclonic storm. In case of cyclonic storm Phailin, IMD issued pre-cyclone watch on 9 October 2013 (http://202.54.31.51/bias/NWP-REPORT-PHAILIN-2013.pdf).

The second stage warning known as 'cyclone alert' is issued at least 48 hours in advance of the expected commencement of adverse weather over the coastal areas. It contains information on the location and intensity of the storm, likely direction of its movement, intensification, coastal districts likely to experience adverse weather and advice to fishermen, general public, media and disaster managers.

The third stage warning known as 'cyclone warning' is issued at least 24 hours in advance of the expected commencement of adverse weather over the coastal areas. Landfall point is forecast at this stage. These warnings are issued at 3-hours intervals giving the latest position of cyclone and its intensity, likely point and time of landfall, associated heavy rainfall, strong wind and storm surge along with their impact and advice to general public, media, fishermen and disaster managers.

The fourth stage of warning known as the 'post-landfall outlook' is issued by the officers concerned at least 12 hours in advance of expected time of landfall. It gives likely direction of movement of the cyclone after its landfall and adverse weather likely to be experienced in the interior areas.

Different colour codes (yellow for cyclone alert, orange for cyclone warning and red for post-landfall outlook) are being used since post-monsoon season of 2006 in the different stages of the cyclone warning bulletins as desired by NDMA.

During disturbed weather over the Bay of Bengal and Arabian Sea, the ports likely to be affected are warned by officers concerned by advising the port authorities through port warnings to hoist appropriate 'storm warning signals'. The department also issues 'fleet forecast' for Indian Navy, coastal bulletins for Indian coastal areas covering up to 75 kilometres from the coastline and sea area bulletins for the sea areas beyond 75 kilometres. Special warnings are issued for fishermen four times a day in normal weather and every three hourly in accordance with the four-stage warning in case of disturbed weather.

The general public, the coastal residents and fishermen are warned through state government officials and broadcast of warnings through All India Radio and national television (*Doordarshan*) telecast programmes in national and regional hookup.

Uttarakhand Disaster versus Very Severe Cyclonic Storm Phialin

Uttarakhand experienced very heavy rainfall during 15–18 June 2013 resulting in floods and landslides. It caused unprecedented disaster causing more than 5000 deaths. Although IMD issued heavy rainfall warning 48 hours in advance and communicated to state authorities and media, the state administration treated it as normal warning and did not initiate any action. Post-disaster analysis showed that the state did not have disaster preparedness and management plan. State administrators and disaster managers did not comprehend the impact of the warning and the gravity of the situation. At the same time some of the organisations and individuals who followed the warnings were able to take timely safety measures. There was also an absence of coordination between the state and central agencies. It is generally observed that most of the states/cities do not pay attention to risk assessment, RC and disaster preparedness. It is only after major disasters such as Bhuj earthquake (2001) and Odisha super-cyclone (1999) that states of Gujarat and Odisha embarked upon systematic risk assessment, RC and disaster-preparedness programmes, which helped Odisha government to undertake safe evacuation of record one million people during the subsequent events of natural disasters. It is an excellent example of the success of accurate early warning system and RC resulting in effective disaster management.

New Challenge in Risk Communication: Climate Change Risks

Climate change is now a reality. The Earth's surface temperature has risen by more than 0.8 °C in the 20th century and by about 0.6 °C since mid-1980s. Some of its impacts are already visible in higher frequency of extreme weather events. Climate projections indicate an increase in extreme weather events such as floods, droughts, heat waves and intensity of rain and storms. Climate change is likely to have adverse impact on agriculture, water and health sectors, and the worst affected will be the poor countries. Sea level rise will lead to migration from small island states and low-elevation coastal zones.

RC in the context of climate change is to address the problem at two levels: immediate one related to mitigation and long-term one to adaptation. With effective communication of risk associated with climate change, most of the countries have realised the need to initiate mitigation process by reducing greenhouse gases and adopting clean and renewable energy. There is broader agreement among scientists that successful implementation of climate change mitigation strategies can provide co-benefits in terms of public savings, air quality and associated health benefits.

While it is not yet possible to predict with precision how climate change will affect each particular place or sector, whether by downscaling global climate change models or extrapolating from past data, there is general consensus that extremes are going to increase in future. Risks are not going to be uniform and will exhibit spatial variability. Climate services planned under Global Framework of Climate Services (GFCS) of WMO are aimed to strengthen the production, availability, delivery and application of science-based climate prediction and services to assist communities to adapt to climate variability and change. RC related to climate change needs to incorporate climate services for taking climate-smart decisions.

Summary

1. Natural hazards pose great threats to lives and property in vulnerable areas. Hydo-meteorological disasters account for 90% of all natural disasters.
2. While there is an increase in economic damages from hydrodynamic disasters world over, there has been significant decrease in deaths since mid-1960s.

3. Disaster risks arise from the combination of hazards and vulnerabilities at a particular location. Poor countries or people are more vulnerable to disasters.
4. Vulnerability maps available as of now are based primarily on frequency and intensity of natural hazards. There is a need to undertake comprehensive vulnerability mapping by including economic, social and physical vulnerability.
5. RC has played an important role in mitigating adverse impacts of disasters. RC has evolved from unidirectional, linear, technocratic top-down approach to two-way interactive process taking into consideration social, economic, educational and cultural aspects of community.
6. RC is essentially a preventive activity and contributes to disaster preparedness.
7. Early warning systems form the core of RC. To be effective, warnings have to be location and time specific, and credible.
8. Improvement in observing and forecasting system has led to improvement in the skill and credibility of warnings of different hazards.
9. Examples such as cyclonic storm Nargis and Uttarakhand disaster bring out that in the absence of effective RC and preparedness, even accurate forecasts and warnings are not useful for the public at large.
10. Risk managers have to understand the complexities of risks associated with climate change and incorporate them in their communication for mitigation and adaption.

Acknowledgements Author is thankful to L.S. Rathore, Director General of Meteorology and M. Mohaptra, Director, Cyclone Warning Division for their help. Thanks are also due to M.K. De and Kiran Pawar for secretarial assistance.

Further Readings

BMTPC (2015). http://www.bmtpc.org/topics.aspx?mid=56&Mid1=334

Covello, V. T., & Allen, F. H. (1988). *Seven cardinal rules of risk communication.* (Pamphlet). Washington: U.S. Environmental Protection Agency.

Gray, P. C. R., Stern, R. M., & Biocca, M. (Eds.). (1998). *Communicating about risks to environment and health in Europe*. Dordrecht, Netherlands: Kluwer Academic Publishers.

Hoppner, C., Buchecker, M., & Brundi, M. (2010). Risk communication and natural hazards. CapHaz_Net WP5 Report, Swiss Federal Research Institute WSL.

IMD 2015 http://www.imd.gov.in/Welcome%20To%20IMD/Welcome.php, http://202.54.31.51/bias/gfsproducts.php

Kasperson, R. E., & Stallen, P. J. M. (Eds.). (1991). *The evolution of attempts in communicating risks to the public: International perspectives*. Dordrecht, The Netherlands: Kluwer.

Kollmuss, A., & Agyeman, J. (2002). Mind the gap: Why people act environmentally and what are the barriers to pro-environmental behaviour. *Environmental Education Research, 8*(3), 239–260.

Lindell, M. K., & Perry, R. W. (2004). *Communicating environmental risk in multi ethnic communities*. Thousand Oaks, CA: Sage Publications.

Renn, O. (2005). Risk governance: Towards an integrative approach. White paper no.1 Geneva: International Risk Governance Council, 156.

Renn, O. (2006). Participatory processes for designing environmental policies. *Land Use Policy, 23*, 34–43.
Seeger, M. W. (2006). Best practices in crisis communication: An expert panel process. *Journal of Applied Communication Research, 34*(3), 232–244.
Tyagi, A., Bandyopadhyay, B. K., & Mohapatra, M. (2010). Monitoring and prediction of cyclonic disturbances over north Indian Ocean by regional specialized meteorological centre, New Delhi: Problems and prospective. In Y. Charabi & S. Al-Hatrushi (Eds.), *Indian Ocean tropical cyclones and climate change* (pp. 93–104). Dordrecht, The Netherlands: Springer.
World' s 35 Deadliest Tropical Cyclones: http://www.wunderground.com/hurricane/deadlyworld.asp

The Challenges of Earthquake Risk Communication to Public

R.K. Chadha

Abstract Earthquake is a natural phenomenon, which, through human intervention, acquires dimensions detrimental to the very existence of the mankind. The common hazards associated with this phenomenon, namely ground failures, subsidence, liquefaction, lateral spreading, landslides or tsunami become life threatening if not addressed scientifically. A lot of progress has been made to understand earthquake source and processes, but predicting this phenomenon to safeguard the lives of people has remained a distant dream. Any claims of successful predictions, whatsoever, have fallen short of scientific scrutiny. The advancement of knowledge in the field of earthquake engineering and its application in the real world has helped in mitigating earthquake hazards to a great extent, but still people die in large numbers when moderate to large earthquakes strike. It is observed that in the countries adhering to strict practices of earthquake-engineered structures the causalities are much less in numbers compared to others—for example, magnitude (M) 7.9 Sichuan earthquake in China in 2008 killed more than 80,000 people in comparison to M 8.0 Hokkaido earthquake in Japan in 2003 where there was no loss of life in spite of the higher magnitude. There are several reasons for such a high number of deaths in the developing countries. One of the reasons, which is emerging, is the lack of proper awareness and communication of earthquake risk to the general public. Because earthquakes are low-probability high-risk phenomena, there is greater challenge to effectively connect with people to reduce the impact. This chapter presents earthquake-hazard scenario in the country and the challenges to effectively communicate earthquake risk to public.

Keywords Natural hazards · Low-probability high-risk phenomenon · Earthquake prediction · Earthquake communication · Natural hazard communication · Seismic zoning map of India · Seismic hazard

---

Views expressed in the chapter are the author's own and not necessarily shared by the editors.

R.K. Chadha (✉)
Council of Scientific and Industrial Research (CSIR),
National Geophysical Research Institute (NGRI), Hyderabad 500007, India
e-mail: chadha@ngri.res.in

P. Bagla and V.V. Binoy (eds.), *Bridging the Communication Gap in Science and Technology*, DOI 10.1007/978-981-10-1025-5_7

Introduction

The communication of earthquake risk to public is fraught with great danger and should be handled carefully. Unlike hydro-meteorological phenomena like cyclones, storms, floods and weather, which are open to direct observations, and hence, can be forecasted, earthquakes occur few kilometres below the earth surface and are not open to direct observations of what actually happens prior to or during the occurrence of an earthquake. Whatever the knowledge we have gained today about the earthquake phenomenon is through inferences drawn from the recordings of the seismic waves at different locations or laboratory experiments on rock failure mechanism. Therefore, predicting earthquakes is not possible with the current state of knowledge. Now and then one finds statements in newspapers claiming successful prediction of an earthquake after it has occurred. Unfortunately, these claims are made by self-seekers claiming expertise in the fields ranging from astrology to cosmology to weather forecast, including few even from the scientific community, to gain publicity. None of these claims withstand scientific scrutiny, as they are not based on sound scientific principles that are understood as of now. Hence, one has to be very careful while conveying earthquake risks to the public with the present state of knowledge about the phenomenon. Scientists who are involved in seismology research have to come forward to spread awareness about the earthquakes and convey the information on earthquake risks in the language suited to the type of audience they are addressing. While addressing the general public they have to be extra cautious to convey the scientific information in a simple language, understandable by a layman. Failing to do so will lead to repercussions beyond imagination. The incident associated with L'Aquila earthquake in 2009 in Italy is an eye-opener to the entire scientific community about the seriousness of the issue of communication with pubic.

In October 2012 six scientists of the Italian National Commission for the Forecast and Prevention of Major Risks were convicted and sentenced to six-year imprisonment on charges of multiple manslaughter for downplaying the likelihood of a major earthquake six days before its occurrence in April 2009 in L'Aquila, Italy. This verdict by an Italian court shook the scientific community world over and created a fear psychosis among the earth scientists of facing legal action over statements that are inherently uncertain. It is a strange verdict in that it is inexplicable how one could downplay occurrence of a major earthquake without even understanding the phenomenon. This is definitely a case of either miscommunication on the part of the scientists or misunderstanding on the part of the judiciary. The scientists were accused of giving inexact, incomplete and contradictory information about the danger of tremors that occurred prior to the main quake on 6 April 2009 and claimed 309 lives. Although this stray incident condemned by the entire earth science community was seen as an error in judgement, it brought out the importance of spreading awareness of the earthquake risks in the right perspective and its limitations in the cross section of society through sound and innovative communication channels.

Earthquakes are low-probability high-risk phenomena. Since predicting them in terms of its location, time and magnitude is beyond the current competency of science, the earthquake hazard is described in terms of probabilities. Although several studies are being carried out to identify regions of future earthquakes or understanding the mechanism and earthquake precursors, these studies are still in the realm of research. On a global scale earthquake research since mid-1960s has brought considerable insight into the patterns of earthquake occurrence and has particularly provided fairly good idea of the occurrence of major and great earthquakes. For example, tectonic plate boundaries exemplified by subduction zones and continent-to-continent collision regions, with some exceptions, are found to be the locales of earthquakes with $M > 8.0$. This chapter presents the earthquake-hazard scenario in the country and suggests ideas to minimize the impact through various mechanisms including proper communication of the earthquake risk to the public.

Earthquake Hazards in India

Seismic zoning map of India (BIS 2002) broadly shows seismic hazard in the country. The map is prepared based on the earthquakes that occurred in the recent past and in historical times (http://www.bmtpc.org/DataFiles/CMS/file/map%20of%20india/eq-india.pdf). According to the map, most hazardous regions lie in the Western Himalayas, north-east India and Kutch region of Gujarat in which great damaging earthquakes occurred in the 20th century—for example, $M > 8.0$ in Shillong (1897), Kangra, Himachal Pradesh (1905), Bihar–Nepal Border (1934) and Arunachal Pradesh (1950) and M 7.7 in Bhuj (2001). The 2600-kilometre stretch of the Himalayas from west to east and the Burmese Arc are some of the most seismically active regions of the world with earthquakes of magnitude between 4.0 and 7.0 occurring frequently (Bilham et al. 2001).

During an earthquake maximum damage to buildings occurs close to the epicentre. However, from the study of the earthquakes of recent past, it was observed that buildings could also collapse at distances exceeding 300 kilometres from the epicentre of the earthquake. For example, during the Bhuj earthquake in 2001, buildings collapsed in Ahmedabad and in Surat at distances of about 350–400 kilometres from the epicentre of the earthquake in Bhuj. Similar observations were made earlier during the 1985 M 8 earthquake in Mexico. The city of Mexico suffered very heavy damage to multistoried buildings located more than 350 kilometres from the epicentre of the earthquake in the Pacific Ocean. The reason for severe damage at such far-field regions was found to be the presence of soft sediments on which these buildings were located. The soft and unconsolidated sediments amplify seismic waves inducing stronger shaking leading to collapse of buildings.

Observations from the studies of the past earthquakes and our own experiences, the most vulnerable regions identified in India, if a major earthquake occurs in the Himalayas, are in the Indo-Gangetic Plains. The Indo-Gangetic Plain is a foredeep basin that runs parallel to the trend of the Himalayas and is filled with sediments varying in thickness from 500 metres to 6 kilometres. In addition to the presence of such a thick layer of sediments that can amplify seismic waves, there is another danger of liquefaction of soil if it is saturated with water (Srinagesh et al. 2011). Liquefaction happens when water-saturated soil is subjected to very strong shaking during an earthquake. The soil which looses cohesion flows like mud and affects the foundations of the buildings leading to collapse. This phenomenon has been observed worldwide during strong earthquakes, and in India during the $M > 8.0$ earthquake at Bihar–Nepal border in 1934 and the 2001 Bhuj earthquake in Gujarat. Research has shown that seismic waves can get amplified several times in certain regions in the Indo-Gangetic Plain, if a major earthquake occurs in the vicinity of the Himalayas. Hence, the major cities located on the Indo-Gangetic plains, such as Chandigarh, Dehradun, Delhi, Lucknow, Kanpur, Allahabad and others are exposed to severe earthquake hazard in case of a major earthquake in the Himalayas. Similarly, city of Guwahati in north-east India that lies in the plains of Brahamaputra valley filled with thick sediments is equally prone to severe earthquake hazard. The density of population in these major cities makes it a worst-case scenario.

Due to rapid urbanisation in India, several major cities even in seismic zones III and II are under threat from moderate-size earthquakes. For example, in 1993 an earthquake of M 6.4 in Killari, Maharashtra killed more than 10,000 people, the 1997 Jabalpur earthquake claimed more than 750 lives and the 2001 Bhuj earthquake killed more than 30,000 people. The reasons for so many lives being lost due to even moderate earthquakes are the poor construction practices, population density and lack of awareness about the impact of earthquakes in low seismic zones. The earthquake risk in India, thus, is real particularly to major cities, if sound practices of house construction are not followed taking into account the seismic codes for the region (BIS 2002).

In the year 2004 a great earthquake of M 9.3 occurred off the coast of Sumatra. This earthquake ruptured a fault length of about 1200 kilometres from Sumatra to northern Andaman Islands and generated a huge tsunami in the Indian Ocean which claimed more than 280,000 lives in several countries. In India, more than 16,000 people lost lives in Andaman and Nicobar Islands, eastern coast of India from Odisha to Tamil Nadu and parts of Kerala. Tsunami is generated when an earthquake occurs below the ocean displacing the ocean floor in a vertical direction creating a tsunami wave which then propagates in the open ocean and travels thousands of kilometres affecting countries in far-off places. Extensive research was conducted on tsunami, and it was found that there are two earthquake sources in the Indian Ocean where earthquakes along thrust faulting can generate giant tsunamis. These sources are along Andaman–Nicobar–Sumatra region in the Bay of Bengal and the Makran coast south of Karachi in the Arabian Sea. Earlier in 1945 a major earthquake along Makran coast generated a tsunami which reached the shores up to Goa along the western coast. Several thousand scenarios have been created using numerical modelling to predict tsunami run-up heights and inundation along the Indian coast in case of a major tsunami in either of these two tsunamigenic sources in the Indian Ocean.

As a response to 2004 tsunami in the Indian Ocean a Tsunami Warning Centre was created in 2007, which can now provide information on tsunami at least 90–120 min before tsunami hits the Indian coast. This warning centre that is located at Indian National Centre for Ocean Information Services (INCOIS) at Hyderabad has demonstrated its effectiveness during the past few years when major earthquakes occurred in the Indian Ocean by providing effective and accurate warnings.

Earthquake Risks

Earthquake risk is commonly misunderstood with earthquake hazard, even though they are fundamentally different. Qualitatively, earthquake risk can be defined as:

$$\text{Seismic Risk} = \text{Seismic Hazard} \times \text{Vulnerability}$$

where 'Seismic Hazard' is a physical phenomenon, namely strong ground shaking, fault rupture, landslide, ground subsidence or failure, soil liquefaction or tsunami caused by an earthquake, and 'Vulnerability' is the potential for damage to the built-in environment which is the primary cause where people get killed. Thus, earthquake risk becomes zero if a great earthquake occurs in some remote corner of the world with no built-in environment and hence, no vulnerability. Seismic zoning map of India (2001) provides information on earthquake hazard in the country in a broader sense but gives no clue on the earthquake risk.

During an earthquake elastic strain that has gradually accumulated across a fault is suddenly released in the process of elastic rebound. The stored elastic energy along the fault drives the movement that generates dynamic strong ground motions lasting for few seconds to few minutes, constituting a primary seismic hazard. The elastic rebound also generates static deformation of the ground. In the epicentral region, the damage is mostly due to static deformation of the ground and strong motions but in the far field, the damage is only due to amplification of strong ground motions in the presence of loose or unconsolidated sediments. Peak ground acceleration (PGA) is the most commonly used parameter to describe strong ground motions representing measure of ground shaking.

Two approaches are used for seismic hazard analysis, namely deterministic and probabilistic. The fundamental difference between these two analyses is that deterministic analyses do not consider the probability associated with the earthquake hazard, whereas probabilistic analyses incorporate the hazard probability. The deterministic approach assumes a seismic fault that can affect a structure at a known distance. An earthquake event of a specified magnitude, associated with an estimated return period, is assumed to occur on this faults and this event is used as the basis of the loss analysis. However, it does not provide a gauge of the likelihood of this loss occurring, nor does it consider the combined effects of multiple faults that may affect the site. As a consequence, this analysis provides an inconsistent measure of earthquake risk when evaluating areas with different levels of earthquake activity. A probabilistic analysis accounts for the full range of possible earthquakes, their location and frequency of occurrence and the size and the propagation of the earthquake motion from the rupture zone to the site of interest. Uncertainty in each of these elements and in the damageability of the buildings is taken into account. This provides a more complete and 'realistic' evaluation of the potential earthquake losses.

There are different roles performed by seismologists and geologists on one hand and civil and structural engineers on the other to provide safer environment for human dwellings. Seismologists and geologists perform seismic hazard analysis and provide values of PGA, peak ground velocity (PGV) and the response spectrum to be utilised by engineers. Further, engineers carry out geotechnical investigations to estimate shear strength of the soil, engineering bedrock and liquefaction potential of a site. Based on all these analyses seismic-design criteria are evolved for safer structures. Built-in environment is also studied with respect to the vulnerability of the existing structures to suggest measures like retrofitting to reduce earthquake risk.

Communicating Earthquake Risk in India

Communicating earthquake risk to common people is the greatest challenge an earth scientist faces—especially, in India where there is a vast urban–rural divide it is difficult for people to appreciate the limitations involved in earthquake prediction. Very often, seismologists are faced with a common question as to whether they can predict an earthquake. The moment the answer is given as *no*, the interest of the people wanes. The greatest challenge is, how one should convey the uncertainties to the public. It is a great challenge to keep the attention of people for a long time on such issues because earthquakes are not very frequent and people's memory is short. After every damaging earthquake there is a short-lived interest in the phenomenon both from the government and from the public. We should devise ways to keep people's interest intact by engaging them frequently with the support from the government or private stakeholders. It is important to convince the people that earthquake prediction is not a possibility, but one can still live safely with them. The secret is to construct safe structures with the existing state of knowledge in earthquake engineering. Examples should be quoted from countries such as Japan and USA where people feel safe when they are indoors during an earthquake unlike India where we advise them to come out of their houses.

There have been a few sporadic attempts to communicate earthquake risks to the public through ministries and National Disaster Management Authority (NDMA), but their efforts have not shown desired results for various reasons. There is an urgent need to develop a long-term strategy on this issue by encouraging formal or informal scientists–public interactions. Some of the points to be pondered for effective communication of earthquake risks in the country are as follows:

- The stress should be on *earthquake preparedness* rather than prediction of time of the impending earthquake because the current competency of science does not allow it.
- One major barrier to effective risk communication is the lack of interaction between professionals and community residents, and public officials and community groups. Most of the research done by scientists and engineers is often geared for the professional groups to advance knowledge and improve engineering practices rather than to motivate public concern or reshape public policy.
- At present there is a total disconnect between the existing scientific information on earthquake hazards and risks and the information needs of the public and stakeholder groups.
- For effective communication of the risks, information should be packaged in ways that make the information accessible and immediately usable by different audiences.
- Different layers have to be developed to disseminate risk information according to the requirements. A single-model approach to communicate with all stakeholders stands to fail. For example, legislators, policy makers, private sector and the general public differ in their information needs.
- Effective use of multiple media to organise press conferences, radio and television programmes, public announcements, newspaper and television features, targeted awareness programmes and even public meetings.
- Organising scientists–students meets at school and college levels to bring awareness on earthquake hazard and risk aspects. The knowledge imparted to schoolchildren at a young age will develop into a culture and will bring about a generational change to look at such hazards more scientifically.
- Involvement of nongovernmental organisations to develop a constituency of people who can continuously promote the efforts to minimise earthquake risks in vulnerable regions.
- Involvement of reputed scientists who can clearly communicate scientific information in a simple language to laymen about the earthquake risks in the country. This will increase the credibility of information and confidence among the people.
- Steps should be taken to discourage claims of earthquake prediction after the events by self-seekers, which creates confusion in the minds of the people.

Why Now?

Data over the past 110 years indicate that we may be passing through a 15-year cycle of a seismic activation period which began in 2004 with an M 9.3 earthquake in Sumatra.

Although there is no change in the average numbers of earthquakes in the magnitude range between M 7.5 and M 8.4, the earthquakes exceeding $M \geq 8.5$ seem to cluster in time (Michael 2011). The first cluster comprising six earthquakes is observed during 1905–1920 and the second from 1950 to 1965 with seven earthquakes. The third clustering seems to have started from 2004 M 9.3 Sumatra earthquake with six earthquakes having already occurred in this cycle, the last being the 2012 M 8.6 earthquake in the Indian Ocean. There have been several reports of increased seismic activity worldwide at local levels after the Sumatra earthquake in 2004. Even in India, increase in seismic activity has been observed. The question being debated now is about the possibility of great earthquakes triggering seismic activity globally. If the 15-year cycle were true, then there is still a three-year period (2016–2019) left in which an M 8.5 earthquake could occur. If this happens, the most likely places would be either in subduction zones or the continent-to-continent collision boundary where earthquakes of such magnitude occur. The Himalayas present one such continent-to-continent collision boundary where earthquakes of $M > 8.0$ have occurred in the past, and hence, it can be a candidate for a future one too. However, again, this observation is still a hypothesis and has to stand the statistical rigour, before definite conclusions are drawn.

Communication of this scenario is again a big challenge. This kind of information should only be confined to the people who are responsible for making policy decisions on disaster preparedness, should such an eventuality occur. We should avoid situations like that of L'Aquila earthquake which is an example of horribly bad risk communication and not of any technical mistake in earthquake observations.

Lastly, there is every possibility that no large event would occur during the period 2016–2019 suggested in the preceding paragraph, but earthquake preparedness is still a worthwhile exercise, which should be continued as efforts in this direction will not go waste.

Recommendations to Bolster Science Communication

(1) Science communication can be very effective if made simple and interesting. It should pursue truth and accept limitations. In our country where the literacy percentage is still not very high, science is considered something very complicated and beyond the understanding of the common man. Science communication need not be in formal sense only. The following case study proves the point. During the 2004 tsunami, several people perished in Vishakapatnam and other places when they went to the seashore to see the unusual receding of seawater. Lori Dengler of Humboldt State University (*New Scientist* Magazine, 15 January 2005) said after the 2004 tsunami, "Even without a warning system, even in places where they didn't feel the earthquake, if people had simply understood that when you see the water go down, when you hear a rumble from the coast, you don't go down to investigate, you grab your babies

and run for your life, many lives would have been saved." On the other hand, in Sri Lanka, Victor Desosa saved the village of Galbokka because he knew what to do when the water receded. Death in his village was limited to only one inhabitant, whereas casualties in the nearby villages were 70–90%.

(2) It is very essential to build scientific temper in the country by encouraging rational thinking from primary education levels. Children at the primary school level and youth at higher and college level should be targeted. It will surely take time, but such a generational change alone can be permanent.

(3) Cases of success stories in effective science communication in the country should be replicated in other parts. There are several isolated cases where excellent work has been done in science communication, which should serve as good examples.

(4) Accurate media reporting on science matters needs to be strengthened greatly in the country. We need to increase the number of good professional science writers and communicators to report complex matters in simple language. Media is the most important vehicle of communication on which scientists can reach greater audience.

(5) Information is distorted less when it reaches the target directly from the source. Hence, direct interactions between the scientists and the general public should be encouraged by utilising different platforms.

Acknowledgements The author gratefully acknowledges and thanks all the researchers whose work has been referred to during the preparation of this chapter. The author wishes to place on record his deep sense of gratitude to V.S. Ramamurthy, Director, National Institute of Advanced Studies (NIAS), Bengaluru for providing this opportunity which is a very different experience. The author thanks Director, National Geophysical Research Institute, Hyderabad for permitting him to communicate this chapter.

References

Bilham, R., Gaur, V. K., & Molnar, P. (2001). Himalayan seismic hazard. *Science, 293*, 1442–1444.

BIS. (2002). *IS 1893–2002 (part 1): Indian standard criteria for earthquake resistant design of structures, Part 1—general provisions and buildings*. New Delhi: Bureau of Indian Standards.

Michael, A. J. (2011). Random variability explains apparent global clustering of large earthquakes. *Geophysical Research Letters, 38*, L21301.

Srinagesh, D., Singh, S. K., Chadha, R. K., Paul, A., Suresh, G., Ordaz, M., et al. (2011). Amplification of the seismic waves in the central Indo-Gangetic basin, India. *Bulletin of the Seismological Society of America, 101*, 2231–2242.

The Queer Case of Communicating Risks Associated with Use of Mobile Phones and Neighborhood Mobile Towers: Are People Contracting More Brain Cancers?

K.S. Parthasarathy

Abstract The phenomenal growth in mobile phone services along with unbridled mushrooming of base stations and antennae caused health concerns across the whole country. There were concerns that cell phone radiation exposure to brain may induce cancer in the tissue. Invariably, in public perception, radiation and cancer are closely linked. The fact that cell tower/phone radiation has a million times lower energy than X-rays and gamma rays that can make structural changes in cells in the human body was not appreciated by the public. Government's prescription of radiation standards that are ten times lower than the level recommended by the International Commission on Non-Ionizing Radiation Protection led to more concerns because some vendors who sold radio-protective accessories demanded still lower values. Only specialists could appreciate the methods of arriving at conservative recommendations in the face of conflicting views. However, seemingly different views among specialists fanned the fire. The spread of scary information through Internet and television interviews was not helpful. Often people sought and obtained information from sources other than responsible agencies such as the World Health Organization (WHO). Government agencies attempted to clear the air, but those efforts were drops in the ocean. Few commercially interested individuals with academic credentials spread disinformation. They misinterpreted radiation standards, organised seminars, published blogs and newsletters and offered interviews always highlighting one-sided opinions. Risk communication in this context must include collecting accurate scientific data and conclusions from agencies such as the WHO and interpreting them transparently by including all stakeholders as partners in the dialogue. Risk communicators should declare conflict of interest, if any. They must keep themselves up to date with all developments.

Views expressed in the chapter are the author's own and not necessarily shared by the editors.

K.S. Parthasarathy (✉)
Department of Atomic Energy, Atomic Energy Regulatory Board, Mumbai, India
e-mail: ksparth@yahoo.co.uk

P. Bagla and V.V. Binoy (eds.), *Bridging the Communication Gap in Science and Technology*, DOI 10.1007/978-981-10-1025-5_8

Keywords Risk communication · Radiation · Mobile phone · International Commission on Non-Ionizing Radiation Protection (ICNIRP) · Radio-protective accessories · Radiation standards · Cell tower radiation

Introduction

As of May 2013, India had 921 million telephone connections including 891 million wireless telephone connections. Seventeen companies provide teleservices. The phenomenal growth in mobile phone services led to unbridled mushrooming of base stations and antennae. This was annoying to the public. Telephone companies vied with each other to secure locations for towers on the top of buildings in crowded urban areas in metropolitan cities.

Invariably, in public perception, radiation and cancer are closely linked. There were concerns that cell phone radiation exposure to brain may induce cancer in the tissue. These concerns spread widely primarily because it was a topic of discussion in popular television shows such as Larry King Live, an American talk show. Often, these shows became sensational; they were seldom unbiased.

Cell tower radiation has a million times lower energy than X-rays and gamma rays that can make structural changes in cells in the human body. Human body consists of cells; cells form tissues; several tissues join together to form organs. When energetic radiations such as X-rays or gamma rays interact with body, they are able to make structural changes in the cell. Large part of radiation impinging on the body may go without any interaction. However, some part of the radiation may be absorbed by the cells. It may kill a few cells. Radiation energy may partly damage few of them. Partly damaged cells may suffer loss of the precisely set multiplication and growth-control mechanisms. These cells may become anarchic with no rules at all or rules of their own! These damaged cells multiply rapidly and proliferate uncontrollably to cause cancer. Cell tower/phone radiations are non-ionizing and are incapable of starting such disruptive action in the cell because they lack energy.

Cell Tower/Phone Radiation Standards

In 2008, Government of India adopted the guidelines developed by the International Commission on Non-Ionizing Radiation Protection (ICNIRP) for electromagnetic radiation from mobile towers. The values chosen for the permissible power density are 4.5 W/m^2 for 900 MHz and 9 W/m^2 for 1800 MHz transmissions.

Another important restriction is on the specific absorption rate (SAR) from mobile sets. From 1 September 2013, government permitted the manufacture of, or import into India, only mobile handsets with revised SAR value of 1.6 W/kg averaged over one gram of human tissue.

Earlier based on media reports and public concerns, the government had set up an Inter-Ministerial Committee (IMC) of specialists on 24 August 2010. The committee examined the environment- and health-related concerns and indicated that most of the laboratory studies were unable to find a direct link between exposure to radio frequency (RF) radiation and health, and that the scientific studies as yet have not been able to confirm a cause-and-effect relationship between RF radiation and health.

Based on the recommendations of the IMC the government accepted the electromagnetic field (EMF) exposure limits of one-tenth of the then existing prescribed limits and made the new limits applicable from 1 September 2012.

The cell phone/tower radiation standards adopted by India are among the strictest. Cell tower radiation levels measured at thousands of locations indicated that the standards could be effectively implemented. The standard accepted by India has a safety factor of 500.

Availability of Information

Internet carries all types of information. However, we must trust only the information provided by responsible and authentic agencies such as the World Health Organization (WHO), the US National Cancer Institute (NCI), American Cancer Society (ACS) and the UK Health Protection Agency (HPA). Far too many sites go for sensational reporting.

The erstwhile National Radiological Protection Board of UK published an excellent review of the health effects of RF EMFs. The UK HPA updated the information in April 2012; its report RCE-20 is freely available from HPA's website (http://www.hpa.org.uk/webc/HPAwebFile/HPAweb_C/1317133827077; accessed on 1 December 2015).

This 333-page report describes the interaction mechanisms of electromagnetic radiation and dosimetry, and reviews cellular, animal and human studies including studies of neurocognitive effects as well as cancer and non-cancer effects.

The report concludes:

In summary, although a substantial amount of research has been conducted in this area, there is no convincing evidence that RF field exposure below guideline levels causes health effects in adults or children.

Reasons for the Spread of Health Concerns

In spite of extensive studies and reassuring conclusions, the health concerns on cell tower/phone radiation grew disproportionately. The reasons for this extraordinary development need more detailed study.

Few companies selling protective accessories against electromagnetic radiation set up shops. Seemingly different views of specialists on the effects of radiation and various uncertainties became a serious issue for public. Vendors of protective accessories made virtually a carpet-bombing effort to exaggerate the effects. They used cleverly designed programmes of disinformation by mixing myths with technical jargon. They deliberately propagated wrong concepts, misinterpreted internationally accepted safety standards and accused telecommunications companies for making profits without caring for the health of the public.

Some of these vendors with impressive academic credentials organised seminars, wrote blogs and issued newsletters. Each of these was biased; they opposed internationally recognised safety standards as ineffective. They exploited the fact that some countries used lower values in their guidelines. They banked on reports such as the Bioinitiative Report (BIR) 2007 and BIR 2012 (http://www.bioinitiative.org/) which advocated this group's own extra-conservative guidelines. Very few individuals noticed that BIRs were nurtured and funded by anti-radiation advocacy groups with commercial motives.

The vendors invariably visited places where 'cancer clusters' reportedly existed. The sound bites excited raw emotions; reporters of most leading newspapers publicised them without cross-checking with real experts or authentic references. One leading national daily started a series of articles in a campaign mode termed *Towering Trouble*.

In the ensuing conflict, there were no umpires! Cellular companies had no credibility as they were accused of conflict of interest. Many non-governmental organisations (NGOs) joined the group—some of them genuinely believing that their cause was noble and worthwhile! Vendors of protective accessories masquerading as experts visited metropolitan cities to carry out well-organised programmes. In an attempt to mitigate this disinformation, Department of Telecommunications published some booklets and issued a few advertisements; nevertheless, they were simply a drop in the ocean.

Cell Tower/Phone Radiation and Cancer and Other Effects

Many agencies have carried out epidemiological studies in this important area. The INTERPHONE study supported by WHO is the largest case—control study of mobile phones and brain tumours conducted to date, including the largest numbers of users with at least 10 years of exposure to cell phone radiation and the greatest cumulative hours of use of any study. Thirteen countries including UK, Sweden, France and Germany collaborated for this study.

In May 2010, The INTERPHONE study concluded that overall, no increase in the risk of brain tumours was observed with the use of mobile phones.

"There were suggestions of an increased risk of glioma at the highest exposure levels, but biases and error prevent a causal interpretation. The possible effects of long-term heavy use of mobile phones require further investigation."–the authors added.

Excellent reviews of the studies carried out in various countries and other scientific facts are available at the following websites:

(1) http://www.cancerresearchuk.org/cancer-info/healthyliving/cancercontroversies/mobilephones/mobile-phones-and-cancer.
(2) http://www.cancer.org/cancer/cancercauses/othercarcinogens/athome/cellular-phones.
(3) http://www.cancer.gov/cancertopics/factsheet/Risk/cellphones (accessed on 1 December 2015).

According to ACS, since cell phones are held near the head when in use, the main concern has been whether they might cause or contribute to tumours in this area. The possibility for the induction of malignant (cancerous) brain tumours such as gliomas, non-cancerous tumours of the brain such as meningiomas, non-cancerous tumours of the nerve connecting the brain to the ear (acoustic neuromas) and non-cancerous tumours of the salivary glands needs to be checked out.

In a Danish cohort study specialists followed up 40,000 phone subscribers who had subscriptions from 1982 to 1995.

ACS asserted:

"Cell phone use, even for more than 13 years, was not linked with an increased risk of brain tumors, salivary gland tumors, or cancer overall, nor was there a link with any brain tumor subtypes or with tumors in any location within the brain."

The NCI observed that "Some case—control studies in Sweden found statistically significant trends of increasing brain cancer risk for the total amount of cell phone use and the years of use among people who began using cell phones before age 20." Some specialists considered this Swedish study as outlier because that particular group alone was consistently reporting similar results.

NCI also noted that, another large case–control study in Sweden did not find an increased risk of brain cancer among people between ages 20 and 69 years.

Thus, a limited number of studies have shown some evidence of statistical association of cell phone use and brain tumour risks, but most studies have found no association.

NCI stated that the reasons for these discrepancies include recall bias, inaccurate reporting, morbidity and mortality among the subjects, participation bias, changing technology and methods of use.

The ACS has compiled the conclusions of the US Environmental Protection Agency (EPA), the National Toxicology Program (NTP), the Food and Drug Administration (FDA), Federal Communications Commission (FCC), Centers for Disease Control and Prevention (CDC), the National Institute of Environmental Health Sciences (NIEHS) and the NCI. All of them agreed that there is no scientific evidence to prove that wireless phone usage can lead to cancer or any other health problem.

Findings of International Agency for Research on Cancer

International Agency for Research on Cancer (IARC), which is a part of WHO, has classified RF fields as 'possibly carcinogenic to humans', based on limited evidence of a possible increase in risk for brain tumours among cell phone users, and inadequate evidence for other types of cancer.

The classification of cell phone/tower radiation as a possible carcinogen by IARC/WHO got undue publicity. Almost immediately WHO issued a press release (Fact Sheet 193) in June 2011, which reassured the worried public:

"A large number of studies have been performed over the last two decades to assess whether mobile phones pose a potential health risk. To date, no adverse health effects have been established as being caused by mobile phone use."

Coffee, Pickled Vegetables, Carbon Black (Carbon Paper) and so on Are also Class 2B Carcinogens

In a review titled, "Are Cell Phones a Possible Carcinogen? An Update on the IARC Report", Lorne Trottier noted that IARC maintains a list of 269 substances in the 2B category (substances that are possibly carcinogenic to humans), most of which are chemical compounds. A number of familiar items are also included in this list: coffee, pickled vegetables, carbon black (carbon paper), gasoline exhaust, talcum powder and nickel (coins).

Trottier's review (2 April 2012) can be accessed at:

http://www.sciencebasedmedicine.org/are-cell-phones-a-possible-carcinogen-an-update-on-the-iarc-report/ (accessed on 4 June 2014).

The IARC ruling was controversial. Trottier asserted that expert groups of most of the world's major public health organisations have taken the same position as the European Commission's Scientific Committee on Emerging and Newly Identified Health Risks (SCENIHR) stating:

It is concluded from three independent lines of evidence (epidemiological, animal and in vitro studies) that exposure to RF fields is unlikely to lead to an increase in cancer in humans.

As comments to Trottier's article, Prof. Kenneth Foster of University of Pennsylvania gave cogent reasons for the confusing results:

"To a large extent it is a consequence of too many exploratory studies in search for small effects (as opposed to hypothesis-driven studies designed to follow up previous findings, or standardised assays). Science is not very reliable when it comes to interpreting the results of fishing expeditions that find tiny effects that are just above the level of statistical significance. The bioeffects literature is filled with such findings."

Final Status

Finally, in September 2013, the WHO issued the following statement:

"While an increased risk of brain tumours from the use of mobile phones is not established, the increasing use of mobile phones and the lack of data for mobile phone use over time periods longer than 15 years warrant further research of mobile phone use and brain cancer risk. In particular, with the recent popularity of mobile phone use among younger people, and therefore a potentially longer lifetime of exposure, WHO has promoted further research on this group and is currently assessing the health impact of RF fields on all studied endpoints."

This truly reflects the current status. It will continue till the ongoing studies are completed over the next few years.

Cell Tower Radiation and Cancer Clusters

A newspaper reported that in a building in Mumbai four cases of cancer were linked to radiation from a mobile phone tower. Based on this, a vendor of protective accessories estimated the power level at the building to be about 0.1 W/m^2 and claimed that the tower was the cause of cancer in "several" people in span of 2–3 years!

To achieve a shock and awe effect among masses, the anti-radiation activists would present the power level of 1 W/m^2 as 1,000,000 $\mu W/m^2$!

Stories on cancer clusters and other diseases near cell phone towers are spread by persons having vested interests, in particular, selling screens, shields and so on against cell tower radiation. Newspapers unwittingly help them by publishing such unsubstantiated stories.

The WHO is aware of scary stories on increased cancer incidence near cell towers and noted thus in its Fact Sheet 304:

> "Media or anecdotal reports of cancer clusters around mobile phone base stations have heightened public concern. It should be noted that geographically, cancers are unevenly distributed among any population. Given the widespread presence of base stations in the environment, it is expected that possible cancer clusters will occur near base stations merely by chance. Moreover, the reported cancers in these clusters are often a collection of different types of cancer with no common characteristics and hence unlikely to have a common cause."

Hypersensitivity to Electromagnetic Radiation

Some people feel that they are hypersensitive to electromagnetic radiation. The topic became so important that the WHO held a workshop on electromagnetic hypersensitivity (EHS) in Prague during 25–27 October 2004.

At the very outset the working group proposed to replace the term EHS with "idiopathic environmental intolerance (IEI) with attribution to EMF" since EHS implies that a causal relationship has been established between the reported symptoms and EMF.

Thus, the workshop concluded:

IEI incorporates a number of disorders sharing similar non-specific medically unexplained symptoms that adversely affect people and cause disruptions in their occupational, social, and personal functioning. The majority of studies indicate that IEI individuals cannot detect EMF exposure any more accurately than non-IEI individuals. By and large well-controlled and conducted double-blind studies have shown that symptoms do not seem to be correlated with EMF exposure.

There are also some indications that these symptoms may be due to preexisting psychiatric conditions as well as stress reactions as a result of worrying about believed EMF health effects, rather than the EMF exposure itself. It was added that IEI should not be used as a medical diagnosis since there is presently no scientific basis to link IEI symptoms to EMF exposure.

Disinformation on Cell Tower/Phone Radiation: A Form of Radiation Terrorism

Individuals with vested interests exploited the uncertainties to achieve their own objectives. They spread disinformation as shown in the following examples.

Disinformation No. 1

They claim that the SAR limit for cell phones—a safety standard limiting power to 1.6 W/kg—is actually for six minutes per-day usage! Thus, they argue, "do not use phones for more than 18–20 min daily."

A cell phone kept near the ear will cause a small increase in temperature in regions close to the phone. Thermoregulatory mechanisms such as blood flow remove the heat, establishing equilibrium in about six minutes. Thereafter, there will not be any increase in temperature. Within this six-minute interval of time the body's defence mechanism restores the thermal equilibrium stabilising the body temperature. Therefore, it is patently absurd to say that using a phone longer multiplies the risk.

Disinformation No. 2

Women run the risk of health problems such as "hormonal imbalances, miscarriages and breast and ovarian cancer, as they tend to spend more time at home and are exposed continuously to these radiations."

There is no scientific evidence to support this statement. No responsible agency such as the WHO drew any such conclusions. Such disinformation is spread by companies selling screens, shields, wall papers and so on for the purported need of protection against radiation. A background search of those who make such statements would expose them as parties with vested interests.

Disinformation No. 3

Cell tower radiation will cause "multiple resonances, localised heating resulting in boils, drying up the fluids around the eyes, brain, joints, heart, abdomen and so on, leg/foot pain, muscle and joint pain." With cell towers erected all over the city, "Mumbai is like a microwave oven."

Exposure to cell tower radiation causes a very tiny increase in body temperature. At the radiation levels recommended by the ICNIRP, the standard-setting body, the increase will be about 0.1 °C; The Department of Telecommunications accepted one-tenth of the ICNIRP values as the safety standard in India. At this level, the increase in temperature would be no more than 0.01 °C! Can such trivial increase in temperature cause the scary effects such as boils and so on?

These 'activists' made presentations containing such misinformation and misinterpretations at medical colleges and other academic institutions in metropolitan cities and small towns. Unfortunately, media gave free publicity to the impressive sound bites, which carried such disinformation.

Report on Cell Tower Radiation

Among the inputs submitted to the Department of Telecommunications was a document titled, "Report on Cell Tower Radiation" authored by a professor from the Department of Electrical Engineering, IIT Bombay (hereafter we call it IIT report for the sake of convenience, although IIT never endorsed it as its report).

Another report also seems to have influenced the decision makers, the "Report on possible impacts of communication towers on wild life including bird and bees" prepared by a committee set up by the Ministry of Environment and Forests. This report is a clone of the IIT report.

"The Birds and Bees" report is, again, one-sided. For instance, while the report refers to the studies on honeybees carried out in India, the committee is oblivious of the following conclusive statement by the US Department of Agriculture, which has been studying the colony collapse disorder (CCD) of honeybees very systematically for several years:

Despite a great deal of attention having been paid to the idea, neither cell phones nor cell phone towers have been shown to have any connection to CCD or poor honey bee health.

Originally, the idea was provoked by the media making a connection between CCD and a very small study done in Germany. However, that study looked at whether a particular type of base station for cordless phones could affect honeybee homing systems. However, despite all the attention that this study has received, it was found that the base station had nothing to do with CCD. Stefan Kimmel, the researcher who conducted the study and wrote an article, e-mailed a communication to *The Associated Press*, to say that there was "*no link between our tiny little study and the CCD phenomenon... Anything else said or written is a lie*."

The US Department of Agriculture report can be accessed at: http://www.ars.usda.gov/News/docs.htm?docid=15572 (accessed on 4 June 2014).

There were few reports on some Indian studies, which were criticised by experts for their infirmities. One of these was published verbatim in two predatory journals!

The IIT report listed symptoms and diseases allegedly caused by electromagnetic radiation. The author mined part of the scary data from 'papers' of Arthur Firstenberg, the founder director of the 'Cellular Phone Task Force' which is 'dedicated to halting the expansion of wireless technology because it cannot be made safe'. Firstenberg filed and lost many suits against the spread of wireless technology.

Wikipedia noted his claim that EMFs from his neighbour's cell phone were destroying his health and that he sued his neighbour seeking USD 530,000 in damages for refusing to turn off her cell phone and other electronic devices! Firstenberg is a symbol of the collective schizophrenia against RF radiation.

The IIT report cherry-picked many references to substantiate such claims. International agencies such as the WHO and national agencies have not accepted such preposterous claims.

This author investigated a cancer cluster story included in the IIT report and found it to be false.

The report uncritically accepted the BIR 2007, the booklet well known for its lack of balance, as mentioned earlier, and further discussed in the following sections.

BIR 2007: An Advocacy Document

The Committee on Man and Radiation (COMAR), a technical committee of the Engineering in Medicine and Biology Society (EMBS) of the Institute of Electrical and Electronics Engineers (IEEE) noted that BIR is an advocacy document. BIR itself conceded that it was written "to document the reasons why current public exposure standards for non-ionising electromagnetic radiation are no longer good enough to protect public health."

Fourteen individuals under the direction of a four-person organising committee wrote the BIR. "Most of its 21 sections are authored by single individual or (in few cases) pairs or trios of authors; the section 'Key Scientific Evidence and Public Health Policy Recommendations' was written by a pair of individuals and appears to reflect their views only." COMAR clarified in an article in *Health Physics Journal*, "There is no indication of how the members of the committee were chosen or how balance was provided in the group of contributors, a majority of whom have public records of criticism of existing exposure standards and guidelines."

COMAR added that after appearing on the Internet in 2007, the BIR has received much media attention; however, more recently, it has been severely criticised by health organisations and scientific groups such as EMF-NET, a coordinating committee of the European Commission's Sixth Framework Programme, the Netherlands' Health Council and Australian Centre for Radiofrequency Bio-effects Research (ACRBR).

These agencies slammed the BIR report thus: "… written in an alarmist and emotive language and the arguments have no scientific support from well-conducted EMF research." "There is a lack of balance in the report; no mention is made in fact of reports that do not concur with authors' statements and conclusions. It is not an objective and balanced reflection of the current state of scientific knowledge." As well as, "as it stands, it merely provides a set of views that are not consistent with the consensus of science."

Bioinitiative Report 2012

The BIR 2012 on standards for electromagnetic radiation is a perfect clone of that published in 2007. According to many responsible agencies it is biased and unscientific. BIR 2012 claimed that the evidence for risks to health from wireless technologies and EMFs has substantially increased since 2007. The studies alleged a link between cell phone radiation and brain tumours contrary to the findings of other responsible agencies.

David Carpenter and Cindy Sage, the editors of the report clarified that each author is responsible for his/her own chapter in BIR 2012; that is, the views are those of individual authors. It is a very unusual practice.

Conflict of Interest

"The great strength of the BIR (www.bioinitiative.org) is that it has been done independent of governments, existing bodies and industry professional societies that have clung to old standards", the editors claim, which is laughable.

A notable weakness of the report is that Cindy Sage, who authored five sections and coauthored one, herself owns SAGE *EMF Design*, a consultancy firm which declares, “Creating low-field lighting for interiors,” and “Remediation: What if your existing home has high EMF?” among its functions. BIR 2012 does not state conflict of interest, if any, of the authors.

Sage stated, “the Report has been written to document the reasons why current public exposure standards for non-ionising electromagnetic radiation are no longer good enough to protect public health”. With the die thus cast, no one can expect in BIR 2012 an objective analysis of the evidence, if any, on the bio-effects of electromagnetic radiation.

Critique of BIR

As mentioned earlier, responsible agencies roundly criticised the report. The European initiative EMF-NET noted that the “Summary for the public is written in an alarmist and emotive language and its arguments have no scientific support from well-conducted EMF research. There is a lack of balance in the report; no mention is made in fact of reports that do not concur with authors’ statements and conclusions.”

The COMAR technical information statement pointed out that BIR has many weaknesses and is a selective, rather than a comprehensive, review of the literature in various topical areas.

The Health Council of the Netherlands highlighted the fact that WHO’s and ICNIRP’s multidisciplinary weight-of-evidence method leads to a scientifically sound judgement that is as objective as possible.

The BIR report did not follow this procedure. The council asserted that “(The report) is not an objective and balanced reflection of the current state of scientific knowledge and does not provide any grounds for revising the current views as to the risks of exposure to electromagnetic fields.”

International guidelines limiting radiation to 4500 mW/m^2 have a safety factor of 50. Indian guidelines include at 450 mW/m^2 a further safety factor of 10. BIR proposes an additional factor of 900!

Although BIR is not based on sound science, cell tower radiation scaremongers selling protective shields and RF measuring instruments (complying with BIR 2007 recommendations) love to uphold BIR values. They can then scare the public further and make hay while the sun shines!

In fact, at least one such company selling protective solutions against electromagnetic radiation uses an instrument calibrated as per the standards followed by BIR. The permissible level of electromagnetic radiation recommended by BIR is orders of magnitude less than those recommended by responsible standard-setting agencies (http://ksparthasarathy.wordpress.com/2013/01/17/health-effects-of-cell-tower-radiation/ accessed on 4 June 2014).

Risk Communication

Risk communication is needed in many fields. In 1997, the Codex Alimentarius Commission (CAC) of the Food and Agricultural Organization adopted the following definition of risk communication: "an interactive exchange of information and opinions concerning risk and risk-related factors among risk assessors, risk managers, consumers and other interested parties."

According to the CAC, "Risk communication has also been described as all those integrated processes and procedures: (a) that involve and inform all interested parties within the risk analysis process; (b) that assist the development of transparent and credible decision-making processes; and (c) that can instill confidence in risk management decisions."

Risk communication process should involve two-way communication.

Broadly, the goals of risk communication will be almost similar in any field.

Food and Agriculture Organization of the United Nations (FAO) identified the following goals:

1. Promote awareness and understanding of the specific issues under consideration during the risk analysis process, among all participants.
2. Promote consistency and transparency in arriving at and implementing risk management decisions.
3. Provide a sound basis for understanding the risk management decisions proposed or implemented.
4. Improve the overall effectiveness and efficiency of the risk analysis process.
5. Contribute to the development and delivery of effective information and education programmes, when they are selected as risk management options.
6. Foster public trust and confidence in the safety of the food supply.
7. Strengthen the working relationships and mutual respect among all participants.
8. Promote the appropriate involvement of all interested parties in the risk communication process.
9. Exchange information on the knowledge, attitudes, values, practices and perceptions of interested parties concerning risks associated with food and related topics.

Cardinal Rules of Risk Communication

The seven cardinal rules of risk communication identified by the US EPA with appropriate modification are well suited for the task in any field.

These rules are:

(1) Accept and involve the public as a legitimate partner. (2) Plan carefully and evaluate your efforts. (3) Listen to the public for specific concerns. (4) Be honest,

frank and open. (5) Coordinate and collaborate with other credible sources. (6) Meet the needs of the media. (7) Speak clearly and with compassion.

This author became interested in addressing the myths and realities on the health effects of cell tower/phone radiation when a senior journalist of a leading daily approached him for his comments on the IIT report. It was two years after the Department of Telecommunications received the IIT report. Newspapers in Mumbai reproduced selected portions from the report frequently whenever the 'harmful' effects of cell tower/phone radiation were discussed. Reporters sincerely reproduced what they were told; they did not cross-check facts as they trusted the credentials of the specialists who purveyed the news.

As mentioned earlier, a leading national daily went overboard and carried a series with the title *Towering Trouble*. This author sent several fact sheets and reviews to the reporters covering the item. However, there was no change in the tone or tenor of the news stories. After having learnt that it was a waste of time to persuade such reporters to a different, more balanced point of view, this author contributed his own articles on the topic in the leading dailies published from metropolitan cities.

The online version of the article titled, "Myths about radiation risks from cell tower" in *The Economic Times* attracted maximum response from the readers. Author received 76 comments online and nearly 30 through email messages. He replied to almost all, including those who used coarse and undesirable language in their discourse. He cannot claim that he has convinced everyone to take a pragmatic approach. But the discussion helped.

In summary, a risk communication programme will have the following steps:

(1) Collect and distribute scientific facts and current status of the field from authentic sources.
(2) List out the concerns.
(3) Present the facts transparently, highlighting uncertainties.
(4) It must be a dialogue.
(5) Identify the conflicts of interest, if any.

Risk communication is a task in which one will never be happy. Whatever you do, you feel you have done only 50% of what you can!

Further Reading

http://www.dailyexcelsior.com/web1/magazine/13jan13/index.htm. Accessed July 14, 2015.

http://www.cancer.org/cancer/cancercauses/othercarcinogens/athome/cellular-phones. Accessed June 26, 2015.

http://www.sciencebasedmedicine.org/are-cell-phones-a-possible-carcinogen-an-update-on-the-iarc-report/. Accessed June 26, 2015.

http://www.ars.usda.gov/News/docs.htm?docid=15572. Accessed on June 4, 2014.

https://ksparthasarathy.wordpress.com/2013/06/03/cell-tower-radiation-and-cancer-clusters/. Accessed June 22, 2015.

www.fao.org/docrep/005/x1271e/X1271E03.htm#ch3. Accessed June 26, 2015.
www.hindu.com/sci-tech/science/biased-unscientific-report-on-electromagnetic-radiation/article4313241.ece. Accessed June 26, 2015.
www.thehindu.com/sci.tech/technology/a-classic-example-of-biased-and-unscientific/study/article3808249.ece. Accessed June 26, 2015.
www.wvdhhr.org/bphtraining/coursescdcynergy/content/activeinformation/resources/epa_seven_cardinal_rules.pdf. Accessed July 14, 2015.
www.cancer.gov/about-cancer/causes-prevention/risk/radiation/cell-phones-fact-sheet. Accessed June 26, 2015.

Atomic Energy: Reaching Out to the People for Perception Management

Swapnesh Kumar Malhotra

Abstract For long-term sustainable development of India nuclear power is an inevitable option. As of 2016 India has an installed nuclear capacity of 5780 MWe and has 5300 megawatts electric (MWe) plants under construction. It has an ambitious plan of attaining more than 60,000 MWe of nuclear power by 2032. Public acceptance is going to be a very crucial rate-determining step in realising this objective. There are perceptions about atomic energy among the general public, which are not based on scientific facts, and these need to be addressed through proper public outreach approach aptly supported by sound rehabilitation and resettlement policy and corporate social responsibility activity. Continuous communication with the stakeholders in a simple language, which is free of complicated scientific and technical jargon, is the basis of a successful public outreach strategy. This chapter elaborates the major public perceptions about atomic energy and the corresponding realities. It emphasises the need for communication and full-time communicators, and also the potential of social media for wider interactive communication with various sections of society, particularly the younger generation. It brings out the importance and role of media for eradicating the myths about atomic energy prevalent in the society. It briefly discusses the need for inclusive growth and seamless development of the neighbourhood areas around nuclear facilities. The chapter also discusses the recent initiatives taken by the Department of Atomic Energy, and finally, outlines the further steps to be taken to create a situation where rather than pushing nuclear energy to the society, a pull for the same from the society can be created.

Keywords Nuclear energy · DAE · Nuclear risk management · Anti-nuclear activities · Nuclear information centres · Communicating nuclear issues

Views expressed in the chapter are the author's own and not necessarily shared by the editors.

S.K. Malhotra (✉)
Department of Atomic Energy (DAE), Public Awareness Division, Mumbai, India
e-mail: swapneshmalhotra@gmail.com

P. Bagla and V.V. Binoy (eds.), *Bridging the Communication Gap in Science and Technology*, DOI 10.1007/978-981-10-1025-5_9

Introduction

During the past almost 120 years, since the discovery of ionising radiation (1895) and radioactivity (1896), atomic energy has established itself firmly all over the world. Its applications can be found in practically all parts of the globe and touches almost every aspect of human life. Today, it is helping the mankind in meeting the challenges of the ever-increasing demand for energy, food and water, and is assisting them in their industrial growth, mitigation of global warming, health-care and environmental issues. In spite of this immense contribution of nuclear energy to the overall societal growth, certain sections of the society are either unaware or are indifferent to these contributions. Some of them even oppose nuclear energy and paint it as a dangerous technology.

If we look back at the past, since the discovery of X-rays and radioactivity, we find that a large-scale 'radioactive quackery' was rampant, particularly in European countries and USA. During this period, numerous products were available in the markets claiming almost magical health effects. Public acceptance of all these products was almost cent per cent, and those who could afford used them. None of these applications were based on any scientific findings, clinical research or trials. The acceptance was based entirely on benefits (risks were not known) as perceived by the people and as advertised by the manufacturers. It was only during 1930s that horrifying results of blatant consumption of radioactive substances and over-exposure to radiation began coming to light. Since 1915, regulatory aspects, particularly in connection with X-rays, started taking shape. Although the craze for these magical radioactive potions died during the 1930s, radio active quackery is still continuing in some parts of the world.

In 1939 nuclear fission was discovered by Fermi and, as happens with almost all major scientific discoveries, fission was also 'usurped' by the human tendency to use it for harming others. As Second World War coincided with this period, rapid development of nuclear weapons took place, and the humanity witnessed the horrific destructive power of atomic energy in the form of atomic bombing of Hiroshima and Nagasaki. This resulted in a deep-rooted fear psychosis in the minds of common public, which persists even today.

Declaration of 'Atoms for Peace' programme by Eisenhower in 1953, subsequent establishment of International Atomic Energy Agency (IAEA) and, of course, marathon efforts by radiation technologists for developing a variety of applications of radiation for societal development slowly but steadily established atomic energy in a wider perspective. Today, these applications have become so much accepted by the society that the total business in these applications all over the world far exceeds the business of electricity generation by nuclear power. This is definitely helpful in acceptance of the nuclear power by society, although the ghosts of Hiroshima and Nagasaki continued to maintain a scare about this technology. The very frequent open-air and underground nuclear weapon tests by many countries are greatly contributing to the negative image of nuclear power. Against this backdrop, the mid-1960s, 1970s and 1980s saw deployment of large number of nuclear power plants (NPPs) particularly in USA, Canada and Europe. After the first major accident in a nuclear power station at Three Mile Island, there was a setback to nuclear energy, particularly in USA, although it continued to flourish in Europe with France as the leader.

When things were appearing alright, Chernobyl happened in 1986 and had a prolonged adverse effect on the global growth of nuclear power. During the first decade of 21st century nuclear power appeared to be making a come back and was being considered as an important technology for mitigation of global warming.

In 2007 the Intergovernmental Panel on Climate Change (established by the United Nations) in its 4th assessment report stated, "Nuclear (Power) has the largest and lowest-cost greenhouse gas (GHG) reduction potential in power generation". Once again when things were looking highly favourable for large expansion of nuclear power sector (even Germany had deferred its earlier decision to close all its NPPs), Fukushima happened and opened up fresh arguments against nuclear power. A BBC opinion poll on nuclear power conducted across many countries during July to September 2011 clearly showed that the dent nuclear power had suffered in its popularity immediately after the Fukushima accident (BBC 2011). In most of the countries hardly up to 20% people fully supported it. The only exceptions were UK, USA, China and Pakistan where about 40% people still wanted to fully continue with nuclear power. In India only about 23% people wanted to uninterruptedly continue with nuclear power—of course as it generally happens in India, a staggering 38% had no opinion at all. Unfortunately, around that time (September 2011), Department of Atomic Energy (DAE) was in the process of commissioning the 1000 megawatts electric (MWe) Unit-1 of the Kudankulam Nuclear Power Project (KKNPP) in Tamil Nadu and witnessed one of the worst and longest anti-nuclear agitation forcing all work to come to a standstill. It was only after DAE convinced the state government about the safety of the reactor (March 2012) and the historical judgement of the Honourable Supreme Court of India (May 2012) that the work there could be restarted. Here it would be worth quoting a paragraph from that judgement: *Nuclear power plant is being established not to negate right to life but to protect the right to life guaranteed under Article 21 of the Constitution. The petitioner's contention that the establishment of nuclear power plant at Kudankulam will make an inroad into the right to live guaranteed under Article 21 of the Constitution, therefore has no basis. On the other hand, it will only protect the right to life guaranteed under Article 21 of the Constitution for achieving a larger public interest and will also achieve the object and purpose of Atomic Energy Act.*

Major Public Perceptions and Realities About Nuclear Power

Although there might be hundreds of frequently asked questions about nuclear power, they can be grouped into the following major perceptions:

1. Atomic energy spreads nuclear weapons (memories of Hiroshima and Nagasaki).
2. Nuclear power is not safe and not good for environment (memories and exaggerations of the Three Mile Island, Chernobyl and Fukushima accidents).
3. Nuclear power is very costly.
4. Presently only India is pursuing it while others are giving it up.
5. Highly radioactive nuclear waste is a problem without solution.
6. Radiation from nuclear reactors is causing cancer to plant workers and the people in the neighbourhood, and would also lead to genetic disorders in future generations.
7. Uranium is an extremely dangerous material (some describe it as an evil spirit), so God has buried it deep in the Earth; but these officers from DAE are trying to take it out, and if they are allowed to do so, it is going to be disastrous.

None of these perceptions is based on scientific facts, and are mostly the consequence of the 'fear of unknown' and—many a times—imaginary exaggerations of figures and facts (NuclearConnect 2015). Briefly, the realities corresponding to these perceptions are as follows:

1. World's first nuclear power reactor came up in 1954 at Obninsk near Moscow—nine years after Hiroshima and Nagasaki. In fact, the first five countries to build atomic bombs did so before moving to electricity generation through nuclear power. Thus, nuclear power reactors never were and are not necessary intermediate steps for making nuclear bomb. In fact, if the world community unanimously decides for a total nuclear disarmament, the nuclear power can help by facilitating utilisation of fissile material from weapons programmes for power generation.
2. Statistical evidence corroborates the fact that nuclear energy is one of the safest methods for large-scale generation of electricity. With about 434 operable nuclear power reactors, the world has more than 14,000 reactor-years of operation with only three major accidents. In India too, with 20 reactors operating and about 350 reactor-years of operation, there has been only one incidence of level-2 on the International Nuclear and Radiological Event Scale (INES) and has been free of any radiation fatalities from nuclear power plants. After the Fukushima accident, thorough safety reviews of Indian reactors have been carried out by both the Atomic Energy Regulatory Board (AERB) and the DAE (BARC 2011). These reviews have established that Indian reactors are well equipped for mitigation of any design-based accident. The fear that a Fukushima type of accident can happen in India is not based on scientific facts. The seismic and tsunamigenic conditions in India are totally different from those prevailing in Japan. Also, during normal operation, the radiation emission from the nuclear reactors is an insignificant fraction of the natural background radiation levels. Nevertheless, not showing any complacency and overconfidence, DAE is taking a number of measures to further strengthen the safety of its NPPs particularly in a scenario of beyond-design-basis accidents.

3. Indigenous nuclear power is very cost competitive on average-cost basis. Nuclear power tariffs are front loaded and tend to reduce with time. The economics of nuclear power appears far more attractive than it did previously. Moreover, nuclear energy becomes economically even more attractive if all energy technologies were to internalise their respective external costs. The cost of safety, waste disposal and decommissioning are included in the price of nuclear electricity. In future, whenever the much talked about carbon tax becomes a reality nuclear energy will become one of the most economic sources of electricity.
4. As of 2016, 73 nuclear power reactors with total installed capacity of about 77,538 MWe are under construction across the world; 172 more nuclear reactors totalling to more than 1,88,000 MWe are planned for construction the world over by 2030. BRIC countries (Brazil, Russia, India and China) constitute barely 13% of global nuclear capacity today, but have over two-thirds of the global capacity in construction and over half of that planned. Meanwhile, early developers (mostly developed countries) are focusing on life extension of their old plants.
5. Nuclear waste rather than being a problem without a solution is the pride of nuclear power. This is because of (a) relatively low volumes of the waste and (b) availability of viable technology for long-term management of the waste ensuring its isolation from public domain. In future, the accelerator-driven subcritical systems (ADSS) will offer the option of incineration of long-lived fission products into lighter short-lived radioisotopes, thus eliminating the need for long-term storage.
6. The radiation levels around the NPPs are an insignificant fraction of the existing natural background radiation. Independent epidemiological surveys of occupational workers and families at NPP sites, carried out by Tata Memorial Centre, Mumbai, have shown no statistical difference in incidence of cancer. Over 40 years of commercial nuclear power in India and more than 50 years internationally have not indicated any increase in cancer occurrence in population around NPPs. Talking about the genetic disorders, there is absolutely no evidence of such disorders even among the next generations of the survivors of Hiroshima and Nagasaki.
7. Uranium mining and processing employs the state-of-the-art technology in the safest manner. People who call uranium an evil spirit fail to appreciate that a variety of societal applications of radiation technology including cancer treatment depend on the availability of radioisotopes that are produced in reactors, and reactors depend on the availability of uranium as nuclear fuel. In comparison to the coal mines in India, particularly the rat-hole mines in Meghalaya, which have inhuman and poor working conditions and where fatalities is a common phenomenon, uranium mines have an excellent safety record.

Need for Effective Communication

In ancient Indian literature it is suggested that for issue management one could opt for a four-pronged strategy based on (i) *Saam* (communication), (ii) *Daam* (payment or monetary compensation), (iii) *Dand* (punishment) and (iv) *Bhed* (stratagem). Given that *Dand* and *Bhed* are not the options in a welfare state, as per the modern-day management techniques, a combination of *Saam* (communication) and *Daam* (payment) is most important for managing the conflicts between the proponents and other stakeholders of any developmental project. What is meant by *Daam* here is paying adequate compensation to the project-affected people as per a sound rehabilitation and resettlement (R&R) policy before the commencement of the project and pursuance of effective corporate social responsibility (CSR) activities. The CSR activities should not be merely few charity acts but should rather be focused towards enabling and empowering the neighbourhood population.

Maintaining a continuous communication with the stakeholders and various sections of the society is a must for success of any developmental project (IAEA 2011). The society can be divided into four groups based on their knowledge about the subject matter and level of their concern: (i) the group with low knowledge and low activity or concern, (ii) those with high knowledge and low activity, (iii) those with high knowledge and high activity and (iv) those with low knowledge but high activity. To begin with, most of the project-affected people fall into the first category, that is, low knowledge and low activity. Professionals generally comprise the second category of high knowledge and low activity. Many anti-nuclear non-governmental organisations (NGOs) and individuals belong to the third category; that is, high knowledge and high activity. They, for the reasons best known to them, often misrepresent the facts and taking advantage of the inactivity of the second group, incite innocent people to move from the first category (low knowledge and low activity) to the fourth category, that is, low knowledge and high activity. The anti-nuclear activities around KKNPP during 2011–2012, continuing even till date, are a perfect example of this. The project had hardly seen significant protest until 2011. During commissioning activities, steam was intermittently being let out through the safety relief valves (as part of testing) often resulting in frightening noise. The anti-nuclear activists took advantage of the situation and frightened the local innocent villagers that such noise will be a regular feature once the plant commences operation. They succeeded in their objective of converting an otherwise sleepy village into an epicentre of one of the largest anti-nuclear movements India has witnessed. This could happen only because of lack of initiative of the second category—the professionals having high knowledge but lack of activity. If we, the professionals from DAE and Nuclear Power Corporation of India Limited (NPCIL), had in a timely manner taken the local community into confidence by properly explaining and communicating with them, probably the whole ugly situation could have been avoided. Majority of Indian media (both print and electronic) normally belong to the first category—that is, low knowledge (about nuclear issues) and low concerns/activity. Nevertheless, in today's world of 'exclusive' and 'breaking' news the media has a tendency to jump into the third group, that is, becoming hyperactive generally without acquiring the knowledge of the field. During the tsunami of December 2004, great damages had occurred in coastal areas around Kalpakkam. So much so that even the residential colony of DAE was badly flooded and few lives were also lost. However, all the nuclear facilities including the two nuclear power reactors—the fast breeder test reactors (FBTRs)—were absolutely safe. The prototype fast breeder reactor (PFBR) project had hardly started a year ago. The large excavated pit where piling work for foundation was in progress was flooded. I recall how, to our amusement, a journalist from a reputed English news channel focused his camera on the pit and gave an 'exclusive and breaking news' that the PFBR building "had collapsed". I recall that during 2007–2008 when hot political debates were going on about the then proposed Indo-US civil nuclear commerce agreement, popularly referred to as the 123 agreement, most of the media was running their own 'exclusive' and 'breaking news' programmes. A self-proclaimed management guru of a heavily advertised management institute suddenly became a nuclear expert and suggested that nuclear technology being "dangerously unsafe", India should rather opt for thorium. Little did he know that thorium is very much a part of the nuclear technology, and it has to be converted into a fissile material in a nuclear reactor before it is used as fuel in another nuclear reactor. Nevertheless, there are many responsible media personnel who are in no hurry to run a spicy story. They first seek to understand the issues (sometimes, better than the subject experts) and prefer to come out with a balanced view which should be rightfully called 'exclusive'. Once again going back to the '123 era', as mentioned earlier, when the media was in general going crazy with the topic, there was one journalist, Pallava Bagla, Science Editor with NDTV and a fantastic science communicator himself, who ran a programme called 'Parmanu ki Pathshala' (meaning, a school for learning about atom) on NDTV that was a fantastically simplified but highly informative tutorial on the nuances of nuclear technology. This really helped the common man and the politicians to understand the need India had for international civil–nuclear commerce and the reasons why India (or any other country) cannot directly use thorium without having set up the uranium- and plutonium-based reactors.

What Needs to Be Done? Some Suggestions

Need for Information Centres and Full-Time Science Communicators

During a visit to France, I saw that at every nuclear facility, they had a well-established information centre outside the security limits of the facility. These centres with their dynamic models of the facility and general information about nuclear energy and radiation are highly interactive platforms for interaction with the general public. Many other countries such as South Korea, Japan, USA and many European countries have networks of such nuclear information centres. DAE has also in recent years initiated action in this direction which needs to be accelerated and multiplied. The NPCIL few years ago established a nuclear gallery at the Nehru Science Centre at Mumbai and is in the process of setting up many such nuclear galleries in various other cities. DAE in its 12th five-year plan envisaged to set up at least a dozen such centres near many of its facilities.

Establishment of information centres would not serve the purpose unless we have a large pool of communicators manning them. Pallava Bagla has at various forums suggested that DAE needs at least one science communicator for every giggawatts electric (GWe) of planned installed capacity. Although DAE has tried to identify a pool of 'part-time' communicators from within its large pool of scientists, engineers and even administrators, that is not going to be sufficient. We need to set up a separate cadre of young professional science communicators just like many other countries namely France, South Korea and so on.

These communicators should be trained not to use complex language and heavy technical terms but instead try to explain the facts in a subtle manner using very simple language. For example, people have a lot of negative perception about the radioactive waste and its management. They claim (rightly so) that the waste from the nuclear reactor is going to be around for thousands of years and that it will therefore constantly endanger the future generations. While talking on this, I normally accept the fact, that yes, the high-level radioactive waste is indeed very hazardous but because of its low volume and technology being available for ensuring its isolation from the society, it is the pride of the nuclear industry. I am sure that this statement would not satisfy them so I try to drive my point by giving a couple of analogies as follows. A lion is definitely more dangerous than a mosquito, but in most cities we are normally isolated from the lion, so mosquitos become a greater health hazard than the lions; governments therefore spend money to protect the lion and kill the mosquito. Similarly, although potassium cyanide is a lethal poison and sodium chloride (common salt) an essential part of our diet, more number of people are suffering because of an overdose of the latter. A visit to a doctor or a hospital will clearly tell us that a large number of people have to be treated because of mosquito bites and excessive consumption of common salt as compared to lion bites and consumption of cyanide. In a similar fashion, although radioactive waste is much more dangerous as compared to carbon dioxide, it is the latter that is endangering our very existence on the planet. At this point, with great conviction I declare that if at all life became extinct on the Earth, it would be certainly because of excessive amounts of carbon dioxide in the atmosphere and not because of radioactive waste. Therefore, if we really want to save our future generations from extinction, we should switch over to nuclear energy rather than continuing burning fossil fuels. This generally has more convincing effect than simply teaching them the complex process of vitrification, interim storage and final placement of radioactive waste in deep geological depositaries.

Inclusive Growth

All nuclear facilities and the townships housing the families of people working there appear like oases of development in otherwise underdeveloped areas. Such facilities become eyesores for the local people who naturally develop a feeling that they have given up their land without getting to enjoy the fruits of development. Such places serve as very good breeding centres for anti-nuclear activities. The isolated acts of charity carried out by the concerned DAE units under its CSR activity are not enough for bridging this gap. DAE should develop such areas with the help of and coordination with the state governments and local administration so that the entire area around a DAE unit/township develops in a seamless manner. I would go to the extent of stating that rather than the current practice of having 'exclusion zones' around nuclear facility, we should re-designate and develop these areas as 'inclusion zones' and promote activities there that would help in boosting the employability and provide avenues of self-employment for the local people. Engaging them in such a gainful manner will naturally help in eliminating the possibility of anti-nuclear activity in these areas. About a decade back, a good beginning was made at Tarapur Atomic Power Station by establishing an experimental agriculture farm in the exclusion zone of the nuclear power station where the local farmers carry out progressive agriculture using the radiation mutant seeds developed by DAE. The entire operation is carried out by a cooperative society which is being managed by the farmers themselves under the guidance and with assistance of scientists, engineers and administrators from DAE. Many more such groups should be formed around other nuclear power stations and facilities, and a good use of a wide variety of technologies developed by DAE should be made through them for the societal benefits. This will definitely result in a win–win situation both for the DAE and for the people living in the neighbourhood.

Interaction with Students and Academics

Teachers and students can be the best messengers for propagation of the right information in the society and thus reducing if not eliminating the fears and misconceptions about safety and environmental aspects of nuclear power. DAE and its units already have interactions with schoolteachers and students through regular seminars, exhibitions, workshops, essay contests, quizzes and other competitions (NPCIL 2015). The number of such programmes needs to be increased manyfold and that should become possible once information centres suggested earlier are in place along with large number of communicators. We should also increasingly use the social media to reach out in an interactive manner to a large number of people, particularly the younger generation.

Media Interactions

Media interaction is one area that probably needs much larger and urgent attention. We normally tend to remember the media only when there is an event that we feel should get media attention or when something adverse about atomic energy or DAE has appeared in the media. The need of the hour is that we must be proactive rather than being just reactive. We must engage media—both print and electronic—by arranging regular media workshops, media visits to nuclear facilities and so on which should result in an intense 'information push' via media into the society.

DAE, all through its existence has not had a culture of advertising. Of course, the trend is now changing. Few years ago, on an experimental basis, we started inserting occasional short advertisements on the national television channel *Doordarshan*. Although a welcome initiative, it may not be the solution in itself to contain the public misconceptions about atomic energy. We may be able to create very informative advertisement clips and may also be able to buy best airtime slots for them, but will not have control on the advertisements preceding or following our advertisements. Many of these commercial advertisements, although very creative, have exaggerated and unrealistic claims, for example, the luck of a person being influenced by the brand of undergarment or a particular brand of body spray resulting in the user being practically mobbed by the opposite sex. The audience, knowing very well the overstatements of such advertisements, develop a tendency of not believing them. A realistic and serious advertisement about atomic energy can become a victim of this lack of faith of viewers in commercial advertisements. Sponsoring informative articles, interviews of experts and interactive discussions and debates in media will definitely be a better option. As already mentioned, we should make good use of the well-established social media network.

Conclusion

For long-term sustainable development, nuclear power is an inevitable option. The public acceptance is going to be an important factor for its growth. The fear psychosis in the minds of people, which originates from certain myths not based on scientific facts, needs to be removed by putting forward accurate scientific information in a language they can understand and appreciate. The best way to engage the public is on a 'one-on-one' basis. It is essential to empathetically address the concerns of each member of the neighbourhood community to build trust and take time to address all their concerns. Here, media has an important role to play. While being thoroughly watchful, they need to act as knowledge facilitators and must contribute their bit to remove the misconceptions about nuclear energy from the minds of the people. In case of any unusual incidence connected with nuclear power or radiation technology, they must indulge in factual reporting after ascertaining the facts from responsible, authentic and knowledgeable sources; they should do so without the unnecessary sensationalisation.

Special efforts need to be put in to continuously engage students and academicians, as they are most influential towards forming a positive opinion among the general public. The communication with them should be interactive and special emphasis should be given to the use of social media.

The people in the vicinity of any developmental project expect that with the advent of the project there would be better roads, adequate water and electricity supply and good educational and health-care facilities. In India, the public view the operator of the nuclear facility as the government and have expectations from it to improve the conditions of their life. Generally, there are huge differences in the infrastructure facilities available within the premises of the NPP sites and infrastructure (or the lack of it) in the vicinity of these sites and the associated residential areas. Therefore, to create trust among the people towards the operator of a nuclear facility, there is great need to take steps to develop the infrastructure and improve the general living conditions around the project site as well.

Finally, I would like to conclude with what Dr. Anil Kakodkar, former Chairman, Atomic Energy Commission often used to say: "The job of the public outreach personnel is as important as that of the person designing a reactor." In fact, with an ambitiously expanding nuclear power programme, gaining public confidence assumes even greater importance. We very urgently need to change the public perception so that rather than opposing nuclear power, the public must begin to ask, "Why not nuclear?"

References

BARC. (2011). http://www.barc.gov.in/egreport.pdf. Accessed June 28, 2015.

BBC. (2011). http://www.bbc.com/news/science-environment-15864806. Accessed June 22, 2015.

IAEA. (2011). https://www.iaea.org/NuclearPower/Downloadable/Meetings/2012/2012-10-09-10-11-TM-NPE/IpsosMori-Knight.pdf. Accessed July 13, 2015.

NPCIL. (2015). http://www.npcil.nic.in/main/PublicAwareness.aspx. Accessed August 6, 2015.

Nuclearconnect. (2015). http://www.nuclearconnect.org/know-nuclear/talking-nuclear/top-10-myths-about-nuclear-energy. Accessed August 6, 2015.

Part III
Old Is Gold: Time Tested Ways of Communication

Current Status of Public Understanding of Science: Results of *Kumbh Mela* Survey Studies

Gauhar Raza and Surjit Singh

Abstract This chapter is a result of 25 years of research on public understanding of science (PUS) in India. The authors have since 1998 carried out PUS surveys during every *Kumbh* and *Ardh Kumbh Mela* (a religio-cultural fair visited by millions of Hindus) held at *Sangam* (confluence of three holy rivers), Allahabad. The analysis carried out and results obtained during this period were reported after every survey. The conceptual model known as 'cultural distance model' and an empirical method to compute cultural distances of various scientific explanations from the thought complex of sampled population were once again put on the anvil of statistical analysis. In this chapter, for the first time, we report and discuss the shifts in cultural distances that have come about since early 2000s. We also have suggested a method to compute the efficacy of media channels using cultural distance model. The analysis has shown that newspaper still remains the most effective channel for communicating scientific ideas. However, like television and radio, it offers a narrow window of opportunity for communicating ideas. This band of opportunity is highly sensitive to cultural distance of a given scientific idea from the thought complex of the target population. The first section of the chapter deals with the history of science communication in India, in brief. The second section reports and discusses the status and shift in PUS and the efficacy of media channels.

Keywords Public understanding of science (PUS) · *Sangam,* Allahabad · Cultural distance model · Science communication · Media channels

---

Views expressed in the chapter are the authors' own and not necessarily shared by the editors.

G. Raza (✉) · S. Singh
National Institute of Science Communication and Information Resources (CSIR-NISCAIR), Dr KS Krishnan Marg, New Delhi 110012, India
e-mail: gauhar_raza@yahoo.com; gauharraza53@gmail.com

P. Bagla and V.V. Binoy (eds.), *Bridging the Communication Gap in Science and Technology*, DOI 10.1007/978-981-10-1025-5_10

Introduction

Scientific ideas, laws, information, data and methods generated by the specialists through what is termed scientific practice follow a complex trajectory to reach the masses (Raza et al. 2002). Even after revalidation and acceptance within the scientific community, a scientific idea takes fairly long time to become part of the cognitive structure of a given cultural set of common people. The notions that are today considered common sense by a majority of citizens all over the world have taken few centuries to become an integral component of the worldview of a sizeable population segment.[1] Revolution of earth, absence of celestial spheres to hold heavenly bodies, diseases caused by microbes such as bacteria and viruse and lightening caused by charged clouds are but few examples of the lag between discovery and its percolation among the lay public. However, since mid-1990s, the speed, spread, efficacy and efficiency of communication channels has increased the speed of propagation of scientific information and acceptance of new technologies among masses (Stamm et al. 2000).

Indian Context

After India achieved independence, the ruling classes of emerging capitalist society were convinced that, in order to mobilise funds for scientific and industrial research, a wider acceptance of modern scientific ideas was necessary. Phrases such as 'scientific temper', 'broad scientific outlook', 'scientific belief system' and 'scientific method' echoed repeatedly in various forms of debate including political speeches (Raza et al. 1995).

This was the period when funds were mobilised for expansion of industry, education, agriculture and health sectors.[2] New universities and institutions were created and old ones inherited from the colonial masters were augmented. The number of institutes within Council of Scientific and Industrial Research (CSIR) rose from two in 1947 to 15 in 1954. Indian Agricultural Research Institute (IARI) opened extension centres all over the country; Defence Research and Development Organisation (DRDO), Atomic Energy Commission (AEC) and Indian Space Research Organisation (ISRO) were established. Besides their core activity each one of them had a mandate to reach out to the people and popularise science. Publication and Information Directorate (now National Institute of Science Communication And Information Resources (NISCAIR)), Indian National Science

---

[1]In India at the dawn of the 20 century, intellectuals were still discussing whether the earth was stationary or moving, and the discourse was deeply religious (Gosling 2007).

[2]The first, second and third five-year plans have been analysed by a number of scholars. For example, both Toye (2007) and Bauer (2011) agree that this was the period when plans emphasised expansion of infrastructure.

Academy and various academies for Indian languages undertook projects for publishing popular science books. Print and electronic media were geared up to propagate science. Museums and planetariums were constructed in the country. Later, two more national bodies dedicated to science communication were created. These were National Council for Science Technology Communication (NCSTC), a funding agency, and *Vigyan Prasar* (meaning, dissemination of science), both funded by the Department of Science and Technology (DST), Government of India.

The history of science communication, although largely a neglected area of investigation in the country, suggests that the first efforts to communicate modern scientific ideas originating in the West were made during the latter half of the nineteenth century. A number of science books were translated from English into Indian languages. Small groups in the form of 'science societies' mushroomed in various parts of the country. Such groups held regular discussions on mathematics and science subjects, but their sphere of activities was confined to the upper 'class' and 'caste' intellectuals (Habib 2000). Their reach in terms of numbers remained quite limited; nor did they make any serious efforts to transform themselves into a large-scale 'science movement'.

The freedom movement was the most potent source and the carrier of modern ideas. Notion of building an egalitarian society in a caste-ridden cultural setting, assertion of gender equality where women were considered as inferior beings, affirmation of education for all where certain sections of society were forbidden from even listening the sacred text (Singh 2005) and above all the idea of building a nation-state in geographical area which was fragmented across many fault lines and littered with small princely states are but few examples of modern ideas that were popularised by the leadership of the freedom movement. The value of science as an agent of social change could not have escaped the attention of the leadership. The phrase *scientific temper* (Kothari 1989) was not coined by a scientist, it was Jawaharlal Nehru, the first Prime Minister of India, who realised the importance of popularising scientific ideas in a society where outmoded irrational beliefs were rampantly prevalent (science comes but wisdom lingers[3]).

However, serious efforts both at the level of the government machinery and non-governmental organisations (NGOs) had to wait for the independence from the British imperialism. The dream of building a modern and industrialised India conceived by the leadership of the freedom struggle required technical manpower to be trained within the country in large numbers for establishing a strong indigenous base for scientific enquiry.

A wider acceptance of scientific ideas in the society was a necessary condition for achieving the two major goals that the leadership had set before the newborn independent India. As stated earlier, phrases such as 'scientific temper', 'broad scientific outlook', 'scientific belief system' and 'scientific method' echoed

---

[3]"What Nehru affirmed in England was his concept of a civilised person, the application of reason to human affairs, tolerance for dissenting opinions and a search for truth, all revealed to him by science'. Available at http://www.garfield.library.upenn.edu/michaelis/title342.pdf.

repeatedly in various forums of debate. Use of newspapers, magazines and electronic media (namely, All India Radio[4] and Doordarshan[5]) began the dissemination of scientific information. During the latter half of the 1950s a large number of new books for teaching science in regional languages were written. The number of popular books on various topics of science and related subjects swelled. Translating English texts posed two major problems: firstly that Indian languages culturally had a limited capability of expressing modern scientific ideas and secondly, there was an acute shortage of standard technical terms. These terms had to be generated; at times this was done mechanically and at others, recourse was taken to using familiar but inaccurate terms leading to obfuscation or in some cases incorporation of ‘anglicised’ terms into the local languages.

Efforts to organise science writers in Kerala began in 1957, but they failed to produce any worthwhile results. In 1962 the *Kerala Sastra Sahitya Parishad* (KSSP) was formed. What was initiated as a small group of 30 science writers and social reformers eager to disseminate science through print medium got transformed into a region-wide science movement in the southern state of Kerala. It is necessary to point out here that similar activities were going on in many other parts of the country. In Maharashtra, Karnataka, Tamil Nadu, West Bengal and Assam, science writers’ associations were involved in publishing popular science books and journals.

Inspired by the success of the KSSP a number of other states saw the emergence of science activists’ groups during the 1970s. The local needs and issues did influence the very nature and structure of these organisations. It was only in the beginning of 1980s that a strong need was felt to create a network of these organisations (Science as Social Activism, KSSP, 1984). For consolidating the formal and informal links between various fraternal groups an all India level activity had to be conceived. KSSP had already prepared the ground for such an event. They had, by the mid-1980s, spread the area of their activities into neighbouring states. In 1985, a workshop of NGOs identified the issues of common concern and interest, which were later to become the basis for long-term cooperation among them. These were ‘self-reliant technology policy of the country’, ‘peace and nuclear disarmament’, ‘Bhopal gas tragedy’ and ‘popularisation of science’. The national coordination committee started calling itself ‘People’s Science Movement (PSM)’, which later transformed into All India Peoples Science Network (AIPSN).

---

[4]Radio broadcasting began in India in 1927, with two privately owned transmitters at Mumbai (then Bombay) and Kolkata (then Calcutta). These were nationalised by the British Raj in 1930 and operated under the name *Indian Broadcasting Service* until 1936, when it was renamed All India Radio (AIR). Available at http://www.answers.com/topic/all-india-radio Last visited on 6 June 2007.

[5]Doordarshan had a modest beginning with the experimental telecast starting in Delhi in September 1959 with a small transmitter and a makeshift studio. The regular daily transmission started in 1965 as a part of All India Radio. The television service was extended to Mumbai and Amritsar in 1972. http://en.wikipedia.org/wiki/Doordarshan Last visited 6 June 2007.

Campaign for Science Awareness

KSSP had evolved a novel method of mass contact programme. Besides publishing books and journals on issues of science and running science clubs throughout the year, they also held 'Science *Jathas*'[6] (science processions) every year. In 1970, for the first time, after the annual conference of KSSP the delegates took out a procession in Ernakulam, Kerala. Subsequently, it became a regular feature to organise public lectures by scientists over a period of about a month throughout the state. Groups of scientists travelled in cars and were received by local reception committees who were responsible for organising the lectures in their area. Often, the travelling '*Jathas*' reached the venue of the public meeting late. Thus, the local committees gradually started producing short plays and songs on various science issues that were performed before the main event. Gradually, a shift in the conceptual framework came about and 'Science *Jathas*' were transformed into '*Kala Jathas*'.

The phrase '*Kala Jatha*' when translated into English language would mean a 'cultural procession'. It is interesting to note that the name itself underlines the nature of the shift in the conceptual framework. Science, in a planned manner, was now communicated through traditional as well as modern art forms, such as songs, street theatre and poster exhibitions. For this purpose, artists, scientists, teachers, students, social workers and unemployed youth were mobilised. Training workshops were held and modules for two-to-three hour performances on socially relevant issues were prepared every year. Groups of these trained activists travelled throughout the state and performed in almost every village where local reception committees, constituted for the purpose, organised cultural programmes. Recalling the experience of 1977 Thomas Isaac in one of his books recorded that the "*Jatha* consisted of a group of science activists who toured (through) the length of Kerala for 37 days, stopping at three to four centres a day." The first *Jatha* performed at 900 places, contacted more than half a million people and sold pamphlets worth more than INR 25,000,[7] thereafter '*Kala Jathas*' became a regular and significant activity of KSSP.

It was against this background that the proposal to stage an all India '*Jatha*' was first mooted by the *Lok Vigyan Sangathana* (meaning 'public science organisation'; Maharashtra State) representatives during a meeting of the 'Coordination Committee of the Campaign for Peace and Against Nuclear War' in 1986. Few PSM groups were also present in the meeting. This suggestion caught their imagination, and they resolved to take the plunge (Ramachandran 2004). Preparatory work started soon after the meeting. The task of nucleating new groups in those parts of the country where science organisations were either weak or did not exist at all was initiated on a war footing. In the process a large number of new groups were formed and old defunct organisations were energised. In about ten months the PSM had put in place the necessary network and begun preparations for

[6]*Jatha* is word of Hindi language, which in English language would mean 'procession'.

[7]INR 25,000 was a substantially big sum during 1970s.

the campaign by holding a workshop for finalising the details of an event which has no parallels in any other country. Twenty-six new and old science NGOs carefully worked out a detailed plan of action in a two-day meeting held in April 1987. A detailed perspective plan was also prepared in the same meeting, which was epitomised in the form of three slogans: 'science for the people', 'science for the nation' and 'science for discovery'.

Workshops were held around conceived themes; lectures, play scripts and songs were written, slide shows and poster exhibitions were prepared; films were produced; and science toys, games, experiments and exhibits were collected and created. Volunteers were drawn from among the artists, writers, poets, scientists, teachers, physicians (medical practitioners) and unemployed youth. In order to cover all the 500 districts of the country five teams each of scientists and cultural activists were constituted and trained to perform three- to five-hour programmes. Five such processions started on 2 October 1987, from five different locations of the country to culminate at Bhopal city on the first anniversary of the Bhopal gas tragedy. Each of these *Jathas* covered a distance of about 5000 kilometres in exactly 37 days. During the period millions of people were exposed to scientific information and issues of social relevance. Thousands of science activists joined the PSM. New local and regional science organisations came into being as a result of this intense mass campaign (Padmanabhan 2004).

The strong national science movement that emerged as a result of *Bharat Jan Vigyan Jatha* (that is, peoples' science procession) gave birth to many regional and national campaigns in subsequent years. National Literacy Campaign, Joy of Learning, Solar Eclipse Campaign and Anti-Superstition Campaign are just few examples (Ministry of Human Resource Development (MHRD), Government of India). Some of the social campaigns that have been cited in national and international forums as successful awareness programmes such as anti-*arrack* (anti-alcoholism) campaign could also be legitimately termed 'spin-off' of the PSM activities (*Bharat Jan Gyan Vigyan Jatha* 2004).

Exploring Public Understanding of Science

A small group of scientists, working on various aspects of science and technology policy who had been associated with the *Bharat Jan Vigyan Jatha* 1987, since the very inception of its idea, undertook the responsibility of organising the northern component of the '*Jatha*' and travelled with other scientists and artist from Srinagar (Kashmir) to the Bhopal (Madhya Pradesh). During interactions with the people it was realised that public did not lack the 'urge to know about scientific issues'. However, when it came to the software modules used in the campaign, the responses were rather mixed. Some of the software packages were very popular throughout the campaign whereas others did not invoke the same enthusiasm. This raised serious questions such as 'what kind of scientific information invokes greater receptivity among the people and why?' Thus, it was decided to frame a questionnaire that warranted strenuous probing. In 1989 the questions for this questionnaire were listed out during an informal workshop of science communicators. A group was soon constituted in National Institute of Science, Technology and Development Studies (NISTADS, CSIR, India) with the objective to carry out research on public attitudes and understanding of science (PAUS).

A review of available literature on the subject proved to be of help, but only to a limited extent. Most investigations on the subject available at that time carried out in the West were directed towards measuring the levels of 'scientific literacy' of the target population (Durant et al. 1991). However, such studies, depending on preconceived scales categorised the populace only into 'scientifically literate' and 'scientifically illiterate' persons. The tools and methodologies developed for carrying out large-scale PAUS survey studies in the Western cultural context were found to be inadequate. The group decided to carry out a survey study (Raza et al. 1991) during the *Kumbh Mela* (Bhattacharya 1956) at Allahabad (a religio-cultural festival) held in 1989, which provided a novel opportunity for such an exercise. In the following 20 years many more large-scale surveys, with the same objectives, were administered.

Shifts in Public Understanding of Science

In order to analyse the shifts in PUS during the period 2001–2013, data collected in 2001 (*Kumbh Mela*), 2007 (*Ardh Kumbh Mela*) and 2013 (*Kumbh Mela*) were analysed. For this, five indicators that marked the area of astronomy and cosmology were used. The data collected were subjected to rigorous statistical tests (Table 1).

The analysis showed that the cultural distance of three indicators, namely shape of the earth, rotation of earth and galaxy had reduced substantially. The revolution of earth or the explanation of formation of seasons as a scientific phenomenon moved farther by 0.6 years. Meanwhile, on the scale of cultural distance, the theory of evolution moved by a factor of 0.2 years. The trends once again validated the efficacy of 'cultural distance model'. It is evident that the scientific concepts could

Table 1 Statistical properties of Indian datasets

Constant	2001 (N = 3484)	2007 (N = 3863)	2013 (N = 4789)
Cronbach's alpha	0.721	0.749	0.756
Standard deviation	5.47	5.15	5.27
Skewness	−1.094	0.755	−0.587
Kurtosis	2.188	2.10	−0.711
Standard error (mean)	0.094	0.084	0.077
Standard error (skewness)	0.042	0.040	0.036
Standard error (kurtosis)	0.084	0.080	0.071

Source Author's work

Table 2 Values of cultural distance computed for Indian populace

Concept	2001	2007	2013
X_{Ci1}	0.7	0.5	−0.07
X_{Ci2}	9.7	8.8	8.1
X_{Ci3}	10.4	10.8	9.8
X_{Ci4}	13.2	12.4	10.2
X_{Ci5}	19.3	19.5	20.5
X_{mean}	10.7	10.4	9.7

Source Author's work

Table 3 Magnitude and polarity of shift in cultural distance

Concepts	Shift (ΔXci) (2001–2007)	Shift (ΔXci) (2001–2013)
ΔXci_{mean}	−0.2	−0.9
What is the shape of Earth?	0.2	−1.4
How do day and night form?	−0.9	−1.6
What causes eclipse?	0.4	−0.6
What is Akaash Ganga (Milky Way)?	−0.8	−3.0
How did humans come to being?	0.2	1.2

Source Author's work

be placed at increasing relative cultural distance on a scale, from the thought complex of the populace sampled during the three *Kumbh Melas* (Raza et al. 2009).

The analysis also shows that the mean cultural distance during the period 2001 to 2013 had reduced by a factor of almost one year (Tables 2 and 3).

Media and Science

In India, it is estimated that over 220 million people have access to print media, and there are 370 million literate Indians (they are seen by industrial houses and multinational companies (MNCs) as large market segments waiting to be explored and exploited) who still do not have access, or do not access, any publication (Newswatch 2007). Put together, this huge market has attracted almost all big publishing international players to India. The sector, to an extent, still remains protected; Government of India regulates the amount of investment that a foreign company is allowed. There are few more protections that have restricted foreign players to enter the regional language printing and publishing industry.

Since the introduction of Internet technology in 1990s its expansion has taken place at a phenomenal rate. In India, in 2006, there were about 21 million people who regularly used Internet. According to a survey this number was expected to increase to 35 million by 2008 (Editorsweblog 2007).

The country's media and entertainment industry revenues rose by 20% in the year 2006. Access to satellite television had increased from 207 million individuals watching in an average week in 2005 to as many as 230 million individuals in 2006. In this century this visual medium has established its supremacy over the print medium. The number of television viewers substantially exceeds the readers in the country. Homes with access to cable and satellite increased by 12% from 61 million to 68 million this year (a snapshot of Indian television history). The cable network infiltrated 61% of all televisions installed in the country. The rise was a massive leap of 56% compared to the previous year. This happened at the cost of traditional, although urban mode of communication and entertainment. The reach of cinema has waned sharply; it is in serious danger in India. The viewership has reduced from 51 million individuals going regularly to the movies (at least once a month) to 39 million. Even in the US the empirical studies reveal that local television news is the most widely used source of news and information and the audience for local television news declined during the late 1990s but appears to have stabilized since 2000 (Miller et al. 2006).

The access to radio, surprisingly, had reduced in rural India from 25% to 23%. However, the increase in listenership in urban centers had resulted in a stagnant 23% of 'national listenership to radio' over the past 3 years (The Voice of India).

The purpose of the discussion was to briefly sketch out the access that Indian public have to various channels in media. The next section deals with the efficacy of various channels in communicating science to the common citizens.

Media and Scientific Information Level

In a survey at *Kumbh Mela* 2013 site in Allahabad all the respondents who were interviewed were asked to tick the modes of communication that they get information through. During the survey 100 trained enumerators interviewed 3836 adults over a period of four days. Besides the questions on the access to media channels, they were also asked to respond to simple questions related to scientific phenomena they encounter in quotidian life.

Some of preliminary results are presented in the following paragraphs. A very large number of respondent 70% reported that they had access to information through television. More than 68% reported that they listened to radio and 64% said that newspaper was their main source of information. About 45.5% said that they received information only through interpersonal mode of communication. They had access to none of the other media channels.

Of all those who were interviewed 68% reported that they had access to both radio and television, 53% said that they received information through television and newspaper and about 49% ticked all three modern channels of information.

In order to analyse the relationship between the access to various channels and the response to questions posed in the area of cosmology and astronomy, a large number of frequency tables were generated. The analysis revealed that those who, along with the interpersonal information exchanges, had access to at least one modern channel of information were scientifically more informed. On the basis of analysis it could be concluded with high degree of confidence that those who read newspapers consistently gave scientifically valid answers in comparatively large numbers. Those who had access to radio and television scored almost equal percentages. The analysis revealed that the efficacy of radio and television in shaping the scientific knowledge base of a common citizen is not quite different.

Cultural Distance and Efficacy of Media

Often it is emphasised by scholars that media does not give sufficient coverage to science issues, and there is a need to increase, in print medium, the column space or, on television and radio, broadcast/telecast time for scientific issues (van Rooyen 2004). Here we argue that mindless increase in coverage will not yield desired results. Selection of content, topics and issues and their communicability through a specific medium are factors that are equally important if effective communication is the desired objective. Ian Hargreaves, Justine Lewis and Tammy Speers in their report on science, public and media observe that "while television is the popular source of news about science, the amount of television news people watch has little effect on knowledge …" (Hargreaves et al. 2003).

It was evident from the frequency tables that as the cultural distance of the scientific phenomena from the quotidian life of the common citizen increased, the percentage of those who gave scientifically valid answers reduced substantially. However, with the progressive reduction of cultural distance, the difference in the percentage of those who had access to various media channels and gave scientifically valid answers also reduced (Fig. 1).

The percentage distribution of scientifically valid responses across questions posed also showed that as the cultural distance of the phenomena increases the distribution curve assumes a bell shape. Therefore, it could be concluded that efficacy of the media channels is high in a region where the cultural distance is neither very small nor very large. Significantly, it is evident that as the cultural distance increases the influence of media channels also increases, and then the curve saturates, reaches a peak at about 7 years in 2007 and 10 years in 2013 of cultural distance for this set of natural phenomena, and then droops to quite low values of efficacy. The analysis makes it amply clear that explanations of all kinds of scientific phenomena cannot be effectively communicated through media channels to a given public. It is a narrow region of cultural distance—sandwiched between two regions of low efficacy—where the media channels could play a significant role in shaping the cognitive structure of the public.

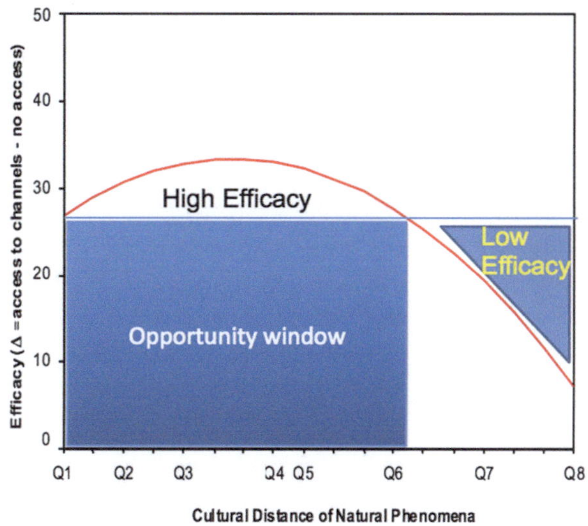

Fig. 1 Efficacy of channels of information versus cultural distance. *Source* Author's work

Table 4 Comparative efficacies of media channels and shift over 12 years

Information channels	2001			2013			Shift in efficiency		
	Peak	Span	Origin	Peak	Span	Origin	Peak	Span	Origin
Cumulative efficacy of all the channels	30.1	14.0	0.5	31.9	19.9	−0.7	1.8	5.9	1.2
Efficacy of newspaper	24.4	12.7	0.5	26.0	14.7	−0.7	1.6	2.0	1.2
Efficacy of television	18.5	10.0	0.5	20.6	10.2	−0.7	2.1	0.2	1.2
Efficacy of radio	18.3	8.8	0.5	20.8	15.2	−0.7	2.5	6.4	1.2

Source Author's work

However, we have computed the span of window of opportunity for a given media channel from the point of origin of graph given in Fig. 1. For instance, see Table 4.

It is evident from the data analysis that the cumulative efficacy of the three channels, newspaper, television and radio, is significantly higher than any one channel. It should also be noted that saturation peak occurs at much higher level and the span or window of opportunity to communicate varied scientific idea is also much wider. The analysis conclusively shows that during the 12-year period the peak had increased (1.8 points) and the window of opportunity had also enlarged (5.9 years) in terms of coverage of cultural distance.

When we look at the computed parameters of newspaper, it scores highest position among the three channels of communication, both in 2001 and 2013. Peak in 2001 occurs at 24.4 percentage points and window of opportunity enlarges from 12.7 to 14.7, an increase of 2.0 years of cultural distance.

Curiously, although the saturation peak of television has increased from 18.5%to 20.6% points, the span of window of opportunity has not increased substantially. It should also be noted that like any other channel origin of the curve has shifted toward the fourth quadrant by 1.2 years. This leads us to the conclusion that window of opportunity that television offers for communicating scientific ideas has narrowed during the past 12 years.

The analysis shows that in 2001 the radio, as a channel of information, on efficacy scale, was a close second rival of television, by 2013, its peak occurred at 0.2% point lower level and the span was shorter by two years compared to television. During the period of 12 years a shift of 2.5% points in saturation peak and an increase of 6.4 years in cultural distance covered by opportunity window, had taken place. The analysis clearly shows that there is a need to use radio as channel of communication by science communicators.

Conclusions

On the basis of data analysis it can be concluded that the understanding of science among the public who visit *Sangam* during the *Kumbh Mela* had increased during 2001–2013. It is also evident that cultural distance of scientific explanations does not increase or decrease evenly. Some of the concepts have moved farther, whereas most indicators show a reduction in cultural distance. Our results also conclude that efficacy of media in communicating science to the public had increased during the period 2001–2013, and each channel of media offers a narrow window, sensitive cultural distance of concept, for communication of science. Only those scientific concepts, information or laws could be communicated effectively, which are placed at a cultural distance that falls within this narrow band.

References

Bauer, P. T. (2011). *Indian economic policy and development*. London: Routledge. (first published in 1961).

Bharat Jan Gyan Vigyan Jatha. (2004). Mass action for national regeneration, December 20 Issue, BGVS.

Bhattacharya, H. (1956). *The cultural heritage of India* (Vol. IV). Calcutta: The Ramakrishnan Mission, Institute of Culture.

Durant, J. R., Miller, J. D., Tchernia, J. F., & Deelen W. V. (1991). Europeans, science and technology (p 7). Presented to the 1991 Annual meeting of the American Association for the Advancement of Science in Washington D.C.

Editorsweblog. http://www.editorsweblog.org/print_newspapers/2007/03/india_more_print_media_boom.php

Gosling, D. L. (2007). *Science and the Indian tradition: When Einstein met Tagore*. London: Routledge.

Government of India, Ministry of Human Resource Development, Department of Education, with National Informatics Centre Compilation on 50 years of Indian Education. (1947–1997). http://education.nic.in/cd50years/home.htm. Accessed June 14, 2007.

Habib, S. I. (2000). Munshi Zakaullah and the vernacularisation of the science in nineteenth century India. In N. K. Sehgal, S. Sangwan, & S. Mahanti (Eds.), *Uncharted terrains*. New Delhi: Vigyan Prasar.

Hargreaves, I., Lewis, J., & Speers, T. (2003). Towards a better map: Science, the public and the media. Economic and Social Research Council.

Kerala Sastra Sahitya Parishad (KSSP). A people's science movement. Available at http://www.kssp.org.in/. Accessed June 4, 2007.

Kothari, D. S. (1989). Nehru aur Jan Vigyan (Nehru and people's science). In S. S. Prasad (Ed.), *Nehru aur Vigyan (Nehru and science)*. Delhi: Parag Publication.

Miller, J. D., Augenbraun, E., Schulhof, J., & Kimmel, L. G. (2006). Adult science learning from local television newscasts. *Science Communication, 28*(2), 216–242.

Newswatch. http://www.newswatch. Accessed May 15, 2007.

Padmanabhan, B. S. (2004). Taking science to the masses. *Frontline, 21*(03).

Ramachandran, R. (2004). The messengers of science. *Frontline, 21*(03).

Raza, G., Dutt, B., & Singh, S. (1991), Scientific attitude among masses. A Report of the survey conducted at Allahabad during Kumbh Mela in 1989, Vol. I. NISTADS-REP-107(AV)/91.

Raza, G., Singh, S., & Dutt, B. (1995). Peoples' attitudes to scientific knowledge: The context of culture. *Journal of Scientific & Industrial Research, 54*, 108–121.

Raza, G., Singh, S., & Dutt, B. (2002). Public, science and cultural distance. *Science Communication, 23*(3), 293–308.

Raza, G., Singh, S., & Shukla, R. (2009). Relative cultural distance and public understading of Science. *Science Technology & Society, 14*, 269–287.

Singh, E. (2005). *Caste system in India: A Historical perspective*. Delhi: Kalpaz. Available at https://www.vedamsbooks.com/no38612.htm

Stamm, K. R., Clark, F., & Eblacas, P. R. (2000). Mass communication and public understanding of environmental problems: The case of global warming. *Public Understanding of Science, 9*, 219–237.

The Voice of India. http://www.hinduonnet.com/mag/2002/08/25/stories/2002082500090100.htm

Toye, J. F. J. (2007). *Public expenditure and Indian development policy 1960-70*. Cambridge: Cambridge University Press. (originally published in 1981).

van Rooyen, C. (2004). Science not a priority for SA. http://www.scienceinafrica.co.za/2004/april/pressscience.htm

Communicating Science for a Better Tomorrow

Hasan Jawaid Khan

Abstract Even the most brilliant scientific discovery cannot and should not remain confined to laboratories or research and technical journals. It needs to be communicated to the civil society to bring about more engagement and increased appreciation of scientific endeavours. An effective way of communicating science is by way of science magazines. Science magazines do not just disseminate scientific information, but can also be very effective means of inculcating scientific temper and, more importantly, generating an interest for science among the young. Taking the case study of the monthly popular science magazine *Science Reporter*, published by the Council of Scientific and Industrial Research, an effort is made here to highlight the important role science communication can play in national development, institutional efforts in the country to popularise science through the print medium, the challenges faced in communicating science through magazines and what needs to be done to make science communication a more potent instrument for raising the scientific awareness of the citizens of the country.

Keywords *Science Reporter* · Science magazines · Popular science · Science communication · Students · Scientific temper

Science Communication: National Responsibility

There is very little realisation, especially in India, that science communication can perform a very important role in the development of the nation. In a democratic setup, where people's power makes and breaks governments and often also decides

Views expressed in the chapter are the author's own and not necessarily shared by the editors.

H.J. Khan (✉)
National Institute of Science Communication and Information Resources (NISCAIR), Council of Scientific and Industrial Research (CSIR), Dr KS Krishnan Marg, New Delhi 110012, India
e-mail: hjkhan1963@yahoo.com

P. Bagla and V.V. Binoy (eds.), *Bridging the Communication Gap in Science and Technology*, DOI 10.1007/978-981-10-1025-5_11

the fate of critical national projects, it is science communication that seeks to empower the citizens with information, thus further strengthening the democracy. India has reached a critical socioeconomic stage where nothing could be allowed to hold it back, least of all lack of information or proliferation of incorrect and mischievous information among its citizens.

At the same time, however, it is for science communicators also to realise that by engaging in science communication they are shouldering the responsibility of a very important societal function. More than the urge to get published and see their name in print, science communicators need to aim at bringing about a change in the understanding of the members of the civil society.

Effective science communication can play a role in further strengthening the functioning democracy that India is by ushering in scientific literacy. A high scientific literacy rate is important for today's technology-driven society. And what does scientific literacy imply? In simple terms, it means that a person should be able to ask questions, or determine answers to questions, derived from everyday experiences. Scientific literacy also entails empowering the members of the civil society to take informed decisions and express positions that are scientifically and technologically informed rather than being swayed by empty rhetoric and faulty arguments.

Today, ill-conceived projects and unplanned and unscientific development of cities and towns could unleash environmental misery on a large scale, as happened during the 2013 Uttarakhand disaster. It is, therefore, equally important that our policy makers, administrators and politicians too be brought up to date with issues such as climate change, genetically modified crops and the like. Science communication can play a big role here.

As Christie Wilcox (2012) says, "When only a small percentage of our populace—including our policy-makers—has a firm grasp on the science behind the debates, we are doomed to make grievous errors in our decisions on a wide variety of issues, from climate change and genetically modified foods to stem cell research and public health and vaccinations."

Inculcating a rational outlook and the need to question unfounded beliefs is yet another important function of science communicators. When seawater miraculously turns sweet or honey starts oozing out of the frame of a deity or for that matter when idols start 'drinking milk' (claims actually reported in the media), it is the duty of the science communicators to dispel such unfounded beliefs with their investigative and unbiased reporting. Science communicators also need to impart to people the ability to seek explanations of natural phenomena rather than merely accepting passed-down judgements or accepting them as some sort of miracles. It is only when the citizens imbibe a culture of science and of scientific enquiry in all walks of life, and are not misled by charlatans, that they can start investing their energy and their intellect in the progress of the country.

There are important environmental issues too that science communication tries to address lest they spiral out of control. For instance, increasing use of polybags, clogging rivers, vanishing species, need to harvest rainwater, growing piles of e-waste, climate change and several such issues would not have come to the fore but for the diligence of science communicators.

Kindling awareness in the young about the techniques of science and encouraging them to engage in scientific pursuits is also a critical responsibility of science communication, especially considering the scenario that the young are fast losing interest in the basic sciences. Since the country needs to have a continuous influx of young scientists, the responsibility of science communicators and popularisers to try and motivate them to take up science assumes a sense of urgency.

Institutional Efforts in Popularising Science

India has a rich tradition of science popularisation through various forms such as folk, puppets, dramas, audio and films. Several efforts have also been made to disseminate and popularise science through the print medium.

The earliest efforts at communicating science through the print medium can be traced to the Asiatic Society set up in 1784 in Bengal, which started publishing the quarterly scientific journal, *Asiatick Researches* in 1788. The journal was so much in demand in the literary and scholarly world that a pirated edition of the first volume came into circulation in England in 1798, and some of the volumes of the *Asiatick Researches* were translated into German as well as into French (http://www.asiaticsocietycal.com/history/2.htm).

Master Ramchandra was a mathematics professor and chief figure in the development of periodicals that were issued from the Delhi College press during 1840s and 1850s. He edited two journals published by the Delhi College in which he made western innovations in science and technology available to the literate public. Ramchandra translated several mathematics textbooks into Urdu and produced two original works in English. One of these, *A Treatise on Problems of Maxima and Minima Solved by Algebra*, first published in 1850, was reprinted in London in 1859, commended by a leading mathematician there, and awarded a prize by the Government of India (Raina and Habib 1989).

Science journalism in Indian languages can be traced to science articles published in the monthly *Digdarshan*, the first periodical from Srirampur (Hooghly), West Bengal in April 1888 in Hindi, Bengali and English. These articles dealt with topics such as flying in a balloon, steam boat and so on. *Vigyan* (meaning *science*) — a monthly popular science magazine in Hindi—has been published by *Vigyan Parishad* (meaning *science council*, a learned society of scientists and academics based at Allahabad) since 1915 (Patairiya 2007).

Several efforts can be cited here of nongovernment organisations (NGOs) taking up the cause of popularisation of science in regional languages. During the 1960s, the Kerala Sastra Sahithya Parishatd (KSSP) took upon itself the mantle of science popularisation through lectures, books and periodicals, songs, *Kala jatha* (meaning *cultural procession*), slideshows and film shows, as well as exhibitions, science festivals and so on. Other efforts notably include those of the Karnataka Rajya Vijnana Parishat (KRVP), the Orissa Bigyan Academy, the Assam Science Society,

Science Popularisation Association of Communicators and Educators (SPACE), All India Andha-shraddha Nirmulan Samiti and several others.

The National Council for Science and Technology Communication (NCSTC), established in 1984, has brought out a large number of books and posters on several scientific topics. Vigyan Prasar established in 1989 also publishes a number of books on popular scienceand also brings out the monthly bilingual popular science magazine *Dream 2047*.

The National Research Development Corporation (NRDC) has been publishing *Awishkar,* a Hindi science monthly magazine and *Invention Intelligence*, an English monthly.

Besides, a number of popular science books have also been brought out by the National Book Trust, India and the publications division of the Government of India.

Role of CSIR

The National Institute of Science Communication And Information Resources (NISCAIR), a constituent laboratory of the Council of Scientific and Industrial Research (CSIR) is also engaged in dissemination of scientific information through research periodicals, monographs, encyclopaedias and popular science magazines.

Since 1950s the Institute has been serving India's scientific community through its 18 peer-reviewed research journals in various disciplines of science and technology ranging from physics to radiophysics, chemistry to chemical technology, experimental biology to biotechnology and intellectual property rights to traditional knowledge. Through its encyclopaedic publication the *Wealth of India,* which documents India's raw material resources, the Institute has been catering to the needs of researchers, farmers and industrialists.

NISCAIR also pursues its science popularisation mandate by bringing out popular science magazines and books. It has brought out a large number of well-illustrated and moderately priced popular science books on topics ranging from cells and genes to computers and artificial intelligence and from atom and materials to space technology and stars. These books continue to be in great demand even 15 years after their publication.

The popular science magazines, however, are the most recognisable faces of NISCAIR. The institute brings out three popular science magazines (Fig. 1): *Vigyan Pragati* (Hindi monthly, launched in 1952), *Science Reporter* (English monthly, launched in 1964) and *Science- ki -Duniya* (Urdu quarterly, launched in 1975).

NISCAIR's popular science magazines have today become the largest selling popular science magazines in the country. They are widely read by people from all walks of life. The magazines, fully illustrated with four-colour photographs and sketches, are being published on a not-for-profit basis and are made available at a

(a) (b) (c)

Fig. 1 Cover pages of science magazines brought out by NISCAIR. (**a**)*Science ki Duniya*, (**b**) *Vigyan Pragati,* (**c**) *Science Reporter. Source* NISCAIR. *Copyright* Author

highly subsidised cost (INR 30.00 per issue for *Science Reporter* and *Vigyan Pragati* and INR 10.00 per issue for *Science ki Duniya*).

Communicating Science Through Magazines

Since the author has been attached with *Science Reporter,* the English monthly magazine, for the past 26 years and has been editor of the magazine since the last eight years, an effort is being made here to highlight salient features and challenges related to editing and overall functioning of the science magazine based on the author's experiences.

Profile of the Magazine

Launched in 1964, *Science Reporter* completed 50 years of its existence in 2014. This is quite commendable in view of the fact that several other science magazines have discontinued publication over the years (including *Science Today, Science Age* and *Invention Intelligence,* to name a few).

Despite not having a very aggressive marketing strategy, *Science Reporter* enjoys a circulation of about 45,000 copies every month. It is available to readers through individual subscriptions (about 12,000) as well as through book/magazine sellers in various towns and cities. The maximum number of copies (reflecting the readership as well) is sold in West Bengal followed by Delhi, Maharashtra and other states. Issues of *Science Reporter* are available online for all subscribers of the print edition (http://www.niscair.res.in/sciencecommunication/popularization%20of%20science/scirep0.asp). However, six-month old issues are available free for all readers.

As of mid-2016, the magazine receives about eight pages of advertisements every month (amounting to an average of about INR 1.00 lakh per month), reflecting the demand for the magazine.

Readership Profile

Since late 1990s, *Science Reporter* has been carrying a prize puzzle every month to be answered by the readers. Almost 800–1000 entries are received every month in response to the prize puzzle. Winners are decided on the basis of a draw of lots from among the correct entries. The prize-puzzle page also carries a questionnaire at its back.

A preliminary analysis of the responses to the questionnaire gives a broad idea of the readership profile of the magazine:

- The age of respondents ranges from 5 to 85 years.
- The maximum percentage of respondents is in the age group of 5–15 years (34 %) and 15–25 years (46 %).
- The maximum percentage of respondents are at the secondary level (55 %) followed by graduates (20 %) and postgraduates (15 %).

Further, an analysis of the profile of authors contributing articles to the magazine shows a variable profile comprising research scholars, scientists, academicians, science administrators, professionals (doctors, engineers, lawyers, and so on) and students (contributing primarily to the interactive columns in the magazine).

Therefore, *Science Reporter* can be said to be enjoying popularity among a wide section of the civil society ranging from school/college/university students to scientists, academicians, professionals and administrators.

Mandate of the Magazine

While there are no records pertaining to the genesis of the publications launched long ago, it may be assumed that while initiating the popular science magazines being brought out by NISCAIR, the CSIR had more than just dissemination of scientific information as its aim. While the purpose purportedly was to reach scientific information to a wide section of the civil society at a cost they could easily afford, a special emphasis would also have been on bringing to light Indian contributions and advances in science, inculcating a rational outlook in the country's citizens, and ensuring wider engagement of the civil society with scientific issues.

Based on the discussion in the preceding paragraphs, the following mandates of *Science Reporter* can be safely assumed:

- Disseminating Scientific Information: The major mandate of *Science Reporter* is raising the level of scientific awareness of the country's citizens by informing them about current scientific developments, providing them insights into controversial scientific issues, giving explanations to complex scientific concepts, bringing views of experts in various fields of science through interviews, giving biographical sketches of scientists with special emphasis on Indian scientists, reporting on scientific events and reviewing scientific publications.
  To add variety to the content and impart dynamism to the magazine, several new columns have been conceptualised and introduced in the magazine from time to time. Most of these columns were written in-house to ensure continuity and flexibility and have been very well received by our readers. Some of these columns are:

  - *Fascinating Animals*: Dealing with interesting facts on some rare and fascinating animals.
  - *IT Trends*: Dealing with issues related to information technology.
  - *Techniques in Science*: Dealing with techniques encountered in various fields of science.
  - *Know Your Body*: Dealing with various organs of the human body.
  - *Medical Innovations*: Detailing innovative medical or diagnostic instruments.
  - *Extinct Animals*: Focusing on animals that have gone extinct.
  - *Strange Plants*: Dealing with plants having peculiar characteristics.
  - *Floral Records*: Dealing with flowers that have set specific records.
  - *Healing Herbs*: Dealing with medicinal properties of plants.
  - *Amazing Adaptations*: Talking about animals that have evolved suitable adaptations to survive in harsh terrains.
  - *EcoTalk*: Dealing with environmental issues.
  - *Living Fossils*: Dealing with organisms classified as living fossils.
  - *For Her*: Talking about science in the kitchen.
  - *Natural Hazards*: Focusing on various natural hazards that wreak havoc.
  - *Health*: Dealing with health issues.
  - *Poison Sleuths*: On interesting forensic science stories.
  - *Entertainment Arena*: Dealing with new gadgets, computer games and educational CDs.
  - *Did You Know?*: Dealing with fascinating facts on various scientific topics.
  - *Ingenious Inventions*: Talking about innovative inventions.
  - *Safe Haven*: Featuring forest reserves, national parks and sanctuaries in India.
  - *Curiosity Corner*: Answering readers' questions on scientific topics.
  - *SciProject*: Covering new scientific projects and experiments.

The magazine also has an added fun and knowledge quotient in the form of science cartoons, fiction stories, puzzles, crosswords and quizzes.

- **Disseminating Information about Indian Scientific Achievements**: *Science Reporter* has always been alive to the need to give coverage to Indian efforts and achievements in the field of science.
  - Since 1960s, the magazine has profiled the achievements of scientific institutions in the country through a column called *Temples of Modern India.*
  - Another column, *CSIR in the Service of the Nation,* talked about the scientific contributions of CSIR laboratories.
  - Perhaps the most popular and well-received column that the magazine initiated and ran for almost two years was *India Can Do It.* The column featured fascinating success stories of Indian science through onsite visits of locations and interviews of experts by *Science Reporter* editorial staff. The stories focused on Indian achievements in the area of space, telecom, agriculture, defence, medicine, road technology, ship building, biotechnology, fisheries and so on.
  - The magazine introduced a column called *Unsung Heroes,* talking about the achievements of some little known Indian scientists. The magazine is currently also running a regular column on *Indian Scientists.*
  - Of course, *Science Reporter* also receives regular contributions from scientists in ICMR, ICAR, DRDO and several other scientific organisations and universities talking about scientific advancements in their respective organisations.

- **Inculcating Scientific Temper**: The Constitution of India provides that it shall be the duty of every citizen of India "to develop the scientific temper, humanism and the spirit of inquiry and reform". Therefore, inculcating rational thinking by demystifying so-called miraculous phenomena through articles and editorials and reporting on events focusing on scientific temper is also a major mandate of *Science Reporter.* With a large percentage of the Indian population still steeped in irrational beliefs, the magazine considers it a bounden duty to talk about such issues.

  In this context, it would be pertinent to point to the tremendous response from our readers in the column *Point Counterpoint* to the question: "*Should media play an active role in eradicating superstitions and faith in miracles so rampant in society?*" Of the hundreds of responses we received from our readers, many of which were published in three consecutive issues (January, February and March 2007), almost 99 % agreed that media did indeed need to involve itself more aggressively and actively in eradicating irrational myths from the Indian society. Many derided the role of the media in dishing out programmes that served no constructive purpose and were, on the contrary, only reinforcing old superstitious beliefs. Most were emphatic in their belief that media being the watchdog of the society was best suited to root out such ills from the society.

- **Ensuring Wider Participation**: While most magazines primarily aim at disseminating scientific information through expert inputs, *Science Reporter* has taken up a conscious position and an added responsibility of ensuring wider participation and encouraging contributions from a wider section of the civil society to bring about greater engagement with scientific issues in the country. Consequently, qualification and experience is no bar for those contributing to the magazine. Novelty, topicality and wide appeal are what count.
  Also, regular contributed columns are not encouraged so as to keep space for the maximum number of contributors. There is also no effort to depend on expert and professional writers; on the contrary, new contributors are encouraged and guided in the nuances of science writing to enable wider coverage of views and opinions and also to bring more people under the ambit of science. However, contributions from experts are indeed sought in case of special topics and issues.
- **Catching Them Young**: A self-imposed mandate of the magazine is also to attract readership and participation from students. As has been shown earlier, although the magazine is read by people from all walks of life, it enjoys a dedicated readership among students throughout the country. This is because it provides latest updates from the world of science, insight into complex scientific projects and phenomena and answers to questions (*Test Your Knowledge* and *Fun Quiz*) from different fields of science at a cost well within their reach—these scientific articles and updates are especially helpful for students preparing for various competitive examinations.
  Some interactive columns have also been introduced in the magazine over the years to enable widespread participation from the youth. These include:

  – *Puzzle Corner*: Since early 2000s, *Science Reporter* has been carrying a prize puzzle every month to be answered by the readers. As stated earlier, almost 800–1000 entries are received every month in response to the prize puzzle, and the winners are decided on the basis of a draw of lots from among the correct entries.
  – *Point Counterpoint*: This column is devoted to garnering views and opinions of readers on various burning issues in science. The column, aimed at helping especially the young readers crystallise their views and thoughts on scientific issues of importance, is receiving enthusiastic participation from students.
    Some of the questions posed in this column include: *Politics and crime are given more coverage in the media than science; Keeping animals in zoos is unethical; A dedicated science channel is necessary to disseminate authentic scientific information and create scientific temper; With the advent of internet, television and computer games, we are now communing with a lifeless collection of microchips, not each other; Can science and technology help tackle corruption?; Is it morally acceptable to experiment on animals to develop products and medicines that benefit human beings; All politicians need to be scientifically aware; Are designer babies acceptable* and so on.

- *Science Fiction and Science Cartoon Competitions*: In the year 2004, a science fiction and science cartoon competition was announced with the belief that science fiction and science cartoons can be an effective way of inculcating interest in science among the younger readers. In view of the tremendous response from the readers, the competitions have become an annual affair with increasing number of entries being received every year. Beginning with about 20 entries in science fiction and 30 entries in science cartoons during the first year, the numbers have now increased to more than a hundred in both the categories with teachers and principals of schools taking active interest in motivating their students to take part in the competitions.

- **Mentoring Young Writers**: At *Science Reporter,* effort is also put into encourage young and budding science writers to display their writing talent. This includes giving them tips on selection of topics, collection of information and presentation of facts; in effect guiding them into the nuances of popular science writing. Consequently, the magazine can today boast of a huge repository of young writers, many of whom have carried forward their love of writing even after joining professional streams. The writing capabilities so gained also, hopefully, give them an edge in their professional life.

Challenges in *Science Reporter*

While communicating science through *Science Reporter* is an extremely exciting and fulfilling proposition, it is quite challenging as well. While some of these challenges are institutional, others are self-imposed, and there are some that are symptomatic of the state of science journalism, and perhaps also reflecting the prevailing general apathy towards science communication, in the country.

In-House Limitations

Science communication as an intellectual enterprise with its intangible outcome only indirectly translating into socioeconomic development has to still vie for trained science communicators. This is perhaps an indication that this area is not considered important enough to be prioritised for intensive deployment of human resource in public policy. Over the past several years the editorial team of *Science Reporter* has been continuously depleting due to retirements and transfers with no recruitments having taken place for the past more than eight years. Passion apart, the author has been single-handedly looking after the editorial aspects of the magazine, taking up alongside other institutional writing, editorial and managerial responsibilities. This has a severe limiting effect on intellectual creativity and functional flexibility of the enterprise. The urgent need to bolster the strength of the science communication community is discussed later in this chapter in the section titled "Increasing Bench Strength of Science Communicators".

Self-imposed Challenges

Publications survive and flourish on the strength, eminence and credibility of their contributors and authors. At *Science Reporter* we have taken it upon ourselves, consciously, the mandate to ensure wider contribution, not just from well-established and expert writers but also from first-time authors and students. This, however, adds much more to the editorial responsibilities since contributions from new authors have to be checked for facts, authenticity as well as plagiarism, and extensive editing is often required to make the articles presentable and appealing for the readers.

Plagiarism

The advent of the Internet has resulted in a drastic increase in instances of plagiarism, more so in popular science writing. Unfortunately, this malaise is not limited to students and first-time authors only but is widely spread across senior experts, academicians and professionals too. Checking for plagiarism is a painstaking and time-consuming task, also not always fool-proof.

Lack of Participation from Scientific Community

While the scientific community in the West is more geared towards presenting their scientific work to the public through lectures and articles, in India this seems to be lacking somewhat. While *Science Reporter* does receive occasional articles on new developments from scientists working in various organisations, in general scientists in the country are still not geared and motivated enough to spend time in preparing articles on their work or on other scientific topics of importance for the general public.

Perhaps there is still a debate and reluctance within the scientific community whether to don the mantle of a communicator or concentrate on their primary scientific work. In fact, not long ago communicating science as an activity was even looked down upon in the scientific community. Says de Semir (2010): "Once upon a time in the world of science, sharing scientific work with the press was heresy.

Journalists, according to the common wisdom, would get it wrong, your research would be distorted, and your colleagues would see you as little more than a shameless grandstander. Science popularisers such as the late Carl Sagan, a master of adroit science communication, were excoriated by some of their colleagues for the questionable practice of trying to make science accessible" (de Semir 2010).

It is accepted wisdom today that scientific knowledge needs to be shared with the civil society to bring about more acceptance of scientific endeavours. In a strong argument for science communication, Strelau (2004) says: "A scientific discovery that does not leave a scientist's laboratory or research field, often remains in oblivion." In the United Kingdom, the Science and Technology Committee of the House of Lords recommended a direct dialogue with the public in the year 2000 (House of Lords, UK 2000).

Besides, in the Declaration on Science and the Use of Scientific Knowledge adopted in the World Conference on Science in 1999, sponsored by UNESCO, it was established that: (1) scientific knowledge should be shared; (2) real cooperation is needed to exist between governments, civil society, the business sector and scientists; and (3) scientists should be governed by the corresponding ethical standards (http://unesco.doc.unesco.org/images/0012/001207/120706e.pdf). Besides, scientists in India are also not very forthcoming with information about their work due to several bureaucratic and administrative hurdles they face in their organisations.

Lack of Participation from Teaching Community

As noted before, *Science Reporter* enjoys wide readership among the secondary level students. Unfortunately, however, our efforts at roping in the teaching community to encourage either reading habits in their students or motivate them to write on scientific topics did not succeed.

In 2004, it was proposed to constitute a *Science Reporter Advisory Panel* with the purpose of making science teachers consultative partners in the magazine. This was with the belief that consultations with teachers could lead to more focused content in the magazine that would be more appreciated by students. Subsequently, letters were sent to principals of schools in Delhi apprising them of the proposed panel's relevance and requesting them to nominate two science teachers each to the panel; for logistic reasons, only Delhi schools were targeted.

About 25 prominent schools from Delhi responded to the request. Subsequently, meetings were conducted, the first with school principals and later with the nominated teachers in batches of 15. However, the enthusiasm of the teachers displayed during the meetings somehow did not translate into the desired action on the ground. While few articles were contributed by some teachers, student participation remained negligible. Eventually, having not lived up to its purpose the *Science Reporter Advisory Panel* was quietly wound up.

Restricted Reach of Science Reporter

The magazine has maintained its circulation figures over the past several years based largely on word-of-mouth publicity despite the advent of alternative media such as the Internet. In fact, *Science Reporter* has in recent years even added to its circulation figures due to its online presence. However, there is still a huge potential for extending the reach of the magazine, which could be brought about through more institutional support as well as networking with stakeholders.

The Way Forward

Spawning Science Publications

Today, it cannot be said anymore that the civil society is scared of science. The public hungers for more information on scientific issues that concern them directly such as health, food, diseases, agriculture and also for issues that interest them such as gadgets and computers, space and astronomy. The public is also curious to know about issues that have the potential to affect them in the future such as nuclear energy, genetically modified crops and foods, climate change, impending environmental disasters and so on.

There is, therefore, great scope for science magazines and science journalists to tell the stories of science in a language the public can easily understand and in a format they can identify with. Unfortunately, most efforts at bringing out high-quality science magazines flounder due to commercial considerations. However, in a huge and diverse country like India there is enough space and scope for more science magazines. Perhaps, the government can take a lead to spark off initiatives and motivate more organisations to bring out science magazines catering to the information needs of the civil society.

Former President of China Hu Jintao once emphasised: "Innovation and science popularization are two important aspects of our science and technology engagement. Science popularization is regarded as part of the accountabilities of science and technology departments, research institutions and associations and societies within the scientific community" (Pitrelli 2005).

Targeting the Young

Since magazines such as *Science Reporter* find a dedicated readership among the student community, they can serve as potent tools to develop an understanding of scientific issues in the young readers and perhaps motivate some to take up science as a career. However, schoolteachers being directly in touch with students need to take an active interest in motivating students to have a passion for reading scientific literature as well as inspire them to research and crystallise their thoughts on contemporary scientific issues and write about them. Publication of such write-ups in science magazines is bound to serve as encouragement for the young students.

Therefore, it would be appropriate to suggest that the teaching community needs to be encouraged to motivate students to think on scientific lines as well as try and give their thoughts tangible outlet through science newsletters or leaflets at the school level so that we have a generation of citizens with an appreciation for the nuances of science communication.

Networking Between Scientists and Journalists

More and more scientists need to be motivated to engage in public outreach of the scientific work they are engaged in. This type of science communication could be further facilitated by networking between scientists and communicators, which would help the scientist understand the journalistic points of view while preparing an article.

According to Gascoigne and Metcalfe (1997), scientists with little or no media experience had far less trust in the media because they thought it trivialised and distorted the science. This group was more likely to judge science stories in their own scientific terms rather than from the effect the stories would have on the readers and viewers.

A networking between scientists and science journalists would not only help scientists gain a journalistic insight into communication, but will also help journalists develop an understanding of the intricacies of the scientific method and how to avoid exaggerating or trivialising scientific endeavours.

Increasing Bench Strength of Science Communicators

Lastly, there has to be a realisation in the state machinery throughout the country that science communication is an instrument that can be employed in a big way in national development. By empowering citizens with scientific information, helping them to take informed decisions and also preventing them from falling prey to false and malicious propaganda science communication can provide a big thrust to progress.

There is, therefore, a need to bolster the strength of the science communication community in the country by providing training opportunities and also creating employment opportunities for science communication personnel. While private companies have realised the need for specialised corporate communication units, there is no such effort in government scientific organisations where communication continues to be regarded as a peripheral and part-time activity. There is a need for all scientific organisations and universities to employ trained science communication personnel to reach out to the public with appealing and palatable science stories.

The relevance and importance of science communication can never fade out. With changing times and changing paradigms, and ever new developments in science, the need to demystify and deconstruct ideas, concepts and breakthroughs will always be there. Communication and propagation of scientific ideas to the civil society will strengthen the democratic fibre of the country and boost progress. If we do not invest in science communication today, we may regret it tomorrow.

References

de Semir, V. (2010). Science communication and science journalism. *Media for Science Forum*. www.mediaforscienceforum.com

Gascoigne, T. H., & Metcalfe, J. E. (1997). Incentives and impediments to scientists communicating through the media—A preliminary report. *Science Communication, 18*(3), 265–282.

House of Lords UK (2000). Science and technology—Third report. http://www.publications.parliament.uk/pa/ld199900/ldselect/ldsctech/38/3801.htm. Accssed June 21, 2014.

Patairiya, M. (2007). *Science journalism in India*. http://www.pantaneto.co.uk/issue25/patairiya.htm. Accssed June 21, 2014.

Pitrelli, N. (2005). The Chinese dream regards science communication. *Journal of Science Communication, 4*(2). http://jcom.sissa.it/archive/04/02/F040201/jcom0402(2005)F01.pdf

Raina, D., & Habib, I. S. (1989). Cultural foundations of a nineteenth century mathematical project. *Economic and Political Weekly, 24*, 2082–2086.

Strelau, J. (2004). *Critical topics in science and scholarship*. Amsterdam: ALLEA Biennial Yearbook.

Wilcox, C. (2012). It's time to e-volve: Taking responsibility for science communication in a digital age. *Biological Bullettin, 222*, 85–87.

Challenges of Communicating Science in Regional Languages: Experiments in Kannada

A.S.K.V.S. Sharma

Abstract Promoting scientific temper among the citizens, a duty enshrined in the Constitution of India, is served well when matters of science are communicated to the public. The plurality of language and other cultural aspects of Indian society largely hinder uniform communication among the public and efforts are on to communicate science in every Indian language. As one of the states that ventured early into modern education, Karnataka demonstrates the possibilities of communicating science in regional languages. Over 4000 books on science have been published in Kannada during the past hundred years. The state has the distinction of hosting seven recipients of national honour the National Council for Science and Technology Communication Awards for science communication in print and electronic media. Rapid changes in the fields of mass media, especially the digital ones, the primary and higher education, and the science and technology are posing fresh challenges. Adaptation of uniform computing platform, access to reliable information resources and problems thrown up by media preferences are some of the issues to be addressed. In addition, manpower sensitive to the world of science and trained in communication is a pressing need to effectively face these challenges.

Keywords Science communication · Vernacular language · Local languages · Jargon · Scientific terms · *Vijnana* · Karnataka

Introduction

Unbiased and rational decision making forms the cornerstone of fair governance. The Constitution of India has given prominence to "develop the scientific temper, humanism and the spirit of inquiry and reform" and has declared it as one of the

---

Views expressed in the chapter are the author's own and not necessarily shared by the editors.

A.S.K.V.S.Sharma (✉)
CSIR—Central Food Technological Research Institute, Mysore 570 020, India
e-mail: sharma@cftri.res.in; kollegala@gmail.com

P. Bagla and V.V. Binoy (eds.), *Bridging the Communication Gap in Science and Technology*, DOI 10.1007/978-981-10-1025-5_12

eleven fundamental duties of the citizen of India (http://lawmin.nic.in/coi/coiason29july08.pdf). Scientific temper or scientific outlook need not be the prerogative of the scientist alone. Progress can be achieved only when the scientific outlook prevails over the innate conservatism. In the case of a society, this outlook acts as an antidote to the evils of prejudice and superstitions (Narlikar 1993). The prejudices need not be traditional alone. The Palampur Declaration (Suboth 2013) argues that without essential support of scientific knowledge base, the rapid pace of technological intrusion could be creating a cognitive gap that is filled in by new forms of religiosity or superstitions that could be detrimental to the progress. Science communication thus becomes inevitable in a rapidly modernising society for developing a scientific outlook over issues affecting it. In India where traditional faith rules and more than 27 % citizens are unlettered, clear and timely science communication is an essential requirement for improving the public understanding of science (PUS).

What to Communicate? And to Whom?

The first question encountered by a science communicator is to communicate what? Or what one should consider as science? Today, when the world of science is fragmenting into more and more specialised branches, the question assumes huge significance. The Royal Society's Committee on the Public Understanding of Science (COPUS) for the purpose defines science very broadly including mathematics, engineering, technology, medicine and all aspects of applied science. The understanding as defined by COPUS includes comprehension of the nature of scientific activity and enquiry (the method of science), and not just knowledge of some facts.

The committee felt that a better understanding of science by the public is essential not because it would help right decisions, but because the decisions made with adequate understanding of issues would be better than decisions made in the absence of such understanding (Walter 2010). For the purpose of PUS, COPUS defines the public as:

(a) Private individuals, for their personal satisfaction and well-being.
(b) Individual citizens, for participation in civic responsibilities and as members of a democratic society.
(c) People employed in skilled and semi-skilled occupations, the large majority of which now have some scientific content.
(d) People employed in middle ranks of management and in professional associations and trade unions.
(e) People responsible for major decision making in our society, particularly those in industry and government.

In other words, everyone other than scientists are the public and the science to be communicated or understood need not be limited to those who form the basis of community decisions. While these definitions suffice for a literate and English-speaking UK, additional parameters of language and literacy are to be included for a pluralist India.

Need for Science Communication in Kannada

Karnataka follows the trilingual policy with Kannada as the official language of the state. Although the literacy rate of 75.36% is encouraging, not all literates could be considered scientifically literate or have adequate command over English to learn about science on their own. In Karnataka, science education is imparted in Kannada up to higher secondary (Class XII) level, and even this is optional. Of the literate, only about 3% are graduates in science or technology (Census of India 2011), while the others migrate to humanities or other streams after their secondary school. On average, only about 20% of those who pass the matriculate examination remain connected with science till graduation. The remaining 80% pursue their education in Kannada medium, thus getting disconnected from both English and science. Therefore, any communication in English in Karnataka leaves out a majority of stakeholders from discussions on science. The situation elsewhere in other languages cannot be much different, and thus, science communication in regional languages like Kannada becomes an obvious need.

Communication through Print Medium

Since the invention of printing press, one of the main modes of mass communication has been the print medium, through books and periodicals. Although science texts on some specialised topics were prepared in Kannada earlier, they were restricted to accounts of traditional knowledge systems. Modern science communication in Karnataka can be traced back to 1880s. The Christian Literature Society in the then Mysore province had published small booklets on topics such as eclipses (1884), description of the cosmos (1884), stars and comets (1891) and description of strange animals of ancient world (1902). These appear to have been intended to evoke interest about the western knowledge among the neo-convert Christians (Somasekhar 2011). Works of missionaries related to botanical or geographical explorations also were rendered into Kannada during the last decade of 19th century.

A conscious effort to bring modern science into Kannada began around the beginning of the 20th century with the establishment of a society called Karnataka Vijnana Pracharini Samithi (KVPS). While earlier efforts were either individualistic or efforts of missionaries, KVPS took up the challenge of introducing the modern concepts of science to Kannada-reading public. The Samithi published *Vijnana*, meaning *science*, the first popular science journal in Kannada. It also organised public lectures and published the lectures in book form similar to the Royal Society, UK. However, there was a difference; the authors of the books of KVPS were enthusiasts of science and not actual practitioners of science.

Science communication in Kannada seems to have become synonymous with publication of books. The Karnataka Rajya Vijnana Parishat (KRVP), a nonprofit organisation that is active in science communication activities, has compiled a bibliography of more than 4000 books published during 1880 to 1998. Based on these, Adyanadka Krishna Bhat, a recipient of NCSTC Award for Life Time Achievement in Science Communication, recognised three distinct periods in the evolution of Kannada science communication (Bhat, A.K., personal communication): the last two decades of 19th century to 1930, 1930 to 1960 and 1960 to 1990. The classification is based on the perceived intent of communication in these publications. The first period is the period of wonderment, the second period is of consolidation and the third is a period of reappraisal when new experiments in science communication and widening influence of media could be felt.

During the period of wonderment, one could see that many of the publications focused on passing on the information and knowledge generated in the West. However, it is during the 1960s that Karnataka saw publication of seminal literature fulfilling the needs of categories (b) and (c) in the previously stated COPUS definition of the public. Noted among these are: *Vijnana Prapancha* (1959–1964), an encyclopaedia of science compiled and published by Jnanpith awardee litterateur K. Shivaram Karanth and *Jnana Gangotri* edited by K. Niranjan and published by Karnataka Sahakari Prakashana in 1968, besides scores of textbooks.

Communication through Periodicals

It is during the period of reappraisal—that is, 1960 to 1990—that science writing came to the streets through writings in daily newspapers and magazines. These developments could be traced to the active support of the government for promotion of science communication in regional languages. With the three-language formula being adapted in 1965, states such as Karnataka, Tamil Nadu, Kerala, Andhra Pradesh and West Bengal promoted education in regional languages providing the necessary impetus for generation of scientific literature in regional languages (Raza et al. 2009).

Karnataka saw a bevy of activities by academic and research institutions during the period. The universities of Mysore and Bangalore (two of the three universities then present in Karnataka) began publishing science magazines. The University of Mysore (UoM) embarked upon an exercise to publish an encyclopaedia including science topics and a voluminous dictionary providing a vocabulary that helped later communicators. While the University of Bangalore published a monthly called *Janapriya Vijnana*, the UoM started a quarterly *Vijnana Karnataka*. The latter had more of academic writings than real popular science.

The period also saw Kannada science literature being influenced by political philosophies. The resurgence of socialist movement saw several Russian popular science books, published by Mir Publishers, being translated into Kannada (the phenomenon also happened in other languages). Whether these books gave any gain to Marxism in India or not, their impact on the youth of the time is recognised.

Connect with the Public

All the efforts described earlier could be called efforts at creating interest in science and resources for science education. Science literature aiming to influence the public opinion began to appear only during late 1960s. It was also during this period that new generation newspapers and periodicals appeared in Karnataka providing a platform for short, informative or critical articles on science. A new breed of writers began contributing educative articles, especially on medicine, agriculture and mathematics. An analysis shows that a majority of early writings, both books and free articles, were related to medicine. During 1980s, environmental issues began getting prominence in the newspapers. The Appiko movement (a localised movement for saving trees in Karnataka similar to the Chipko movement in the Himalayas) gave space to a flurry of articles and books on environment. These efforts in communicating environment issues went beyond just informing, and shaped public movements against several irrigational and power projects to protect environment (Rao 2011).

Another noteworthy development during 1980s was the birth of anti-nuclear movement in Karnataka. A series of articles by Nagesh Hegde on the safety aspects of nuclear energy, environment, nuclear bombs, nuclear winter and others created an unusual interest among the public forcing the Government of Karnataka and the Department of Atomic Energy to organise a public debate on the pros and cons of Kaiga nuclear power project. Although the conclusions arrived at in these science articles are questioned by many, the fact remains that they not only kindled the interest of public in issues related to nuclear energy, but also set a new trend of science writing bordering on sensational in Kannada. The Kaiga debate organised in the Indian Institute of Science (IISc) is considered as the first example of PUS activity in Karnataka.

Late 20th century also saw establishment of institutions like KRVP, especially interested in engaging the public in science. KRVP in association with the Karnataka State Council for Science and Technology (KSCST)—an initiative of IISc and Government of Karnataka—took up popularising Astra stoves across Karnataka. It also organised the Karnataka part of the Bharat Jan Vigyan Jatha (1986)—a science communication movement styled as procession. The Parishat also began publishing a popular science magazine *Bala Vijnana* in an effort to promote interest in science among children. Although it has only limited circulation, the magazine is the only long-standing science magazine in Kannada as of 2016.

The period also saw electronic media being used extensively for developmental communication. Live programmes on stargazing by astronomy enthusiasts were a novel experiment in engaging the public. Prof. G.T. Narayana Rao, A. Krishna Bhat and few others used radio to communicate astronomy to remote places through these programmes. Programmes on environment, evolution and history of mankind were also aired on the radio as part of a national effort in development communication.

Post-1990s Karnataka saw a boom in periodicals publication. Two new science magazines were launched (*Vijnana Sangathi*, Kannada University, Hampi, first published in 1994 and *Vijnana Loka*, Karnataka Science and Technology Academy, Bengaluru, first published in 2007) and several newspapers also began publication. Among the magazines, *Adike Patrike*, a monthly devoted to agricultural technology, stands out as the only private publication with large circulation. All other popular science and science education magazines rely on government funding for sustenance.

The liberalisation of economy which happened around 1990s spurred the growth of periodicals, newspapers and television channels. A welcome outcome of this was the publication of supplements by the newspapers, including those devoted to science. Weekly supplements and weekly columns on science became common among leading newspapers helping a new breed of columnists, some of whom were active scientists and technologists, to connect with the public. Some television channels (such as ETV) have been telecasting applied science programmes, especially in agriculture.

Science Communicators in Kannada

Modern science entered Kannada psyche only after it was introduced in the curriculum. The vocabulary was also not adequate to communicate new concepts of modern science. It is not surprising then that the early communicators were teachers inspired by and interested in modern science (Rao 2011). No wonder the general trend has been of teachers (in higher and lower education) writing science articles. Instructions to authors given in popular science magazines like *Bala Vijnana* still ask authors to target 10th grade pass students! An issue of the magazine carried articles, all written by secondary school teachers (Bala Vijnana 2014). Practitioners of science themselves being science communicators is rare in Kannada, except in the field of applied sciences like medicine and agriculture.

On the other hand, examples of nonscientists venturing into science communication are available in Kannada. Noted among these are K. Shivaram Karanth, a Jnanpith awardee litterateur. As a renowned author in Kannada, he also wrote *Bala Prapancha*, an encyclopaedia for children and *Vijnana Prapancha* mentioned earlier. A staunch critic of nuclear programmes, peaceful or otherwise, his was a forceful voice asking for a public debate on Kaiga nuclear power plant. The other litterateur who was equally popular for his science articles was Dr Poorna Chandra Tejasw. Many of his novels revolved around scientific concepts. His musings on insects, animals and plants were built on a sound scientific platform. Kannada literature also has the distinction of having the only example of a popular science book being awarded the national Sahitya Akademi prize. Prof. B.G.L. Swamy's *Hasuru Honnu*, a book on taxonomy of plants received the Sahitya Akademi prize in literature category! There were also bureaucrats like H.Y. Sharada Prasad (Did You Know? 1947) and T.K. Tukol (Rural Development and Agricultural Development: A Practical Guide, 1937) who wrote science in Kannada. However, these are exceptions.

Popular lectures and mathematics communication have been conspicuous by their rarity in Kannada. KVPS used to organise popular lectures during the beginning of 20th century. The UoM also organised public lectures by eminent speakers on topics of general interest and published these lectures as booklets. As these lectures were organised in schools and colleges they have largely remained away from the public except in the form of published text. There are individualistic efforts that are noteworthy: V.S. Siva Sankara Sastry, a bank official and a mathematics enthusiast, and Prof. M.R. Nagaraj, a retired professor of chemistry have set exceptional examples of organising demonstrations and lectures in homes or social gatherings.

It is generally agreed that there is a dearth of competent science communicators who could effectively convey the principles of science and their implications to nonscience audience. A concerted effort to train science writers was conceived by KRVP, which has now become a national programme coordinated by NCSTC. KRVP has been conducting workshops on science communication since 1990s. The aim is to feed human resource to the field of science communication, including the electronic media.

Current Status

The present seems to be unusually conducive to science communication in regional languages. There is more space available in every medium for science if we could use it. Increasing number of media, both print and electronic, is creating both new opportunities and newer challenges for science communicators. Internet has brought scientific resources closer to communicators. Gone are the times when science communicators had to spend days scouring for information in far off libraries. Digital publication and e-libraries, open source publication, arXives, e-prints and other tools have made recent developments in science and technology available at our own doorsteps.

Columnists Nagesh Hegde, Sudheendra Haldodderi, Kollegala Sharma and Dr U.B. Pavanaja, exploited the opportunities offered by the Internet and electronic publishing to bring news of latest developments to the public at the earliest.

A browsing of their work shows that their writings were not restricted to any specific discipline, but they communicated on every branch of science, although each showed a bias towards some specific field. The reason probably is a dearth of communicators with access to sources of fresh knowledge. It could also be due to the demands of the media.

New newspapers and specialised science supplements bring more opportunities for science communicators. This advantage is offset by the severe competition for circulation among the newspapers, which forces them to always look for 'exclusive' or sensational content even in science stories. Despite such expansion, journalists exclusively covering scientific issues are conspicuous by their absence.

Television provides another opportunity. It is a medium, which is not hindered by the usual hurdles faced by textual communication. One need not look for analogies, can communicate with limited vocabulary and supplement with animations for effectiveness or clarity. However, television programmes demands high cost of production. A via media would be voice overs or dubbing.

Social medium on the Internet provides another opportunity. Blogs, YouTube and e-magazines offer exceptional opportunity for communication. Kannada can fully exploit the potential afforded by social media, if the Unicode standards are adapted in Kannada computing. YouTube, however, has no such limitation.

Science magazines in Kannada are living a precarious life. The circulation of the three magazines (*Vijnana Sangathi*, *Vijnana Loka* and *Bala Vijnana*) taken together does not exceed 20,000 copies per issue, of which *Bala Vijnana* alone contributing to more than 80 per cent of the circulation. Publications of universities of Mysore and Bangalore have ceased to exist or are very irregular.

Conclusions

The present is the best time for giving a greater thrust to science communication in local languages. Several sources to help with technical vocabulary are available. The public is sensitised to the needs of knowing about science (*Kannada Jana Mana* personal communication). Media, especially newspapers are aplenty. The only hitch is the availability of communicators. Science writing workshops appear to have not yielded desired results. The by-lines appearing in the media and on the books reflect that only limited number of communicators are active. If there are any new additions, the number of new communicators is not commensurate with the numbers that have been trained. Another observation of the author as a participant and faculty of several training workshops is the near absence of journalists and researchers as participants in these workshops. More focused training to journalism students would be helpful. Participation by scientists needs to be promoted. Making science communication a part of pre-registration study for Ph.D. students would be a step in that direction.

References

Bala Vijnana. (2014). Published by Karnataka Rajya Vijnana Parishat, Vijnana Bhavan, 24/2, 21st Main Road, Banashankari II Stage, Bangalore.

Census of India. (2011). Chapter III, literates and literacy rates.

Narlikar, J. V. (1993). Need for scientific temper. http://www.iucaa.ernet.in:8080/jspui/bitstream/11007/1979/1/334E%20Need.pdf

Rao, H. R. R. (2011). Kannadadalli Vijnana Sahitya Nadedu Banda Daari. In H. R. R Rao, T. R., Anantharamu, & U. K. Udayabhanu (Eds.), *Kannadadalli Vijnana Samvahane, A collection of essays on science communication.* (pp. 1–19) Kempegowdanagar, Bangalore: Advanced Study Centre.

Raza, G., Singh, S., & Habib, I. (2009). Science, media and the public: Past and present. *Indian Media Studies Journal, 2*, 128–140.

Somasekhar, B. S. (2011). Aarambhada Vijnana Kruthigalu- Bhaashe Svaroopa. In H. R. R Rao, T. R., Anantharamu, & U. K. Udayabhanu (Eds.), *Kannadadalli Vijnana Samvahane, A collection of essays on science communication.* (pp. 21–61) Kempegowdanagar, Bangalore: Advanced Study Centre.

Suboth, M. (2013). A perspective on scientific temper in India. *Journal of Scientific Temper, 1*, 46–62.

Walter, B. (2010). Public understanding of science: The BA, the royal society and COPUS. *Notes and Records Royal Society Journal of the History of Science, 64*, S151–S160.

Challenges Faced by Science Journalists and Communicators Working in Vernacular Languages and Insights Pertaining to Science Communication Courses

Nimish Kapoor

Abstract There is an increasing communication gap between citizens of India who speak vernacular languages and scientists who read and write in English using a lot of scientific and technical terminology. It is up to the scientists and science journalists to bridge the linguistic divide and tackle the challenge of translating scientific jargon into vernacular languages so that people can understand science communication. There is no nodal agency or service of science communication available in India to provide scientific and technological content from laboratories to national or vernacular media. Challenges faced by journalists to communicate science in vernacular media, need of a nodal agency to provide science news, training of scientists and journalists in science communication and utilisation of trained science communicators in scientific organisations should be seriously considered if we wish to reduce the gap. This chapter deals with the approach to bridge the gap between science and society connected by vernacular media.

Keywords Science and technology communication · Science journalism · Vernacular language · Courses in science communication

Introduction

There are several challenges of science and technology (S&T) including environmental issues, public health and bioresources management that need to be communicated by well-informed groups of science communicators and science journalists in vernacular languages. Everyone has a right to know about scientific

---

Views expressed in the chapter are the author's own and not necessarily shared by the editors.

N. Kapoor (✉)
Vigyan Prasar, Department of Science and Technology, Government of India, A 50, Institutional Area, Sector 62, Noida 201 309, India
e-mail: nkapoor@VigyanPrasar.gov.in; nimish2047@gmail.com

P. Bagla and V.V. Binoy (eds.), *Bridging the Communication Gap in Science and Technology*, DOI 10.1007/978-981-10-1025-5_13

developments in his or her vernacular language that will affect their lives (Gopichandran 2013a, b). There are 22 official languages and hundreds of living languages spoken in various parts of India. However, only a small number of science journalists and communicators are involved in science communication through vernacular languages. Scientists are enthusiastically involved in their research and least interested in communicating research findings in vernacular languages. Similarly, media has little interest in science coverage due to unavailability of scientific information in simple and/or vernacular languages and their priorities for nonscience issues. Laypersons including weaker sections of society have many unanswered questions related to health, nutrition, agriculture and environment, which can be addressed by the scientists through vernacular media. In the same way, media can be attracted to create more space and scope for science communication by regular interactions with scientists. There is a need for synergetic efforts between journalists covering science and health issues and experts in those fields, to avoid passing of inaccurate information to the readers. During September 2008 when rumours related to Large Hadron Collider, the world's largest and most powerful particle accelerator, were sensationalised by the media that its activation could spark a *big bang* that would destroy the world, a teenage Indian girl in outskirts of Indore, Madhya Pradesh was reported to have committed suicide after being traumatised by the reports (Telegraph 2008). Efforts have been made by both government and voluntary organisations to enhance the public understanding of science in vernacular languages to establish scientific temper in our societies (Sinha et al. 2011). However, a remarkable gap still persists between scientific knowledge and common people.

Some Indian universities now have a full-time postgraduate course (masters in science communication) for over a decade, and hundreds of trained science communicators are being produced by these universities every year (Patairiya 2007). However, only few opportunities are available for trained science communicators in scientific institutions and media. Attention of policy makers and job providers is required to absorb these potential science communicators for Indian laboratories, other scientific organisations and media as press officers (public relation officers) science communicators or science journalists. Trained science communicators from universities can be helpful to reduce the information/perception gap between science and public.

Science Coverage in Vernacular and Hindi Newspapers: Scientific Terminology and Awareness Are the Major Barriers

English is the language of S&T in India, and nearly all S&T journals are being published in Western languages. Scientific and technical content available in laboratories and scientific institutions is not easy to understand for Hindi- or vernacular-speaking journalists. It is a big challenge before the science communicators and journalists to communicate national or international research to the citizens in their vernacular languages. Journalists writing in vernacular language have to cope with the fact that much of their source information is in English. Scientists are not inclined to translate their research articles or findings into vernacular languages (Nautiyal 2010a, b). It cannot be expected from a vernacular language journalist that he or she be an expert of scientific and technical English. Ignorance of scientists in providing scientific information to the regional media, lack of information on local research in vernacular language and geographical as well as professional distances between science and vernacular media are also identified as barriers.

Coverage of science in vernacular and Hindi media is often superficial because content availability such as research reports or studies are available only in highly technical English. If journalists try to interact with a local scientist, many prefer to comment in English because they usually discuss their research in English and feel uncomfortable in communicating science in Hindi or vernacular language. Many Indian laboratories are working on the issues such as soil salinity, low-cost housing, low-cost toilets, tissue culture, high-yield crops, agricultural technologies and so on that are well connected with vernacular-speaking population or rural India. These themes have much potential to be covered by vernacular media. Due to lack of awareness on these issues and use of English in press releases, journalists in vernacular media avoid carrying science stories or if any interested journalists carry that story, it finds a very small space in newspapers (Gopichandran 2014). Vernacular language journalists experience difficulties in identifying and reaching scientists to cover their research findings in detail, and in obtaining news and information about research because of highly technical communication from research institutions. It has been reported that some institutions never send out any research news while having number of stories to share with media (Nautiyal 2010a, b).

Editors treat science as soft item, and it is considered for a small single column story that does not increase sale or readers' attraction. Editors and journalists have never received any training or orientation for science reporting (Bagla 2002). They prefer ready-made science news stories received from foreign media which may not be suitable for their region (Kumar 2013). The biggest celebration of Indian science —Indian Science Congress or achievements in space science like Mars mission or Mission *Chandrayaan* have never been covered as lead story by Hindi and vernacular press.

A study revealed that science coverage in Tamil print media is hardly 3.5%. They publish articles on new products or technologies for market and consumers' point of view (Arulchelvan 2010). More or less same conditions exist in Hindi and other regional languages. Another study conducted at Indore, Madhya Pradesh to compare science coverage in Hindi and English newspapers of India revealed that on average Hindi newspapers carried only 1.74% as science news while English newspapers covered 2.34% (Kumar 2013). This study revealed that English newspapers cover science-, health-, nutrition- and environment-related news as per the requirement of daily life with research details. Hindi newspapers need more space for better science coverage. *vigyan Prasar*, an autonomous organisation of Department of Science and Technology (DST), Government of India, conducted a survey of science coverage in media during 1999–2000. The study comprised 52 newspapers—31 in Hindi and 21 in English. The study revealed that on average only 3.3% of the items covered by newspapers were science items comprising 4.3% in English and 2.5% in Hindi (Kumar 2013).

Scientists of National Institute of Nutrition of Indian Council of Medical Research, Hyderabad carried out a research study on the quality of nutrition research reporting by leading daily newspapers in India (Maheshwar et al. 2014). They chose six newspapers (three English dailies and three Telugu newspapers) for their research study, namely *Deccan Chronicle* (with a daily net paid circulation of 658,037 copies), *The Hindu* (452,096), *The Times of India* (197,911), *Eenadu* (1,537,086), *Sakshi* (1,338,845) and *Andhra Jyothi* (573,857). This study revealed that one-fourth of the newspaper reports did not convey nutrition research results properly. Almost one-fifth of the reports had no mention of the source, which denies access to the reader to the original report. This study also revealed that more than 70 % of all the science reporting based on health and nutrition across the dailies was taken from secondary sources like foreign news agencies. Less than 30 % were based on the primary source where over two-thirds were published in English newspapers (Maheshwar et al. 2014). This survey clearly indicates that vernacular newspapers largely depend on secondary sources like foreign news agencies for science news. In another study conducted by National Institute of Nutrition it was indicated that percentage of space occupied by health topics in *Eenadu* (Telugu) and *The Hindu* (English) newspapers was 2.30 % (Maheshwar and Rao 2012). Survey also indicated that quantity of health and nutrition information covered by vernacular newspaper was less than that in the English newspaper.

Vernacular newspapers are more popular among the local readers and also reach the weaker sections of society who can understand or read only vernacular language. Efforts are required to provide research content and findings in local languages to the vernacular media and public for maximum dissemination of scientific information at grassroot level. Localising science by vernacular reporting is crucial to take information about science to the wider audience, allowing local communities to learn about scientific developments in a meaningful way.

No Official Agency, Centre or Service Is Responsible to Provide Research Findings or Other Updates on S&T to the Media

As of 2016, there is no responsible and exclusive authentic government (national or regional level) source available to provide science and technological research findings to the press in any language. There is no official source dealing with media for providing authentic scientific information on science, agriculture, environment or energy sector from a single window (Kapoor 2012a, b; Patil and Kokate 2011). Under this circumstance India's regional or vernacular press and Hindi media covering S&T are highly dependent on the science news provided by foreign news agencies. They translate the foreign news items and place them in newspapers, magazines and web portals (Patairiya 2007). Some interested media persons and science communicators involved in science reporting do translate the Indian science news stories into vernacular languages, but they are exceptions.

Vigyan Prasar initiated a science news/article service titled "readytoprint science page" during early 1990s. In this service a group of science writers was engaged to prepare popular articles and ready to print articles and news items were provided to newspapers and magazines. Ready-to-print science page service covered more than 30 editions of 20 newspapers with about 25 lakhs copies (Sinha et al. 2011). At present Science News Service by Press Trust of India (in English) and *Strote* Science Feature Service by *Eklavya* (in Hindi) are quite popular. Nevertheless, there is no service under the supervision of government available to showcase Indian scientific and technological development and provide content or information to the press.

There is an urgent need to establish what could be called a media science resource centre (Salwi 2002) or Indian science news service at national level with its regional centres in collaboration with interdepartmental/ministerial network, scientific organisations, scientists, science communicators and journalists where authentic information from various sources can be collected, reviewed, designed and packaged as per the science news requirement. This centre may act as a nodal agency to coordinate science news from various regions at national level and disseminate it across the country through the regional centres. This will not only catalyse vernacular media to cover the science stories, but also promote scientists–journalists interaction for better science reporting in regional languages.

There Is Limited Circulation and No Mechanism of Dissemination/Marketing of Science Magazines and Books in Hindi and Regional Languages

The popular science magazines and books in vernacular languages have become an important source of knowledge. Continuous efforts for publishing of popular science magazines in Hindi and vernacular languages have been made by science communicators at individual and institutional level. *Vigyan*, monthly published by *Vigyan Parishad Prayag*, Allahabad is the oldest popular science magazine, started in 1915. National Council of Science and Technology Communication (NCSTC) is supporting publication of popular science magazines including *Nirantar Soch* (Punjabi); *Vaigyanik Drishtikon* (Hindi); *Arivukkan* (Tamil) and *Vigyan Aapke Liye* (Hindi). In addition, *Virat Surya* (Hindi, Marathi and English); *Arivial Poonga* (Tamil); *Bigyan Jeuti* (Assamese) and *Vigyan Katha* (Hindi) were also supported by the council. CSIR's National Institute of Science Communication And Information Resources (NISCAIR) publishes *Vigyan Pragati* (Hindi), *Science Reporter*

(English) and *Vigyan Ki Dunia* (Urdu). *Electroniki Aapke Liye* (Hindi) by AISECT, Bhopal; *Aavishkar* (Hindi) by National Research Development Board; *Chakmak* and *Strote* (Hindi) by Eklavya, Bhopal/Hoshangabad; *Vaigyanik* by Bhabha Atomic Research Centre (BARC); *Vigyan Katha* by *Vigyan Katha Samiti*, Faizabad; *Discovery* (Hindi, English and Telugu with the softcopies in Malayalam and Urdu) from Science for You Knowledge Society, Hyderabad are some notable magazines. *Lok Vigyan Parishad*, USA/India publish only soft copy of *Vigyan Prakash* due to lack of funding. *Phal–Phool* and *Kheti* by Indian Council for Agricultural Research (ICAR) and *Kurukshetra* by Ministry of Rural Development are popular magazines designed for farming community. *Vigyan Prasar* publishes *Dream 2047* (bilingual Hindi–English magazine) as a resource material for students and teachers and *VIPNET News* (Hindi/English) for Vigyan Prasar Network of Science Clubs (Sinha et al. 2011; Mochahari 2013; Nautiyal 2010a, b; Patairiya 2007).

Vigyan Prasar, DST, Ministry of Science and Technology Publication Division, Ministry of Information and Broadcasting, National Book Trust and Ministry of Human Resource Development are the leading publishers of popular science books in Hindi and vernacular languages under Government of India. Lack of distribution and marketing strategies in government publication houses is one of the major challenges for dissemination of popular science magazines and books in vernacular languages. Dissemination of popular science books and magazines can be assured through collaborated effort by the publication houses mentioned earlier for wider distribution.

Of the magazines mentioned in the preceding paragraphs, limited copies of only *Vigyan Pragati* and *Kurukshetra* magazines are available at the bookstalls for sale. Other magazines have limited circulation or other mechanism of distribution which targets small number (few hundreds to thousands) of readers. It has been found that research reports or popular articles published in popular science magazines in Hindi and regional languages are generally translations of English research articles. Efforts are required to initiate original popular science writing in vernacular languages (Patairiya 2007).

Lack of Press and Public Relation in Vernacular Languages

The absence of PROs or press officers and trained science communicators in most scientific organisations who could communicate science in vernacular language is another challenge to promote science in vernacular press. Absence of media interaction in laboratories leads to distrust among the research community about the media and misconceptions about how the media works. Scientists also think that journalists misreport, misrepresent or sensationalise the facts during translation in vernacular language. This kind of misunderstanding prohibits scientists to share their findings with the vernacular press (Osama 2012; Clayton and Joubert 2012).

Many initiatives of research and development (R&D) that are designed in public interests in the areas of S&T are not being disseminated due to unavailability of PROs. Post of a PRO or extension officer must be made mandatory in a scientific organisation. A trained science communicator with the enthusiasm to communicate science to the public and understanding of vernacular media of the particular region can be an asset for any scientific organisation. Study reveals that the lack of full-time, dedicated communication staff is a major drawback in scientific institutions as well as district level establishments such as *Krishi Vigyan Kendras* (*KVKs*; agricultural science centres) of ICAR, especially designed to communicate agricultural science to farmers in vernacular language. Studies indicate that to manage and run *KVKs* or science communication work as per its mandate, self-motivated young science communicators or extension workers are required who are committed to work with stakeholders with the clear understanding of local problems and issues to be communicated in vernacular language. For this purpose priority must be given to self-motivated trained science communicators who can speak local dialects with good communication skills (Chauhan 2011; Patil and Kokate 2011).

In a survey titled "The Need for an African Science News Service" conducted by UNESCO and SciDev.Net revealed that "many of the scientific organisations agreed that there need to be more opportunities and training for scientists to communicate with the media, as well as the creation of press officers at research institutions to advise researchers on how to get stories to the media, and to provide a link with journalists … we need to train our scientists and researchers on how to communicate better, and what are the advantages of communicating through the media" (Clayton and Joubert 2012).

Scientists and Journalists Are Not Trained for Science Communication

Many scientists get no media training as part of their professional development and have absolutely no experience or sense of how to communicate their work to a lay audience or media. Scientists involved in the solutions of regional problems do not communicate in simple or plain language. Scientists must be trained in science communication as well as in public and media relations. Science communication training is also required for journalists (Gopichandran 2014) so that they can go 'beyond the press release' to develop science stories and feel confident in talking to scientists. More than 55,000 newspapers and periodicals are published in India, but there is an insignificant contribution of S&T content in print media.

There should be proper training for scientists and science journalists or communicators for development of their skills in translating jargon into vernacular and explaining scientific research in simple manner. Strong and effective group of science communicators is an important concern in R&D organisations and science communication institutions, as well as media establishments. Opportunities are needed to enhance scientists' communication skills such as dealing with media, popular writing for newspaper and answering stakeholders' questions in simple language. Scientists and journalists must be trained in the techniques to select relevant information from research and scientific developments for masses, popularising technical information into concepts easily understandable to nonscientific audience and engaging with the media (Bagla 2002; Osama 2012; Patil and Kokate 2011).

The literature reveals that there is a lack of trained science journalists in media establishments—newspapers and television channels. Generally, a reporter or writer who reports or writes political stories or covers campus issues is given the responsibility of science coverage (Bagla 2002). Without proper training, it is difficult for a reporter or writer to understand science, translate scientific jargon and repackage technical information into a form that is interesting and relevant to readers, listeners or audience. To meet these challenges, science communication training workshops must be organised regularly by various institutions associated with science and media (UNESCO 2007). These workshops will be helpful to scientists, journalists and other stakeholders to update them with scientific and technological developments and equip them with the necessary skills and practices to impart key messages to be communicated for variety of audience. As a result of these workshops new generation of science communicators may be developed (Salwi 2002; Nautiyal 2010a, b; Clayton and Joubert 2012).

Regular Interaction between Scientists and Journalists Is Absent

To ensure the effectiveness of science communication, the information must be delivered on time with consistency (Gopichandran 2013a, b). When a journalist approaches a scientist, he or he has to go first to the head of the institution for taking permission to interact with the scientist. This process often takes considerable time; very often, due to delay in getting permission, the reporters cannot meet the deadline of their assignment. Each and every scientist in scientific organisations must have reasonable freedom to interact with journalists, especially on the issues that are already in public domain and directly related with public welfare and societal benefit. Permission may be restricted to sensitive issues only. An interactive platform is required for journalists working in vernacular languages and scientists to meet and to break down the communication barriers they currently face. Regular interaction between scientists and journalists is required where scientists can understand the requirement of media (for instance, what kind of press notes and press briefs are required) and journalists can learn how to communicate factual scientific information with the scientific message to the target groups (UNESCO 2007).

Studies indicate that journalists lack skills and expertise in science reporting, including handling of academic information sources. Journalists try to get science news from various sources where role of scientists is very crucial to check their information and sources from where the information or facts are being gathered. Here, the big problem is that scientists are not trained to communicate their research findings to the media and masses (Bagla 2002). On the other hand, scientists can explain their research findings in seminars or in group discussion with other scientists. Sometimes, it has been found that the need for science to be understood by the public is not appreciated by many scientists or institutions. Science communicators are generally not appreciated by other scientists who are involved only in research. In the context of science communication in vernacular languages, scientists of various regions should come forward and share their R&D information with local media.

No or Very Less Space Is Available for Research Articles in Hindi and Vernacular Languages

Our research journals have no space for vernacular language research articles. As per norms, scientists write research articles in English and Indian journals publish articles only in English. For decades, researchers in Russia, Germany and other European countries have contributed significant research in their own languages. Indian contribution in research is not satisfactory in terms of quantity and quality. Indian scientists know that if they wrote in a language other than English, it would not be helpful for their promotions with least chance of their work to be quoted in any other journal. However, the moot point is that if an article on method of 'nuclear fusion on the table or method of getting gold from coal' is published in a journal, a layperson would definitely want to read it in his/her vernacular language. Similarly, articles written in regional languages on public understanding of nuclear energy or genetically modified crops will attract science journalists working in that region. It will give an opportunity to the journalist to cover that research in local language (Nautiyal 2011, 2014).

We should create some space for research articles to be written in vernacular languages by scientists or research scholars to protect the right of the common public to understand science in their regional languages. If done, then it will be a great effort to attract large number of students as well as readers from various regions of the country who have command over their vernacular language, and they would love to read scientific development stories in their language. Some space can be created in prestigious journal such as *Current Science* for a research article in vernacular language in each issue.

Some institutions did start research journals in Hindi language, for example, *Bhartiya Vaigyanik evam Audyogik Anusandhan Patrika* (BVAAP) by CSIR-NISCAIR. *Vigyan Parishad Prayag,* Allahabad also publishes a journal *Shodh Patrika* in Hindi.

Translating Technical Jargon and Scientific Language into Vernacular Language(s) Is a Challenge

Reporting science effectively in vernacular press requires a dictionary or glossary in that particular regional language which could provide suitable words to report science. Translation of technical/scientific content from English in vernacular languages is a challenge especially for majority of media persons having no science background. This indicates the need of preparation of dictionaries of scientific and technical words in vernacular languages and to standardise the usage of technical terminology in the particular language. It will help vernacular press to simplify technical information and transform it into a more interesting and relevant format from the viewpoint of local stakeholders. Generally, technical terminology created in vernacular languages and Hindi is very difficult to adopt by the laypersons. The challenge of translating the technical jargon and scientific language into the vernacular language(s) is different from literary translation. Meanings can be changed or lost in translation as the process involves additional tasks of interpretation with the knowledge of the subject (Osama 2012). Technical terms can be used as actual, or simple vernacular words can be created (Sinha et al. 2011; Nautiyal 2011, 2014).

Starting 1980s, the Council of Scientific and Technical Terminology (CSTT), Government of India has created glossaries/dictionaries of scientific terminology in Hindi and other regional languages (Sinha et al. 2011). Unfortunately, the technical terms coined by the CSTT are tough and difficult to get accustomed to. Scientific and technical words translated or coined in vernacular languages or Hindi are not easy to pronounce and memorise. Only government departments adopt and use that terminology from *Takniki evam Vaigyanik Shabdawali* (meaning list of technical and scientific words). Majority of media houses including vernacular press prefer to use common international terms in original as accepted by the target audience.

Standardisation of technical terminology in vernacular languages is an important task to communicate science for local stakeholders. Many technical words in regional languages derived from Sanskrit are common in more than one language. Selection and popularisation of those words can be a step forward. Continuous institutional and individual efforts are required to adopt a standard vernacular terminology of S&T. Centre for Science Communication, Cochin University of Science and Technology, Kochi, Kerala has developed a scientific glossary in Malayalam language with the support from Kerala State Council for Science, Technology and Environment. Similar efforts must be made to popularise science communication at regional level and among the vernacular media with availability of vernacular science dictionaries on the Internet.

Aspects Associated with Effective Science Communication Strategy in Vernacular Languages

An effective science communication strategy in vernacular languages may be summarised in the following questions which must be asked by a scientist, science journalist or science communicator when he or she is engaged in science communication or extension: What are we really trying to achieve? What is the actual problem of the particular area that needs to be solved? Are we trying to provide knowledge to the local people, change their attitudes or alter their behavior? Do we want to engage stakeholders proactively with two-way communication in their language, or simply inform them of what we are doing so that they passively support or do not resist it? Are our communication goals clear, realistic and achievable? Are our stakeholders too ambitious or too weak?

Identification of target audience and sharp and clear communication of messages as per the audience's knowledge level are other important aspects of science communication in vernacular. When elaborating a message, it is necessary to use language that is accessible and attractive to the target audience and to keep the message as simple as possible. Science communicator should avoid jargon, acronyms and unnecessary detail. Communicator must put himself/herself in the place of general public and give them what they need. The message must be realistic and focused on behaviour or action that are relevant and that they can actually pursue. Only when we are confident about our audience and our messages, can we determine the best way to transmit our message. Evaluating the effectiveness of a communication effort can be extremely difficult. However, we may get an idea with some key questions like "Has our message been received, understood and acted upon by the audience?" Feedback on these questions from the target group can be obtained through follow-up interviews or survey and from close observation of changes in behaviour. Such feedback is useful for future communication. Guidelines for scientists on communicating with the media prepared by the Social Issues Research Centre, Oxford, United Kingdom and Meta-Review by Media for Science Forum, Madrid, Spain are worth considering for sharp and clear communication by the vernacular science journalists and communicators (SIRC/ASCoR 2001; Media for Science Forum 2010).

Use of Web and Social Media in Vernacular Languages Can Penetrate Science News in Remote Areas

Increasing access of social media in remote areas clearly indicates its potential as an effective communication tool. It would be a wonderful tool for vernacular science journalists also to be informed and to build science stories. Social media in vernacular press could be useful if scientists of various regions are ready to share their findings in regional languages on social media. It will provide a platform to access information in regional languages, and journalists will be motivated to publish those stories without any difficulty.

Journalists complain that scientists or researchers publish their findings mainly in research journals and it does not matter to them that journalists cannot access these publications. Hence, if there is an avenue like social media, it will go a long way in enhancing and encouraging science journalism in vernacular languages. Web platform of social media may provide sufficient information of particular institute and other information on research, scientists' contact details and so on. Sharing of information with social media must be mandatory in each and every research institute, and local scientists may volunteer to communicate the work in vernacular language (Rohn 2011; Shipman 2012).

Now text can be shared in vernacular languages on social media like Twitter and Facebook. This has become very easy by using vernacular language typing sites/software like www.quillpad.in. It is one of the easiest typing software that picks up Hindi and other Indian languages, namely Tamil, Telugu, Bengali, Oriya and so on. Typed messages (in the form of news, feature or any form of popular article or information) in vernacular languages can be placed on Twitter or Facebook.

NISCAIR initiated a promising web programme entitled 'CSIR-NISCAIR Tube'. This web platform provides large number and variety of audiovisual programmes in public domain. There is enough scope in this initiative to focus on vernacular language science films and load these films on CSIR-NISCAIR Tube. This site can be converted into Indian science films' repository for showcasing India's contribution towards science films in various regional languages. Science Bloggers Association of India is also motivating science communicators to initiate science blogs in Hindi and vernacular languages.

Collaborative Efforts and Networking of Potential Science Communication Partners Working in Vernacular Languages Are Desirable

Formation of communities of science communicators, that is, networking of science communication agencies and individuals is an important action to be taken by the nodal agencies of science communication (Gopichandran 2013a, b). This will catalyse the collaborative efforts in science communication in vernacular languages through experiential sharing and motivate new generation of science journalists and communicators towards collective work at regional level. In India, there are many organisations working for science communication, especially for science writing or reporting in vernacular languages. All India People's Science Network, New Delhi; Tamil Nadu Science Forum, Chennai; *Kerala Sasthra Sahithya Parishad*, Thrissur;

Vigyan Parishad Prayag, Allahabad; *Marathi Vigyan Parishad,* Mumbai; Indian Science News Association, Kolkata; Assam Science Society, Guwahati; Swadeshi Science Movement/*Vijnana Bharati,* Kochi; *Jan Vigyan Vedika*, Hyderabad; Indian Planetary Society, Mumbai; Bangalore Science Forum, Bengaluru; Centre for Science and Environment, Ahmedabad; National Bal Bhavan, New Delhi; Pondicherry Science Forum, Puducherry; Science Centre, Gwalior; Madhya Pradesh Vigyan Sabha, Bhopal; Indian Science Communication Society, Lucknow; Karnataka Rajya Vigyan Parishad, Bengaluru and *Eklavya,* Bhopal/Hoshangabad are few of them (Sinha et al. 2011; Maheshwar and Rao 2012). Besides these organisations many universities offer science communication degree or diploma and produce science communicators. A strong networking among the science communication organisations and universities involved in science communication courses is desirable. This network can be promoted for SWOT (strengths, weaknesses, opportunities and threats) analysis of science communication in regional languages; reality checks can be initiated with the help of student science communicators to understand science and media relationship in vernacular languages. This network can be used to initiate science media news centres at regional level.

Vigyan Prasar has developed a database of science communicators in India recently, which can be helpful to establish a network of individuals and institutions working in the area of science communication.

Review of Science Communication Academic Courses Is Required

During the 100th session of Indian Science Congress in 2013, noted agricultural scientist Dr M.S. Swaminathan suggested in his address, “Our universities should help in developing science communicators who can explain to the general public in local languages the significance of important scientific discoveries. Similarly, issues of biodiversity, biotechnology, nuclear technology and nanotechnology need priority attention in efforts designed to bridge the scientist–society perception gap” (Kapoor 2014). To bridge this gap, some Indian universities have started an academic discipline of ‘Science Communication’ in higher studies. NCSTC, DST, Government of India initiated and University Grant Commission (UGC) recognised postgraduate courses in science communication in the Indian universities during mid-1990s. The idea behind these courses is to promote science communication and public understanding of science through a group of trained science communicators (Kapoor 2012a, b; Rajput 2008). Nevertheless, unfortunately, majority of decision makers associated with job creation in S&T and media industry do not know about these courses. It shows the unresponsive attitude of the system in which these courses are being conducted since early 2000s.

Science communication courses have been started with the aim and objective to produce professionals in the area of science communication. The broad objectives of the academic courses of science communication are to impart the basic science communication skills and techniques to the students to communicate effectively through different media with a view to stimulate scientific outlook, to promote science communication in media industry, R&D centres and corporate houses and to provide them hands-on experience to communicate critical issues of S&T at grassroot level. Science communication courses are very well designed with the theoretical and practical aspects of science communication and train students to become successful communicators in S&T. Students learn about the need, significance and role of S&T communication, S&T policies, history of S&T, emergence of modern science, landmark achievements of great scientists and science journalism in India. Science reporting and writing on health and environmental, water and sanitation, media and disaster management, peace journalism, corporate communication, rural communication, agriculture extension and issues related to climate change, global warming, genetically modified crops and nuclear technologies have been incorporated in the courses. Students also learn development communication, advertising, public relations, publication of science newsletters and media management in the light of science communication (Kapoor 2014).

Indian universities conducting these courses include University of Lucknow, Lucknow; Devi Ahilya University, Indore and Anna University, Chennai. National Council of Science Museum (NCSM), Ministry of Culture, Government of India in collaboration with Birla Institute of Technology and Science (BITS), Pilani also conducted master of science course in science communication. Few other universities namely Madurai Kamaraj University, Madurai and Makhanlal Chaturvedi National University of Journalism and Communication, Bhopal also started post-graduate diploma courses, which were discontinued due to number of reasons. Faculty members and students pointed out that university administration and academic councils did not support the courses due to lack of trained teachers of science communication; absence of direct employment as science journalists or PRO in scientific institutions or media industry; unavailability of standard books, manuals and systematic study material of science communication; inadequate financial or infrastructural support within the university system; absence of any scheme or policy to attract right candidates for these courses and lack of branding and inadequate publicity of the courses. These reasons for discontinuation of courses are based on interaction of the author with the faculty and students. As of 2016 no further research or review study has been reported.

Courses of science communication must be promoted and reviewed by leading institutions such as NCSTC, National Institute of Advanced Studies (NIAS) or Vigyan Prasar. UNESCO and World Conference of Science Journalists have also designed science communication courses (UNESCO 2007). Representatives from these agencies can be invited as the collaborator to review Indian academic courses of science communication to know their efforts. Reviews of science communication courses are required for betterment and modifications in accordance with the current trends of science communication in the country. Most of the curricula of science communication courses were designed during mid-1990s. Now the time has come to revisit the curricula as per the current needs of public understanding of science with more practical orientation and effective linkages with laboratories as well as media. There are some issues to be reviewed—such as creation of regular posts and 'special chairs on science communication' for faculty of science communication in universities; recognition of science communication as a subject for National Eligibility Test conducted by UGC; development of books and study material as per the current requirement of scientific institutions for science communication or public understanding of science; employment generation for science communicators in scientific institutions and media; and publicity and branding of these courses to attract suitable students as well as employers.

Employment Generation for Trained Science Communicators from Universities Is Necessary

Trained science communicators from these courses can be employed as PROs, press officers or science communicators by Indian scientific institutions. A large number of students have been trained as science communicators by the universities, but very few of them are actually working as science communicators as no such jobs or vacancies are available in our scientific institutions. Only few trained science communicators from these courses are working in various institutions of DST, DAE and DRDO. However, majority of them are employed with the media, corporate communication and public relations sector where they have limited scope of science reporting or communication.

Science communication units/departments can be established in our S&T research centres/laboratories related to space, health, agriculture, biotechnology and so on with the proper utilisation of the talent of trained science communicators. Trained young science communicators can contribute their services in various research centres and organisations where science communication is required especially in vernacular languages (Kapoor 2014). Efforts are also required to establish science reporting units in newspapers and television news channels where students from science communication courses can be employed. NCSTC has initiated *Swarna Jayanti* (silver jubilee) fellowship to motivate young science communicators for science writing in newspapers. Our policy makers and job providers may create regular vacancies for science communication where these science communicators can be employed (Kapoor 2012a, b, 2014; Rajput 2008). Appointment of trained science communicators would help bridge the gap between scientists/experts who work at the forefront of science, and the citizens who enjoy the benefits of science or have a right to know about scientific and technological breakthroughs/misadventures.

Recommendations

1. Establishment of a nodal agency, say, 'Science Media Resource Centre' or 'Indian Science News Service', to initiate a sustainable dialogue between science and media. Such an agency may provide research findings and information on development work to the national and vernacular media from a single window. Variety of content, contact details of scientists and press releases issued by organisations related to various scientific fields such as health, environment, agriculture, space, atomic energy and biotechnology under the various departments and ministries can be made available through this nodal agency. This agency would also catalyse regular interaction and training for scientists and journalists focused on science communication in vernacular languages on local and global issues.
2. Establishment of science communication and public relations units in various S&T research centres/laboratories with the employment of trained science communicators from universities as press officers or science communicators. Such offices created in few S&T organisations like DRDO and DAE need to employ qualified science communicators in greater numbers and substantially enhance their contributions to science communications in vernacular languages.
3. Creation of new jobs in science communication for R&D centres and media; development of content (course materials and manuals) as per the current requirement of science communication.
4. Regular practice of sharing of information with social media in vernacular language in each and every research institute, if necessary, with the help of local scientist volunteers.
5. Networking of potential science communication partners working in vernacular languages for SWOT analysis and reality checks of science communication to understand science and media relationship in vernacular languages. This network can be used to launch science media news centres at regional level.
6. Preparation of online scientific glossary for standardisation of technical terminology in vernacular languages.
7. Promotion of research articles in vernacular languages in Indian research journals; space for one research article in vernacular language per issue can be reserved.

Acknowledgements The author is grateful to V.S. Ramamurthy, Director, NIAS and Pallava Bagla, NDTV Group and Visiting Professor, NIAS for giving the opportunity to share his views on science communication. The author is also thankful to his mentors Dr. R. Gopichandran, Director, Vigyan Prasar and Dr. C.M. Nautiyal, Sr. Scientist, BSIP and well-known science communicator, for their encouragement and guidance for this chapter.

References

Arulchelvan, S. (2010). Science and technology dissemination through Tamil newspapers: A study. *Indian Journal of Science Communication, 9*(2), 3–9.

Bagla, P. (2002). Good science journalism-and barriers to it in India. Science and Media: An International Workshop, Tobago, West Indies.

Chauhan, N. M. (2011). Aptitude of the programme coordinators of Krishi Vigyan Kendras of India. *Indian Research Journal of Extension Education, 11*(3), 19–24.

Clayton, J., & Joubert, M. (2012). *The need for an African science news service*. Report Prepared on behalf of SciDev.Net for the UK National Commission for UNESCO. http://www.unesco.org.uk/uploads/The%20Need%20for%20an%20African%20Science%20News%20Service%20May%202012.pdf

Gopichandran, R. (2013a). Some important facets of science communications. *Dream 2047, 15*(10), 35.

Gopichandran, R. (2013b). Some important insights regarding the objectives and process of S&T communication: A synthesis. *Dream 2047*, 15(6), 39.

Gopichandran, R. (2014). Classic insights from Jean-Yves Le De'art Foreword. *Dream 2047, 16*(4), 43.

Kapoor, N. (2012a). A science information resource hub for sustainable science communication in India. In *Proceedings of International Conference on Science Communication for Scientific Temper* (pp. 98–103). New Delhi: Communication and Information Resources (NISCAIR), CSIR.

Kapoor, N. (2012b). Science communication as a discipline in higher education: Promotion of scientific temper. In D. P. Singh & N. S. Singh (Eds.), *Issues and challenges on higher education* (pp. 273–283). New Delhi: Words Worth India.

Kapoor, N. (2014). Career and courses in science communication. *Employment News* (weekly), *XXXIX*(2), 1, 48.

Kumar, M. (2013). Comparison of science coverage in Hindi and English newspapers of India: A content analysis approach. *Global Media Journal—Indian Edition*, Summer Issue (4), 1.

Maheshwar, M., Gavaravarapu, S. R. M., Venkaiah, M., & Rao, D. R. (2014). The quality of nutrition research reporting by leading daily newspapers in India. *Journal of Media and Communication Studies, 6*(6), 92–98.

Maheshwar, M., & Rao, D. R. (2012). Quantitative analysis of nutrition and health messages in Indian print media. *Public Health Research, 2*(2), 28–31.

Media for Science Forum. (2010). Meta-review: The crisis of media, the relocation of the journalists' world and the decline of science sections in the context of the Internet communicative and social revolution. Media for Science Forum, 12–13 May 2010, Madrid, Spain. Available on http://www.mediaforscience.com/

Mochahari, M. (2013). Revisiting India's science communication and journalism: Issues and challenges. *Global Media Journal—Indian Edition, 4*(1).

Nautiyal, C. M. (2010a). Role of scientists in science communication. *Indian Journal of Science Communication, 9*(2), 10–16.

Nautiyal, C. M. (2010b). Science and science communication in India. In S. Priest (Ed.), *The encyclopaedia of science and technology communication* (Vol. I, pp. 381–388). Newbury Park, CA: Sage.

Nautiyal, C. M. (2011). *Hindi mein Vigyan Lekhanki Chunautiyan; Kyon se Kyon Nahin Tak.* In A. Sinha, S. Mahanti, & N. Kapoor (Eds.), *The Souvenir/Proceedings of the Seminar at IIM Lucknow, Noida Campus* (pp. 122–128). Noida: Vigyan Prasar.

Nautiyal, C. M. (2014). *Hindi men Vigyantatha shod lekhan ki vyavharik samasyaen. Hindi Garima*-2014. Souvenir of the seminar on '*Rajbhasha Hindi ka sankraman kaal aur bhavishya'* (pp. 26–27) at BSIP, Lucknow on March 23, 2014.

Osama, B. (2012). How to report science in local languages. SciDev.Net.

Patairiya, M. (2007). *Science journalism in India*. The Pantaneto Forum Home Page, January 25, 2007. Available on www.pantaneto.co.uk/issue25/patairiya.htm

Patil, S. S., & Kokate, K. D. (2011). Training need assessment of subject matter specialists of Krishi Vigyan Kendras. *Indian Research Journal of Extension Education, 11*(1), 19–22.

Rajput, A. S. D. (2008). Science communication: Careers and courses in India. *Current Science, 95*(11), 1513.

Rohn, J. (2011). *How scientists can reach out with social media*. SciDev.Net.

Salwi, D. (2002). *Science in Indian media*. Noida, India: Vigyan Prasar.

Shipman, N. (2012). *Scientists: Social media is not necessarily a waste of time*. SciLogs in association with Nature.com available on www.scilogs.com

Sinha, A., Mahanti, S., & Kapoor, N. (Eds.). (2011). *Hindi me vigyanlekhan: vyaktigat evam sansthagat prayas*. In The Souvenir/Proceedings of the Seminar at IIM Lucknow, Noida Campus Organised by Vigyan Prasar, Noida.

SIRC/ASCoR. (2001). *Guidelines for scientists on communicating with the media*. The Social Issues Research Centre, Oxford, United Kingdom and Amsterdam School of Communications Research, Amsterdam, The Netherlands. Available on www.sirc.org/messenger/

The Telegraph (2008). www.telegraph.co.uk/news/worldnews/asia/india/2796536/Large-Hadron-Collider-fears-prompted-Indian-suicide.html

UNESCO. (2007). Developing a science journalism course for developing countries: A report on the UNESCO workshop and questionnaire conducted in association with the World Conference of Science Journalists. Prepared by Econnect Communication in association with the University of Queensland and Boston.

Part IV
The Bold New World: Using Television, Internet and Social Media for Communicating Science

Enhancing Science Content on Indian Television: Status, Issues and Way Forward

T.V. Venkateswaran

Abstract Media, in particular television, significantly influence the public sphere and construct what constitutes our common sense in the modern world. If we desire science and technology to be part of culture, then science needs to be present on television as a routine. On this premise and drawing from the international experiences and empirical studies on science on television, particularly in Europe, this chapter argues that in addition to regulation various measures are required in contemporary scenario to enhance airtime for science in India.

Keywords Science programmes · Indian television · Doordarshan · Prasar Bharati · Science shows · Vigyan Prasar

Scientific enterprise plays a monumental role in contemporary society. In fact, it may be modern era's defining feature. We are surrounded by valuable results of scientific pursuits (artefacts such as Internet, refrigerators, vaccines and hybrid seeds) or understandings (such as harmful effects of smoking, genomic unity of humankind, natural evolution of life on earth and so on). Some of these understandings are beyond mere accumulation of facts about the world and have great explanatory and often predictive power, which can be used to anticipate, control and organise our environment in newer ways.

As all aspects of modern life are permeated by science and technology (S&T), for a vibrant democracy, citizens should be scientifically literate so as to appreciate and make technological choices and to embrace a scientific world view. Citizens run the risk of being powerless if they fail to understand science as a complex social activity; and both uncritical adulation and alarmed rejection would severely dent the development and social justice agenda.

Views expressed in the chapter are the author's own and not necessarily shared by the editors.

T.V. Venkateswaran (✉)
Vigyan Prasar, New Delhi, India
e-mail: tvv@vigyanprasar.gov.in

P. Bagla and V.V. Binoy (eds.), *Bridging the Communication Gap in Science and Technology*, DOI 10.1007/978-981-10-1025-5_14

Formal education is perhaps the best nurture for scientific literacy. However, as most of us are consumers and beneficiaries, rather than participants of S&T, beyond school media is the only source of our knowledge about S&T. All developments in rapidly changing scientific and technical fields, as well as information about the implications of these changes for our lives are largely communicated by the media (Nelkin 1995).

Pivotal Role of Television

Newspaper, periodicals, radio, television, Internet and mobile phones are some of the widely used media platforms for access to information and entertainment, including information that shape our appreciation and perception *on* and *about* S&T. Even as we have entered the brave new world of Information and Communication Technology (ICT), television seems to dominate the Indian scenario. The arrival of Internet was seen as a *spectre* that would haunt television; even while the practice of downloading YouTube recordings of favourite soap operas to watch at leisure is on the ascent, the reach of television especially in countries such as India is unchallenged at least for some time to come (Neilsen 2010).

As on March 2012, of the total 247 million estimated total households, 148 million had television sets at home and of these an estimated 94 million had cable and/or satellite television access in their television sets. In contrast, of the total population of 1270 million, only 243 million had any sort of Internet access (Internet And Mobile Association of India—IAMAI, Businessline 2014).

A recent survey has shown that television remains the primary source (57%) of information in the country, and is almost five times as popular as newspapers. Close to three-fourths of urban households rely on television for information, as do half the rural households. Indeed, even educated people rely more on television than on any other medium. In the case of postgraduates, for instance, 65% rely on television as the primary information source compared to just 27% for newspapers (Shukla 2005). In a survey conducted at *Ardh Kumbh* (a religious–social fair attended by multitude of Indians) in 2007, 70% reported that they accessed information through television (Raza and Singh 2009).

Despite the common myth, television seems to be the dominant medium even among youth. The National Youth Readership Survey indicates that "television remains the most popular source of information for the youth followed by newspapers." "Medium is the message," so the saying goes; the survey also shows that 54% youth view television for entertainment and 22% for current affairs and news; 63% of newspaper readers do so to gather news and information and only 10 % for entertainment, reflecting that perhaps television in India today is largely entertainment, whereas even today newspapers have largely not deviated from the current affairs and news focus (Shukla 2010).

Television has become the window to the world. A survey (Shukla 2010) observes that literate youth spend as much as 98 minutes daily viewing television,

32 minutes reading newspaper and 70 minutes surfing the net. Another study indicates that an average viewer spends around 130 minutes per day watching television (Deloitte 2013). A micro-study in a rural area notes that both young men and women watch over 120 minutes of television per day (Vasan 2010). As of March 2013 there are 824 permitted private channels operating in India. Even after many decades of opening the television sector for private players, we have Doordarshan (the state-run national television channel) as the largest television network in the world with 31 channels, 66 studios, 1413 transmitters and access to 128 million households (Thomas 2010).

Science on Indian Television

Science on television immediately brings to our minds the fond memory of the '*Cosmos*' of yesteryears or evokes an appreciative nod of its Seth Macfarlane's remake '*Cosmos: A Space Time Odyssey*', featuring astrophysicist Neil deGrasse Tyson and produced with Hollywood film values, jaw-dropping computer-generated imagery (CGI) and the music of a thousand violins, tubas and trumpets. BBC's *Horizon* series, David Attenborough's grandiose science documentaries, historic science series such as *Quest*, *Bharat ki Chhap* (a series on science in Indian history) or *Turning point* would be staple representations.

Truly, whatever be the criticism, "Doordarshan can be credited with occasional awakening to the cause of science on small screen" (Kala 2002). As stated, Doordarshan (popularly known as DD), part of the national public broadcasting agency Prasar Bharati, is one of the largest networks with its own production units. Apart from covering science news events in its news bulletins at the national level it produces a five-day a week health communication programme '*Swasth Bharat*' in 19 Indian languages and 17 dialects broadcast through 30 stations. Intermittently it commenced science programming; only to be off-air after some time. In 2002, for example, a serial '*Imaging Science*' anchored by film director Mahesh Bhatt was telecast once a week on DD News, DD Metro and DD National channels. In 2012, collaborating with Deutsche Welle (DW) DD National launched a weekly serial science magazine *Manthan* (meaning churning) focusing on science, technology, environment and medicine, with special focus on the latest developments in Germany and Europe. In addition to the efforts at the national level, some of its regional centres produce weekly science magazines, for example, '*Ariviyal Aayiram*' (meaning myriads of science nuggets) in Tamil regional channel DD Pothigai; '*Sasthra kauthukam*', science magazine programme anchored by famous cine star Nedumudi Venu from DD Malayalam and *Bigyan Prasange*, a weekly science magazine show produced from DD Bangla.

Public Service Broadcasting Trust (PSBT) is a unique public–private initiative established jointly by Prasar Barathi, Films Division of the Ministry of Information and Broadcasting and Public Diplomacy unit of the Ministry of External Affairs.

It has supported more than 350 independent film makers to produce 475 documentaries so far. Of these, 10 pertain to agriculture, 57 are on wildlife and environment, 16 on health and 7 on HIV/AIDS. Most of the films on agriculture, health and environment deal with the science–society issues, and wildlife films largely are documentaries on nature.

The National Council for Science and Technology Communication (NCSTC) of the Department of Science and Technology, Government of India has supported production of science serials and documentaries since its inception in mid-1980s. Initially it supported production of short (30–60 minute) documentaries on various topics such as Safe Use of Pesticides (Puducherry Science Forum) and Athisiyam Arputham (on explaining miracles; CDIT) and biographies such as The Enigma of Srinivas Ramanujam and CV Raman—The Scientist and His Legacy. Thereafter it commenced production of such television serials as Bharat ki Chhap, Kudratnama (1997), a 27-episode science quiz programme telecast by Star Plus, Zara Si Baat, a serial on physics in daily life, Hum Honge Kamyab, Ignited Minds and Challenge Chuttny involving young people in the genre of 'science in backyard'.

Vigyan Prasar, a national institution for science communication, has garnered more than 28 slots (of about 30-min duration) per week, that is, 728 h per year in various public broadcast channels. It is this solid telecast space that gives Vigyan Prasar the scope to explore wider variety in genres and audiences. Such serials as *Aisa Hi Hota Hai* and *Baatein Raaz ki* (on science in everyday life), documentaries *Jeethe Raho* (exploring some of the most common medicines we use, say, paracetamol), *Taron Ki Sair* (on astronomy and cosmology), *Nano ki Dhuniya* (on nanotechnology), *Kahani Darathi ki* (story of planet Earth), *Experiments That Shaped the World, World of Chemistry*, animation serials such as *Our Celestial Neighbour* (CGI animation film serial on Sun and solar system), science magazine programmes such as '*Science Watch*' and science quiz-based shows including *Jigyasa* and *Chamatkar* were produced. Many of these serials, initially, were produced in association with Development and Educational Communication Unit (DECU) and Indian Space Research Organisation (ISRO), and subsequently, are being produced by Vigyan Prasar independently. Yet the limitations of funding, institutional mechanisms for identifying quality producers and absence of adequate capable science producers seriously limit the choices.

Private channels do occasionally transmit science programmes. NDTV (New Delhi Television) has a weekly show '*Science in spotlight*' an interview-based talk show. Many channels have phone-in type health shows wherein audience can discuss their ailments with medical experts. Sun TV, a popular private cable and satellite (C&S) Tamil channel, produced a weekly science show explaining everyday science in simple yetelegant manner. Surya TV, a private Malayalam channel, conducted a regular '*Sasthra Veedhi*' a science magazine genre programme. The Energy and Resources Institute (TERI) had been producing highly acclaimed *Terra Quiz* for telecast initially on DD, then on Nat-Geo and Discovery channels for 14 years.

Why Is Science Scarce?

While quantifying the extent of airtime for science on Indian television is arduous for want of data, it is not too difficult to visualise the minuscule space occupied by 824 channels and a combined 7,218,240 h of airtime per year. Why is science scarce on Indian television?

When the poor nation invested public funds in television way back in 1959, development communication was the stated *raison d'etat*. This spirit is best captured in an article, "Television for Development" presented by Vikram Sarabhai at the Society for International Development Conference at New Delhi in 1969. He argued that "backward countries can and should tap the most advanced communication technologies including television for leapfrogging into rapid economic growth and social transformation" (Joshi 1985). Television was seen as a tool for development communication, and consequently educational programmes and agricultural programme *Krishi Darshan* for farmers dominated the airtime.

The advent of satellite television during 1990s changed all these, and impelled by the television rating point (TRP) wars and competition for advertisement revenues private channels found it a luxury even to pay lip-service to public broadcasting commitment of the media. It became much of entertainment-driven, low-quality soaps and trivialisation of news at the centre stage (Page and Crawley 2001; Singhal and Rogers 2001; Nikhil 1998). Even while the satellite television revolution has increased choice for the consumer, it is located in urbanity, responding largely to the needs and desires of urban middle class (Mehta 2008; Thomas 2010).

No wonder, the aim of education and development through television is made to seem outdated, and rural India is no longer a viable constituency to cater to. Little or no space is devoted to issues related to how the other half of India lives/perishes; consequently, science communication for all as a public good, hardly finds priority.

While the pressures of privatisation and ethos of neoliberalism could explain near absence of public good, including science in the contemporary private television landscape, the experience of DD has not been exemplary too (Herman and Mc Chesney 2001; Mehta 2008). Regrettably, the tradition of public service broadcasting of DD was informed by an ethos of paternalism, mired in a top-down approach. The rhetoric of expertise that it promoted sought that the writers or programme makers would engage with contemporary reality and experience and see deeper, beyond the immediate and the obvious, and that it was the duty of broadcasting to bring these insights and ways of looking to the widest possible public. But in reality, often, under bureaucratic and political control, 'expertise' was reduced to a voice of those in power. Public broadcaster was in reality government media.

With the formation of *Prasar Bharati*, DD has become *Janus-faced*. As a public broadcaster, it is expected to uphold the broadcast code and fulfil its commitment for public good. However, as it is expected to raise revenue by sale of airtime and seek advertisement to meet part of its expenditure and compete in marketplace, it is forced to follow the path set by the markets. It is well known that "if public service broadcasters are forced to compete openly in the commercial marketplace, they will inevitably come to resemble private broadcasters with whom they compete, thereby abandoning their mandate and undercutting their very raison d'être" (Mendel 2000). Thus both private channels, due to market pressures, and DD, for lack of clarity, fail to meet public-good needs and, as a consequence, do not provide adequate airtime for science.

"Science has few takers" is the often stated refrain, but facts prove otherwise. It is reported that 35 % youth are interested in science and technology and 34 % in environmental pollution. About 75 % literate youth evinced some interest in science—25 % being very much interested and 50 % fairly interested. The majority of urban youth and more males agreed that global warming, biodiversity and environmental pollution were serious issues (Shukla 2010). Even if one takes the much disputed TRP as the measure, DD's *Turning Point* (programme) had consistently high viewership and in recent years Vigyan Prasar's science telecast had TRP of 3 (and in state segments like Uttar Pradesh as high as 8), yet not many advertisers showed interest.

One can only conclude that science has few takers in media, not necessarily so among the audience. An anecdotal experience, perhaps, is not out of place here; the author discussed with a reputed private production house to explore harnessing of C&S channels for science telecast. Her immediate riposte was "*our children*, who after school go to private tuitions, come home exhausted, seek entertainment and not instruction from the television." By the qualifier *our* she meant children who are *their* target; not the multitude of children in rural households and urban slums, but children of urban middle class. Private C&S channels talk of 'catching eyeballs', but the fact is that they are interested only of the eyeballs with deep pockets and purchasing power.

Is Regulation an Option?

Even the whiff of the word 'regulation' results in tantrum and is seen as a throwback to 'licence-control *raj*'. Changes in the television system and massive privatisation did not occur in India alone. There was a worldwide trend during the 1980s towards commercialisation of television. Perhaps it would be educative to look at how these countries met the challenge of ensuring that public good is delivered in a private-greed scenario.

World over there has been concern with the ascent of private C&S channels, exploring new ways of satisfying the need for an alternative to commercially driven programming. As one possibility many countries imposed public interest obligations on private broadcasters as part of licensing condition. It is argued that airwaves are public property and hence for its exploitation private player needs to pay 'rent', which is imposed partly as the compulsion to telecast programmes of public good.

In the United States, way back in 1946, the Federal Communications Commission issued a directive 'Public Service Responsibility for Broadcast

Licensees', which required commercial radio and television broadcasters to meet their commitment to public service. In France, the notion of even commercial broadcasting as a public service rather than simply a commercial exercise is still very strong. The main elements of the public service mandate are contained in the 1986 law, which seeks to ensure that all broadcasters promote quality and diversity of programming. The British Communications Act 2003 requires even commercial channels to provide airtime for 'high-quality and diverse programming'. This includes 'programmes of an educational nature and other programmes of educative value'. Other western countries such as Germany, Canada and South Africa also have laws that require private operators to fulfil certain public service obligations (Mendel 2000; Ndlela 2007; Page and Crawley 2001). Such obligations have been imposed with or without providing funding for selected areas of programme production. Some countries, in addition, have established national funds to support independent film producers to prepare documentaries on public-good themes including science, rather than funding private channels directly. Page and Crawley (2001) in their analysis of South Asian broadcasting systems noted that the state has to play a more creative regulatory role, if public interest is to be safeguarded in the commercial media environment.

In 1995, the Hon'ble Supreme Court of India in the Cricket Association case held that "airwaves or frequencies are public property" and that "their use has to be controlled and regulated by a public authority in the interests of the public and to prevent the invasion of their rights." It advocated establishment of an independent authority to regulate various aspects of broadcast media. Following this judgement, a draft bill (2007) was prepared which envisaged "to encourage broadcasting services to be responsive to the educational, developmental, social, cultural and other needs and aspirations of people and include in their programming public service messaging and content."

Regulation Is Only a Part of the Answer

Enabling regulation may go a long way in making televisual space available to enlighten and empower public instead of allowing it to be hijacked by the forces of commercialisation. A multination empirical study in Europe (Lehmkuhl et al. 2012a) indicated the positive impact of regulation regime on airtime for science on television—meaning, it is necessary, but not sufficient. The study also points out that (a) the segmentation/fragmentation of public television markets, (b) existence of mid-sized commercial channels which as part of their niche marketing tend to present more science programmes and (c) the weight of market forces on public service channels influence the volume of airtime for science significantly—that is, regulation may be a necessary step, but it would be insufficient to ensure quality airtime for science in prime time television.

The study reveals paradoxically that more the fragmentation of the television market (that is, more the number of television channels that reached a combined market share of 85 %) the more the airtime for science. In general, commercial channels, in particular big ones, did not contribute substantially to the provision of science programmes in Europe (only 20 % of all science programming was by big commercial channels), but the presence of mid-sized commercial channels (with market share of 5–9 %) increased the chances of airtime for science. The study also found a strong positive correlation between the number of public broadcasters in a country and the airtime devoted to science in each country. Further the study found that less airtime was given by public broadcaster to science, when the said public broadcaster depended more on commercial revenue.

Given that the commercial market is much more fragmented and that there are no mid-sized commercial channels in India, hope from that sector is not that bright, at least for some time to come, unless regulations are put in place. On the other hand, the emergence of new public broadcasters such as Lok Sabha TV, Rajya Sabha TV and noncommercial channels such as DD Bharati gives hope. It is in this context that Vigyan Prasar had formulated a programme to 'enhance science content in television' in 2005. It entered into a dialogue with the national public broadcaster, Prasar Bharati, and a joint committee was constituted to examine the proposal. Subsequently, a memorandum of understanding (MoU) was executed between Vigyan Prasar and Prasar Bharati to meet their respective mandates of nurturing scientific temper and communicating science. Under this MoU Vigyan Prasar is to produce video programmes and Prasar Bharati is to provide transmission slots. Since then Vigyan Prasar has been producing regular science shows/serials to feed the slots allocated by Prasar Bharati. Subsequently, public broadcasters such as Lok Sabha TV and Rajya Sabha TV also came forward to join the initiative.

Although studies have shown that science items can be regularly found in television news (León 2008; Cheveigné 2006), Lehmkuhl's study (2012a, b) shows that thematically specialised public channels within a media system are keys to the presence of a specialism in science news on television. This raises the question of dedicated Indian science channel or working out a mechanism of apportioning significant 'branded' airtime from one or two public broadcasters (just as MTV took branded time from DD Metro in yesteryears) as a viable way forward in Indian scenario.

Why Is Science Absent? Some Structural Issues

Science being public good, deriving commercial benefits from science programmes may be a limitation. Yet science is not totally absent from Indian television landscape. When the *Mangalyaan*, Indian Space Research Organisation's Mars Orbiter Mission, was launched, Indian television was not found wanting. Its coverage was huge and extensive. Many channels do provide airtime for health-related shows (often in the phone-in format) and interviews with experts including scientists. Studio-based format with table-top science demonstration has also found space in many channels. A 27-episode science quiz programme *Kudratnama* was telecast by Star Plus, a private C&S channel in 1997; and one may be able to list few such examples.

Phone-in health programmes or interviews with experts are most popular and widely used formats because they are easy from the production point of view. Not much editorial inputs or expertise on the part of the production team is required. They can be produced in a short time and typically have ready audience. Even if they do not attract prime time television, private channels could resort to telecasting them without hesitation at least to infuse variety.

However, science serials, such as documentaries, are another matter; these are expensive to make, specifically if they need to target a large audience, have high production value and sleek in presentations. As today's audience are exposed to high-quality visuals and editing patterns, an acceptable level of aesthetic value is required to ensure mass appeal.

Further, in Indian context, when funds are limited, such documentary productions with higher budget make sense only if they have higher shelf life and are easy to render into other Indian languages for multiple transmissions. The very need to dub implies interviews are to be avoided and one has to resort to 'voice of god' narration taking us through the film narrative. Longer shelf life implies that 'timeliness and currency' factors are difficult. Most Vigyan Prasar productions fall into this category, as many of them are typically produced initially in Hindi and dubbed into 10 other Indian languages. On the other hand, typically edutainment programmes have 'characters', and dubbing makes them a bit dull. On the other hand, the magazine type would typically have a 'presenter' linking a few short films with, voice of god, commentaries; therefore, they easily lend themselves for dubbing as a desirable format. Thus, the transmission requirements such as ease of dubbing limits the formats of the shows that are produced, and often popularisation documentaries or magazine types are favoured.

These programme formats typically require long preparation time and committed involvement from the side of scientist consultants, science communicators and the television professionals such as script writers, directors and so on for a sustained period of time during production. They require intense editorial input, and organisations such as Vigyan Prasar are able to provide institutional mechanism for bringing in diverse expertise to produce such serials. In the absence of assurance for continued funding, possibility of nucleation of such groups in private commercial realm is remote. It is essentially these structural issues that make science appear and disappear from the television landscape.

Being Part of Television Routine

Television has emerged as the primary source of 'common knowledge' in the present era, including that of science (Nucci and Kubey 2007, 2010). It serves as a widely shared pool of information and perspectives from which people create their understanding of self, world and citizenship (Miller 2013). Televised images provide an opportunity for the viewer to observe the actions, attitudes and thoughts of others, which in turn can influence the viewers' action, values, thoughts and behaviour (Contractor et al. 1988; Brown 1990). Television is the "theatre of discourse … that enters the life of the entire public" (Price 1995). It is through the presence without absence in television, that politics, economics (share market), gadgets (cell phone, motorbikes, cars and so on), sports (largely cricket) and even life style (food, travel, leisure and so on) become part of our everyday discourse. If science has to be part of culture and quotidian life, among other things it has to have a place in the media, in particular television, on a regular and sustained. Nevertheless, today science, if at all is seen, is only at the margin, once in a while, off-prime time and hardly in news.

To capture space in media on a regular basis it must have a 'news value'. Selection of coverage in mass media including television is guided by 'eyeballs' it will attract, which in turn is determined by the informational or entertainment value of the programme, as well as relevance of the information to the recipient. A statement is informative only if it is 'new' and 'novel'. Only that, which is informative and relevant, can gain attention. Media experts usually list eight factors that make the news value of an information: impact, timeliness, prominence, proximity, bizarre, conflict, currency and human interest. In short, for science to be regular and sustained on television, it has to be 'television science', and for that we need to look beyond the genres such as documentaries, magazines and edutainment programmes. Usually it is absence of adequate news value, among other reasons, that restrict space for science in media. As emphasised earlier, when India launches *Mangalyaan*, or a controversy over genetically modified organisms (GMOs) spreads, media is not found wanting.

Beyond documentaries and phone-in shows, we need science in news, both regular in prime time news and specialised in science news segment, reporting on latest developments, discoveries and innovations. We require 'advice programmes', having takeaways on topics such as consumer health. In modern societies government controls a lot of what goes on in science, medicine and environment. It makes many laws and drafts regulations, ideally to make people's life better and safe. One needs to look at the political and social implications going beyond science instructions, and meet the specific need, such as the social need of environmental protection, by exploring these regulations or absence of it. Advocacy programmes that look at the emerging news and its connect with science including regulation issues are required.

These formats have to be relevant and timely, and hence produced in a short duration. If topic selection, research, scripting and productions have to be carried out in short duration, a dedicated team is a prerequisite. The journalists assigned to science beat must monitor current events in the national science research system and report on the latest developments as a routine. Only a dedicated science editorial team would be able to monitor the developments as a routine and have adequate expertise to interpret science messages and journalise them. For an effective and balanced advocacy programme one needs experts who would be willing to join a panel with representatives from adverse advocacy groups along with presenter/producer with sensitivity and certain level of familiarity with science and technology involved.

Thus, the problem of science on television is not merely about availability of slots. While willing public broadcasters, regulations and provision of airtime could facilitate, science on television also requires nurturing of specialised and capable editorial units that can be sustained over a period of time (Lublinski 2011). Assigned science editorial units, if present on a regular basis, over time accumulate expertise, develop efficient routines to find relevant news items, and sift relevant from irrelevant news items and accumulate knowledge about research progress in particular fields of science they observe (Lehmkuhl et al. 2012a). Science in media must be understood as outcome of an organised production of informative and relevant messages on science (Boon 2014).

However, even a cursory glance at Indian television industry shows stark absence of specialised units, except one or two. It also does not seem practical that many of the private channels would invest resources for a science unit from their own revenue. It is in this context that institutions such as Vigyan Prasar assume significance, and their potential for establishing such unit(s) and providing science feature service to media needs to be explored.

Way Forward

Based on the preceding discussion, the following specific action points are suggested:

(a) Formulate and implement regulations requiring even private channels to provide public-good airtime in consonance with the Supreme Court ruling.
(b) Establish a public trust to fund independent film producers for making science, technology, environment and science–society documentaries.
(c) Increase allocation for science communication budgets with contribution from all S&T-related ministries, departments and agencies.
(d) Appoint PRO or media outreach officers in every science institution to give information on latest developments to media on a regular basis, such as the initiatives taken by Defence Research and Development Organisation and Department of Atomic Energy in the recent years.
(e) Establish a centre on the lines of Science Media Centre that can support television networks in India with regular supply of science features and video clips, and facilitate media–scientist/expert interaction.
(f) Institute a regular intensive in-service course for people working in media on science reporting.

References

Boon, T. (2014). 'The televising of science is a process of television': Establishing horizon, 1962–1967. *The British Journal for the History of Science, 48*, 1–35.

Brown, W. J. (1990). Prosocial effects of entertainment television in India. *Asian Journal of Communication, 1*(1), 113–135.

Cheveigné, S. (2006). Science and technology on TV news: The creation of content. In J. Willems & W. Göpfert (Eds.), *Science and the power of TV* (pp. 85–100). Amsterdam: VU University Press & Da Vinci Institute.

Contractor, N. S., Singhal, A., & Rogers, E. M. (1988). Metatheoretical perspectives on satellite television and development in India. *Journal of Broadcasting & Electronic Media, 32*(2), 129–148.

Deloitte. (2013). Technology, media & telecommunications India predictions 2013.

Herman, E. S., & Mc Chesney, R. W. (2001). *The global media: The new missionaries of corporate capitalism*. London: Continuum.

Joshi, P. C. (1985). *An Indian personality for television: Report of the working group on software on Doordarshan*. New Delhi: Ministry of Information and Broadcasting, Government of India.

Kala, L. D. (2002). Imaging science. *Indian Journal of Science Communication, 1*(1), January–June.

Lehmkuhl, M., Karamanidou, C., Mörä, T., Petkova, K., & Trench, B. (2012a). Scheduling science on television: A comparative analysis of the representations of science in 11 European countries. *Public Understanding of Science, 21*(8), 1002–1018.

Lehmkuhl, M., Mörä, T., Karamanidou, C., Boyadieva, K., Cunningham, Y., & Team, A. V. S. A. (2012b). Understanding television's science audiences: An analysis of viewer rates in 10 European countries. In M. Bucchi & B. Trench (Eds.), *Quality, honesty and beauty, in science and technology communication* (pp. 270–274). Florence: Proceedings of the PCST-network.

León, B. (2008). Science related information in European television: A study of prime-time news. *Public Understanding of Science, 17*(4), 443–460.

Lublinski, J. (2011). Structuring the science beat: Options for quality journalism in changing newsrooms. *Journalism Practice, 5*(3), 303–318.

Mehta, N. (2008). *India on television: How satellite news channels have changed the way we think and act?*. New Delhi: Harper Collins.

Mendel, T. (2000). *Public service broadcasting: A comparative legal survey*. UNESCO.

Miller, T. (2013). *Television studies: The basics*. Abingdon: Routledge.

Ndlela, N. (2007). Broadcasting Reforms in Southern Africa: Continuity and Change in the Era of Globalisation. *Westminster Papers in Communication and Culture, 4*(3).

Nelkin, D. (1995). *Selling science: How the press covers science and technology*. New York: Freeman and Company.

Nielsen (2010, August) How people watch: A global Nielsen Consumer Report.

Nikhil, S. (1998). Doordarshan, public service broadcasting and the impact of globalization: A short history. In M. E. Price & S. G. Verhulst (Eds.), *Broadcasting reform in India: Media law from a global perspective* (pp. 22–40). New Delhi: Oxford University Press.

Nucci, M. L., & Kubey, R. (2007). "We begin tonight with fruits and vegetables" Genetically modified food on the evening news 1980–2003. *Science Communication, 29*(2), 147–176.

Nucci, M. L. & Kubey, R. (2010). How TV shape public understanding of science, *Television Quarterly, XXXIX*, 33–37.

Page, D., & Crawley, W. (2001). *Satellites over South Asia: Broadcasting culture and the public interest*. New Delhi: Sage Publications.

Price, M. E. (1995). Free expression and digital dreams: The open and closed terrain of speech. *Critical Inquiry, 22*(1), 64–89.

Raza, G., & Singh, S. (2009). Cultural distance and science communication: Few lessons from the past and present. *History and Sociology of South Asia, 3*(2), 310–336.

Shukla, R. (2005). *India science report: Science education, human resources and public attitudes towards science and technology*. Delhi: NCAER.
Shukla, R. (2010). *Indian youth: Demographics and readership*. New Delhi: National Book Trust and National Council of Applied Economic Research.
Singhal, A., & Rogers, E. M. (2001). *India's communication revolution: From bullock carts to cyber marts*. New Delhi: Sage Publications.
Thomas, P. N. (2010). *Political economy of communications in India: The good, the bad and the ugly*. New Delhi: Sage Publications.
Vasan, A. (2010). *Films and TV: Viewing patterns and influence on behaviours of college students. Health and population innovation fellowship programme*. Working Paper No 13. New Delhi: Population Council.

SCIDEV.NET: Advent of Online Science Journalism in India

T.V. Padma

Abstract The Science and Development Network, or *SciDev.Net* has emerged as a pathbreaking Internet-based initiative to promote science journalism in developing countries, especially in places where science journalism either did not exist in the form recognised in advanced countries, or was struggling to find a place in the wider world of national and international journalism. 'SciDev.Net' is an open-access, not-for-profit, international science website focusing on development-related science with an aim of empowering individuals from both developed and developing countries to participate in the dialogue between science and technology and society. This chapter illustrates how Internet challenged the traditional 'linear' model of flow of science information, in which science journalists were the intermediaries who relayed information faithfully 'from privileged sources to diverse publics', in the light of the story of *SciDev.Net*, as a successful science communication website.

Keywords SciDev.Net · Science journalism · Science communication · Developing nations · Science websites · Science news

SciDev.Net: The Seed of an Idea

The Science and Development Network, or SciDev.Net (URL: www.scidev.net) emerged as a pathbreaking Internet-based initiative to promote science journalism in developing countries, especially in places where science journalism either did not exist in the form that is recognised in advanced countries, or was struggling to find a place in the wider world of national and international journalism.

Views expressed in the chapter are the author's own and not necessarily shared by the editors. This chapter was written when the author was working with SciDev.Net.

T.V. Padma (✉)
SciDev.Net, New Delhi, India
e-mail: tvpadma_10@yahoo.co.in

P. Bagla and V.V. Binoy (eds.), *Bridging the Communication Gap in Science and Technology*, DOI 10.1007/978-981-10-1025-5_15

> SciDev.Net, a website on science and world development, is a relatively rare example of a journalist-driven service that demonstrates the power of the (web) medium for rich communication. (Trench 2009)

By the time SciDev.Net was formally launched in London in December 2001, as a website for, by and of science journalists in developing countries, with focus on development-related science issues, its time had come.

The Internet had begun to influence science journalism (and indeed journalism in general) in myriad ways.

The Internet challenged the traditional 'linear' model of flow of science information, in which science journalists were the intermediaries who relayed information faithfully "from privileged sources to diverse publics" (Trench 2009).

The Internet era led to the emergence of a nonlinear model of flow of information with multiple sources of information, and a wide access to them. The reader could decide what she or he wanted to delve into further (Hermida 2010a, b).

SciDev.Net, meanwhile, was conceived to challenge another model—the dominant Western model of science journalism, in which reporting high-end science from developed countries, and sometimes from few emerging economies, was the norm. To focus only on science and development, and that too from often little-known developing countries, was considered unconventional by some, and foolhardy by the uncharitable.

To reach out to a myriad of developing countries, in at least three or four international languages, and to readers all of who may not have the means to pay to read the information, needed an out-of-the-box solution. Thus was conceived the idea of an open-access not-for-profit international science website focusing on development-related science. There was, and still is, the question of varying Internet access in developing countries, but the Internet has come to stay globally, and it was left to the organisation to tap its enormous potential to meet its objectives.

SciDev.Net hoped to tap the potential of the web to attain its objectives.

> In industrial economies, innovation is crucial. In democracies, journalism is essential. So in democratic industrial economies, journalism dedicated to covering innovation is essential. (Nordfors 2004)

The Rationale

By then, there was a growing realisation among some sections of science journalists that the key task of science communication was to empower individuals to participate in the dialogue between science and technology (S&T) and society (Dickson 2004).

As SciDev.Net's founder director, the late David Dickson reasoned, social progress in countries, developed or developing, requires effective and accurate communication of S&T, and this information should be accessible. Dickson (2004) argued, "in the specific context of developing countries, the application of scientific knowledge is at the heart of social progress and of the development process. But such knowledge will not reach the heart of development unless it has been effectively and appropriately communicated to those who can put it into practice. It is also imperative for science communicators to ensure that positions on controversies on science-related issues are grounded in the current state of scientific knowledge."

Increasingly, several analysts were looking at the role of science communication in the interlinked sectors of science, technology and innovation (ST&I), which led to certain conclusions.

The first was that ST&I are essential to address the "triple helix of development", as Dickson described: meeting the three challenges of achieving economic growth, reducing poverty and ensuring environmental sustainability.

The second was that communication of accurate and accessible information about S&T, and the contributions of both to innovation, are essential to achieve these objectives.

The third was that policy makers at various levels need accurate, timely and reliable information about how S&T can contribute to development.

And that public understanding of S&T, and engagement with it, ensures an open and democratic debate about the promises they offer—as well as their potential risks.

Communication in general, and research/science communication in particular, represents the critical fourth strand to the development helix (Dickson 2008).

Perhaps what proved fortuitous for SciDev.Net was the emergence of the Internet as a powerful medium to communicate information.

The crucial role of science communication in framing sound policies based on evidence was also highlighted in a 2008 Working Paper of the UK-based Overseas Development Institute (ODI). Their analysis showed that in developing countries, often there is weak institutional capacity for framing policies based on evidence. Hence, merely disseminating research results does not suffice to bridge the gap between S&T and innovation research on the one hand and policy making on the other. Researchers and policy makers need to interact with each other, and S&T information must be aligned not just to the needs of policy makers, but also when and how the different players use the information (Jones et al. 2008).

In addition, those engaged with policy making and development prefer a set of policy-relevant recommendations to accompany scientific findings, and they favour more public participation in ST&I-related policy debates, and improved public understanding of ST&I.

ODI also concluded that there is a need for 'intermediary organisations' that serve as 'knowledge brokers' at the interface between science and development policy.

SciDev.Net: The Origins

The origins of SciDev.Net lie in a website set up by staff of *Nature*—with financial aid from the UK's Wellcome Trust—to report on the World Conference on Science held in Budapest in 1999. According to SciDev.Net, "The positive response to this website led to discussions about creating a permanent website devoted to S&T and development", and proposals for the initiative were endorsed at a meeting at The World Academy of Sciences (TWAS, then known as the Third World Academy of Sciences) in Trieste, Italy, in October 2000 (www.scidev.net).

The UK's Department for International Development (DFID) financed a six-month planning stage, starting November 2000, "to produce a prototype design of the website, a business plan for a two- or three-year development stage, and financing strategy" (www.scidev.net).

Subsequently, DFID, the Swedish International Development Cooperation Agency (SIDA), International Development Research Centre (IDRC) and Rockefeller Foundation provided funds for staff recruitment and the opening of a permanent office in London in August 2001. A preview version of the website went live in November 2001, and the website was launched in December 2001. Its "goal was to help individuals and organisations in the developing world make informed decisions on S&T-related issues that impact on social and economic development." The organisation aimed to "provide reliable and authoritative information" on these issues, by "operating a free-access website containing news, views and analysis about science and technology in the developing world" (www.scidev.net).

Dickson, who was the news editor at *Nature* before founding SciDev.Net and had also worked as correspondent with *Nature* and *Science*, effected a unique trilateral arrangement with *Nature*, *Science* and TWAS to support SciDev.Net. Under this arrangement, both *Nature* and *Science* agreed to have a direct linkage to development-relevant papers they published in the two journals.

SciDev.Net: The Journey across the Globe, and to India

As in other news organisations, the website contained news, features, opinion articles, notices (including lists of meetings, grants and jobs), book reviews and letters to the editor.

The initial version (Fig. 1) also had some other features and detailed dossiers on important topics, for example, malaria, biodiversity, policy briefs and e-guides on science communication.

The news, back then, was in three languages—English, Portuguese and Spanish for the Latin America 'gateway' (as regional pages were called then, now called 'edition') that was launched in 2002. The South Asia gateway was launched in November 2004, and the China gateway, with Chinese content, in June 2005.

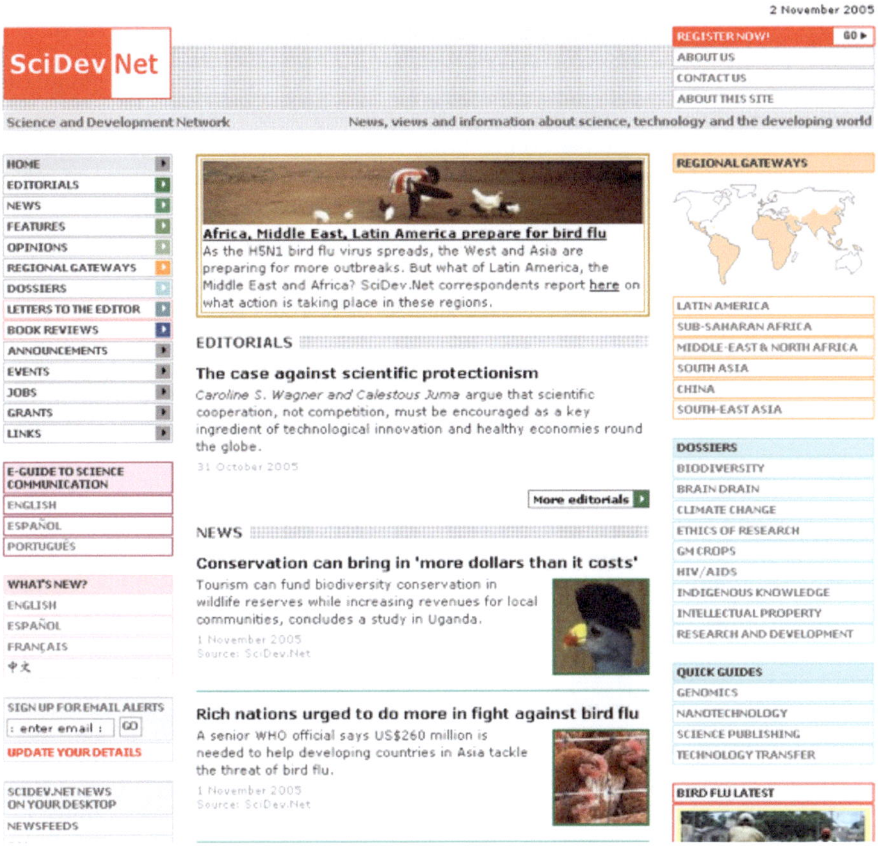

Fig. 1 SciDev.Net's initial version, when it came to India

There have been additions and some deletions along the way, both in regional gateways and arrangement of the content. For example, the Portuguese version was scrapped as majority of the countries in Latin America worked in Spanish. A French version was added for French-speaking countries, and the launch of the Middle East and North Africa gateway saw the introduction of content in Arabic. However, the China gateway was scrapped.

Meanwhile, as the number of topics for dossiers increased, the second version of the website regrouped them under six umbrella 'topics' and 'subtopics'—for example, 'health' as a topic and 'malaria', 'tuberculosis' and so on as 'subtopics'.

All news articles appearing on the website can be reproduced by any printed media or website free of charge, and the only conditions are acknowledgement to SciDev.Net and a link to its website.

The website provides a free news feed. In addition, it partners with regional networks of science journalists and communicators and institutions who share its commitment to improving S&T communication, by way of organising capacity-building workshops and other events in the developing world.

SciDev.Net made a formal bow in India—although it was already beginning to be known informally in some circles—with the launch of the South Asia gateway in November 2004 in New Delhi. Since then, it underwent two revamps, and it now comes in mobile-friendly versions, has blogs and social media accounts and offers audio slide shows and podcasts.

SciDev.Net: Tapping the Potential of the Internet

However, when SciDev.Net came to India, there was still scepticism in several quarters about the credibility of the web as a medium for well-researched, analytical information, on any topic, leave alone science.

Nevertheless, the winds of change were already blowing across the world of journalism and would have reached India sooner or later.

A November 2006 report of the Pew Center, USA on ‘The Internet as a Resource for News and Information about Science’, based on a survey in the USA, found that the Internet was the ‘primary source’ of science news for 40 million American adults—that is, 20% of all Americans. It was second after the television that was used by 41%, while 14% got their science news from newspapers and magazines.

The Pew survey found that the Internet was the most popular source for science news for adults under 30 years, who had broadband facility at home. In this group, the Internet was the main source of science news for 44%, while 32% turned to the television and just 3% relied on newspapers—an indicator of where the future was heading towards.

The survey also found that people often are not hunting for a specific science news item, but often stumble on it on the web, when they are searching for some related information.

As Alfred Hermida, associate professor at the Graduate School of Journalism, University of British Columbia, observed “if the future is online, then the way journalism is practiced must change to take account of this new medium. Behind the hyperbole of the transformative potential of the Internet lies a pressing reality of how changes are taking place in the way news is produced, distributed, and consumed. Online information dissemination has transformed the way that news and science reach the public. The Internet allows the audience to have greater control over information, in terms of where they get the news, when they get the news, and how they get the news” (Hermida 2010a, b).

Hermida observed in a keynote address at Future Directions of Science Journalism Conference in 2007, at the University of British Columbia, that unlike the traditional print or broadcast model that relied on committed readers or viewers to come to the publishing/broadcasting outlet, in an online world, “one should not expect people to come to you out of habit; rather one should create more opportunities for people to stumble upon your content” (Hermida 2010a, b).

"There is still a need for professional journalistic skills—sourcing, researching, storytelling and editing—but the way these skills are used has to change to create new forms of journalism and reach new and, invariably, younger audiences. Mainstream news organisations are increasingly exploring the idea of news as a conversation with the audience, offering more ways for readers to participate, such as allowing comments on stories or soliciting photos and videos from the public" (Hermida and Thurman 2008). Online journalism is characterised by the ability to use multiple media to provide varying textures, combining elements from print—text and graphics—with those of broadcast: sound, music and video. It also offers the scope for interactivity (Hermida and Thurman 2008).

Brian Trench, former senior lecturer at the School of Communications, Dublin City University, noted that "the Internet provides the means to present new scientific developments in ways that promote dialogue and conversation" (Trench 2007). Additionally, he observes, "new opportunities, media, and genres for reporting science have emerged, that challenge established modes of science journalism." The web has tremendous power to connect scientists, policy makers, government officials and civil society organisations with science journalists and communicators.

The web's facilities for interactivity offer yet more possibilities for communicating science in richer, more textured ways than traditional dissemination or transmission models of science journalism can encompass. Through e-mail, forums, and weblogs, as well as other services, the Internet facilitates three-way communication—from producer to consumer, consumer to producer and consumer to consumer (Trench 2007).

Fahy and Nisbet (2011) observed that the new science media ecosystem in the USA and the UK, which they studied in detail, is "a mostly online environment that is deeply pluralistic, participatory and social." And that, "online science journalists have a more collaborative relationship with their audiences and sources and are generally adopting a more critical and interpretative stance towards the scientific community, industry, and policy-oriented organizations" (Fahy and Nisbet 2011).

Also, say Fahy and Nisbet (2011), "in the new media ecosystem, instead of journalistic authority resting heavily on being first with the news, a new kind of credibility and authority has arisen for all reporters, based on a 'competitive logic' … one of knowing more, knowing better, knowing comprehensively, and knowing in as much depth or extent as readers would wish."

The advent of the web has also changed the professional lives of science journalists and democratised access to publications. It has cut time in communication, and helped journalists access research abstracts and articles quickly; laptops, mobile phones and Internet have revolutionized information gathering and reporting. Journalists across the world receive embargoed e-mail alerts on research published in the world's premier peer-reviewed journals, to help them have sufficient time to read the scientific article, interview the authors and independent experts in the field and prepare their report. Webinars and live streaming of press briefings mean journalists whose organisations cannot afford to send their reporters across the globe can still know what happened at a press briefing and can even send their queries through e-mail or Twitter.

So have the ways in which this information is accessed and disseminated changed—it is now routine for organisations and individuals to maintain an official Facebook and Twitter account from where even entire reports can be downloaded.

Breaking news is now available on the web, and the advent of Twitter has further quickened dissemination of short pieces of information and data from various official and nonofficial sources (Hermida 2010a, b).

"These days, science journalists do not have to chase stories, the stories are chasing you", observed Jayaraman, veteran Indian science journalist and correspondent with *Nature*, in a public lecture at Hyderabad in 2007.

The web has also made scientists more accessible. Most, despite their busy schedules, reply to e-mails. Also, there are others who maintain blogs—although scientists in India are yet to take to blogs or Twitter in a big way. "The informal nature of blogs allows for a more conversational approach to science, and provides a platform to explore the process of science rather than just the published findings", says Hermida (2010a, b).

Blogs are a platform native to the web, and thus they share one of the defining characteristics of the Internet: the use of hyperlinks. The ability to create a network of ideas through the use of links, whether in blogs or other forms of online storytelling, marks a major departure from other forms of journalism. In traditional journalism formats, such as print or broadcast, the aim is to retain audience, to keep them reading, listening or watching. Links in online content invite audience to explore further and construct their own narratives. Hyperlinks shift control to the audience and are part of the nonlinear nature of the web (Hermida 2010a, b).

The web has also opened newer platforms to disseminate a science journalist's work—the reports often appear during Google searches on a topic. Web technology (online versions of articles, magazines and so on) gives stories a wider reach, and they are subjected to some sort of a peer review, as readers and fellow journalist post their comments.

The Internet has also redefined who exactly is a producer of science news. Trench (2007) says, "there are many more people producing science news than there are people who can justifiably define themselves as science journalists." For example, "institutional websites frequently use mass media journalism formats, such as daily or weekly news updates. In this way, they make what may have been intended as peer communication publicly accessible."

"The plethora of Internet-based media that have developed for the dissemination of scientific information to wider publics include versions of services already provided via print and broadcast media, but also new media formats, such as portals, e-zines, forums and weblogs (blogs). Publishers of science information include higher education and research institutions, established scientific publishers and scientific societies, but also science centers and museums, public education initiatives, individual scientists, interest groups, hobbyists, lobbyists, and many more." "Also, open access scientific publishing facilitates sharing of knowledge between scholars. But it also means that members of non-specialist, but interested, publics have access to information prepared by professionals for professionals. Some sites maintained by scholarly societies and scientific journals require only that users register by name, and parts of such sites are freely open to any passers-by. Access to the web has opened up many aspects of scientific research previously hidden from the general public" (Trench 2007).

"And, many publicity services have been developed on the web with journalist audiences in mind and some, like the European press release distributor, AlphaGalileo, restrict access to reporters and correspondents who specialize in science. However, much of the material available through that service is also available directly—perhaps after some delay—from the original institutional sources, and can be accessed by any web user. There are more such information distribution services that operate without restriction" (Trench 2007).

SciDev.Net: Taking to the Web Medium

Some of the emerging, exciting developments on the Internet are reflected in SciDev.Net too as it evolved over the years. A newer version of the website was launched in 2008, which offered conference blogs and social media. In addition to regular news reports, the website started publishing conference blogs as additional, interesting bits of information from, and personal reflections by reporters covering key conferences. The conference blogs offered developing countries' perspectives that were often missing in other international blogs. SciDev.Net's blog from the UN Climate Change Conference COP-15 in Copenhagen in December 2009 found mention in Columbia Journalism Review as something to watch out for among the likely interesting blogs from COP-15.

Similarly, SciDev.Net's conference blog from the conference on the UN Convention on Biological Diversity (CBD), held in October 2012 at Hyderabad, attracted a huge number of readers.

SciDev.Net by then had begun to publish 'practical guides' (http://www.scidev.net/global/content/practical-guides.html) for science journalists, science communicators and even scientists wishing to communicate to the public or policy makers.

The practical guides cover a range of topics—for example, how to report on clinical trials, stem cell research, how to communicate risk and uncertainty to policy makers, guidelines on data journalism, audio slide shows, investigative science journalism and so on.

The organisation's latest and current strategy guided by its present director reflects the keenness to tap into the potential of multimedia to present information in newer formats such as audio slide shows and podcasts and the social media to disseminate information in newer ways.

It has also begun to run columns on development issues, such as poverty, migration and gender. The new strategy aims to take science to nongovernmental organisations—which most scientists would welcome, undoubtedly.

It is also keeping up with the trend of news outlets making their content as widely accessible as possible, through syndication, social bookmarking or technologies such as RSS—Really Simple Syndication. This variety of available routes to the news is particularly important for science journalism, as often people happen to come across science news online.

As a SciDev.Net editorial in 2013 notes, as the digital revolution sweeps the globe, it is becoming ever clearer that 'readers' want to become multimedia consumers of information. They enjoy a mix of media, relishing the fact that information can be more visual and does not have to stay still.

It says, science stories can generate striking images "that are magnets to the eyes, attracting readers who might otherwise have shunned their content", and multimedia can draw users in and engage them emotionally (Janowski 2013).

An example—just one; there are plenty in SciDev.Net and the world of online science journalism—is the topic of climate change. While websites published both news and opinions on the Intergovernmental Panel on Climate Change (IPCC) reports, a series of stunning visuals from South Asia captured the ground realties of climate change and its impact on the lives of the poor. Or, the website that produced both a news report and a slide show on Western Ghat's endangered biodiversity.

Therefore, now that online science journalism, in general, and SciDev.Net, have taken root in India, we address the question that the sector faced in the beginning—do readers in India take the web as a source of science information more seriously now? SciDev.Net surveys and questionnaires are showing an encouraging impact of the articles. For example, an 'Opinion' piece by Vineeta Dixit, from the Department of Information Technology, titled "Local needs are the future for IT innovations" (17 November 2011) was noticed by one of the companies mentioned in the article, and they have since contacted the Department of Information Technology, Ministry of Communications and Information Technology in India to find synergies with the Common Services Centre Project, an e-governance project.

As SciDev.Net's example shows, and taking cues from analysis of the future direction of journalism in general, including science journalism, there is a bright future for online science journalism in India, provided its potential is tapped properly.

Recommendations

(1) University and scientific research institutes should have information and links on their published scientific articles, with a press release written in a simple format, to help science journalists access the information on their websites.
(2) Key Indian scientific journals should adopt organised system of embargoed releases on the lines of those used by *Nature*, *Science*, *Lancet*, Public Library of Sciences and so on.
(3) Dedicated press officers should be appointed by scientific organisations who would create e-mail lists and provide details of reports and articles published as well as advance information on key conferences hosted.
(4) An Indian version of 'Eurekalert!' science news service (http://www.eurekalert.org/) for journalists, from the American Association of the Advancement of Science (AAAS), should be launched.

References

Dickson, D. (2004). Presentation at launch of SciDev.Net's South Asia gateway.

Dickson, D. (2008). Presentation at Euroscience Open Forum (ESOF), Barcelona.

Fahy, D. & Nisbet, M. C. (2011). The science journalist online: Shifting roles and emerging practices. *Journalism: Theory, Practice and Critisism 12*(7), 778–793.

Hermida, A., & Thurman, N. (2008). A clash of vultures: The integration of user-generated content with British newspaper websites. *Journalism Practice, 2*(3), 343–356.

Hermida, A. (2010a). Revitalizing science journalism for a digital age. In D. Kennedy & G. Overholser (Eds.), *Science and the media*. Cambridge, MA: American Academy of Arts and Sciences.

Hermida, A. (2010b). Twittering the news: The emergence of ambient journalism. *Journalism Practice 4*(3), 297–308.

Janowski K. (2013). http://www.scidev.net/global/digital-divide/editorials/scidev-net-s-multimedia-future-can-widen-our-reach.html

Jayaraman, K. S. (2007). Public lecture on science communication, University of Hyderabad, Mar 07, 2007.

Jones, N., Jones, H., & Walsh, C. (2008). Political Science? Strengthening science–policy dialogue in developing countries, Working Paper 294 London: Overseas Development Institute. http://www.odi.org.uk/sites/odi.org.uk/files/odi-assets/publications-opinion-files/474.pdf

Nordfors, D. (2004). The role of journalism in innovation systems. *Innovation Journalism, 1*(7), 3–18.

Pew Center, US. (2006). http://www.pewinternet.org

Trench, B. (2007). How the Internet changed science journalism. http://www.doras.dcu.ie/3628/2/internet_science_journalism.doc

Trench, B. (2009). Science reporting in the electronic embrace of the Internet. In R. Holliman, E. Whitelegg, E. Scanlon, S. Smidt, & J. Thomas (Eds.), *Investigating science communication in the information age: Implications for public engagement and popular media*. Oxford: Oxford University Press.

Smart Websites: Insights from the Department of Biotechnology

Archita Bhatta

Abstract Communication is crucial for growth and inclusive development—the very factors that make science useful to society. In an age when the web has become a crucial element of our existence, communication through this medium has also become equally important. Till date most government websites, specially the Indian ones have limited themselves to providing information. This needed a change, specially in the case of science institutions that have an additional responsibility of making science attractive to the public. Department of Biotechnology took the plunge in bringing forth a site that is communicative and projects the dynamic image of the department, which administers world-class science. This chapter captures the experience of how the idea of the new website took shape, and the journey involved in shaping it into effective communication tool—an interface between the public and the government—to build confidence in the government and also to dispel prevailing myths about science and technology.

Keywords Public–government interphase · Government departments · Department of Biotechnology · Science communication · Science and Internet · Participative science communication

Introduction

One of the premiere institutes of the country was into redevelopment of its website. Here is a typical comment from there: "We have survived the past several years without a communicative website and we can do so in the future. Why do we need to spend time and money on a website?"

Views expressed in the chapter are the author's own and not necessarily shared by the editors.

A. Bhatta (✉)
DBT Communication Cell, Vigyan Prasar, Noida, Uttar Pradesh, India
e-mail: architadbt@gmail.com; architabhatta@gmail.com

P. Bagla and V.V. Binoy (eds.), *Bridging the Communication Gap in Science and Technology*, DOI 10.1007/978-981-10-1025-5_16

This reflects the lack of appreciation of the importance of communication and of the web media as a crucial forum for the same. This is where the need arose for a new website for the Department of Biotechnology (DBT). It had in its background the massive gap between the way the younger generation wants to perceive content and the way government websites present themselves. The reflection is seen in the difference between the Government of India websites and the websites of science departments and science institutions in the modern world beyond the Arabian Sea.

It is not that the government or the scientific institutions do not realise this gap. Forward-looking government officials and scientists in several science-based ministries and scientific institutes realise the necessity of a youth-friendly and modern communication mechanism to make science and technology attractive to people. However, while ‘communication’ is understood to be important, it is still not considered a priority. In some quarters, there are even doubts about its necessity. As a result, the obvious constraints in funds are delaying such efforts. Thankfully, the DBT decided to take the plunge.

Why Is Web Communication Important for the DBT

Biotechnology in India is one of the fastest growing knowledge-based sectors and is now poised to be potentially a larger player in the gross domestic product (GDP) growth of the country, more than even the information technology (IT) industry (EBTC 2013). The DBT had been set up in 1986, following the vision of some top scientists, which the then Prime Minister of India, late Rajiv Gandhi endorsed and helped give a shape to.

Between then and now, a lot has been achieved. The latest towering achievement is the cheapest anti-rotavirus vaccine in the world (Bhandari et al. 2014), which entailed cooperation and collaborations at all levels—individual scientists, Indian institutes, foreign institutes and the industry (Madhi and Parashar 2014). Several new funding opportunities have also emerged, which opened up opportunities for more of such collaborations that are becoming increasingly critical to address the crucial issues facing humankind. Parallel to these developments, the DBT felt the need for a globally attractive face that reflected them. This is because at any given moment anyone wanting to check out the Indian biotechnology sector would first look at the website of the DBT, since it is the nodal agency that has a direct or indirect influence on anything related to biotechnology in the country.

As an information portal the website could be an instrument of interface that ensured rapid exchange of information—to and fro. For example, it could be a people-friendly way to find information regarding funding opportunities available for students, researchers and scientists.

It would project all activities of DBT—funding, programmes, policy and regulations—and significantly highlight any new development in the department upfront. This could be about some new research area of a DBT institute; a new scientist who has started working in a new arena of biotechnology; a proposal floated for a new project and so on. Someone, somewhere will relate to any of these 'new' developments, and hence, communication will be enhanced. That means, cutting down on costs and time of getting a message across and speeding up interactivity. One of the crucial outcomes would be a boost to collaborative efforts across the world.

Here let us go back in history a bit and see how it works: Rabindranath Tagore had established Visva-Bharati, now a central university, at Santiniketan. When he was in America, he heard from Albert Einstein about physicist Satyendra Nath Bose. Tagore decided that after his return, he would invite Bose to establish a nuclear physics department at Visva Bharati. It took him months to 'touch the base' as Americans would put it, return home, ask someone to find Bose, who would then 'humbly' request the scientist to meet Tagore (Quantum Indians 2013).

We live in the age where science needs to be fast tracked. Hence, the need for a cutting-edge communication tool, which would not only be an information window about DBT to the world, but also an interface between public and government. The website is a medium to engage the public and build its confidence in the government.

It is this aspect of the website that underscores the necessity of a 'facial' that includes two aspects. First, the dressing of the site. Breaking away from bureaucratic, officious and monochromatic look that government websites usually have, the new website has distinctive images that visitors could immediately identify as a biotechnology site. Second, rewriting of the pages with special attention to perspective and in a lucid language shorn of jargon. For example, here is what the original website page on animal biotechnology programme looked like; it read:

> Efforts continued for the improvement of animal productivity, development of newer animal vaccines and diagnostics, molecular characterization of indigenous breeds of livestock and development of animal byproducts. The brief achievements of the programme are as follows:ET and related areasEmbryo transfer technology was standardized in cattle, buffalo, goat, sheep, camel and equines which includes optimization of in vitro fertilization of oocytes, culture of embryos, micro-manipulation and embryo cloning using nuclear transfer, sexing through PCR, reproductive ultrasonography and endocrine profiles for augmenting fertility.

But the new website puts it differently:

> Livestock sector plays an important role in the economy of our country. The contribution of this sector in national economy was approximately 3.9% in 2011–2012. India ranks first in the world in milk production with an estimated production of 132.4 million ton in 2012–2013. India is also third largest egg-producer in the world, over 69.7 billion eggs were produced in 2012–2013.While the majority of India's animal products are consumed domestically, exports are also growing. India is the top global exporter of buffalo meat, and is also the fourth largest exporter of soybean meal, an important ingredient in commercial feed for farmed animals. In addition, India's leading poultry producers are increasing sales in Asian and Middle East countries. International investment is also expanding.

The basic idea behind this approach to an opening page of a technology-driven site is to go beyond the obvious audience of the scientific community and create an additional target group, which includes science students sitting on the hedge, the general public with 'not too friendly perceptions about science' and a media hungry for sensational stories to excite the public, and bring them closer to biotechnology.

In this milieu, the website can play a crucial role by engaging in public debates to dispel myths about biotechnology and vaccines.

Consequences of Lack of Communication

Information and communication technology (ICT) has been shown to result in economic growth. The period 1990s saw the reflection of the growth in the USA and in other developed countries (Colecchia and Schreyer 2002). The ICT-boosted growth percolated to the developing countries during 2000s. While the world celebrated the triumph of ICT, what remained a whimper against this huge celebration was the role that communication played in this growth.

ICT diffusion and usage depend on the existence of the right organisational framework and conditions (Colecchia and Schreyer 2002). One of the conditions that is necessary is the use of simple and interesting language for communication so that people are attracted to the information.

People still fail to understand the connection between lack of communication, delay in decisions and resultant economic setback.

Lack of communication on the part of the government has led to proliferation of several myths about technology amounting to a factor called 'techno fear' among the public. Aggressive campaigning by pressure groups coupled with silence on the part of the government has led to confusion among the common public about technology, lack of faith in the government and delay in policy decisions on necessary vaccination programmes and technology implementation. The fallout has been obvious: long-term economic, social and health impacts. Effective and timely communication is an essential solution in this scenario.

Why Web Communication

We live in times when web communication has gone much beyond maintaining websites—it has moved to social media. Reading habits have reduced significantly and so have patience levels, especially among the youth (Loranger and Nielsen 2013).

India has surpassed Japan and is today the third highest user of Internet after USA and China (Digital Future in Focus 2013). This is in contrast to situations in 2005 when the country's Internet subscriber base was 0.4%, much less than that of Korea and Malaysia which stood at 58% and 11%, respectively (India Telecom News 2003).

The advent of the Internet protocol television (IPTV) and the YouTube has changed the concept of audiovisual media. The Internet also first brought in the concept of democratisation of information (Bajwa 2003): information flow was not limited to the top-down model any more. The audience have the freedom to ask questions and trash inadequate information. Web-based communication has in fact changed the way in which communication is carried out. To a large extent, it has drastically changed the meaning of communication.

The Web and Participative Science Communication

Traditionally, science communication was a process of scientists explaining concepts to people. Perhaps, that is the reason behind the fear of science that is widely prevalent in Indian society and the perception that science news will not be a 'seller'. These have grown because of the way science has been taught in the schools and colleges through the ages. It has been disconnected with life and not been able to convince people how it is linked with their daily existence.

Yet another reason is the lack of a comprehensive policy on science communication (Obregón and Waisbord 2010). As a result, attempts in science communication have so far yielded limited results in India.

Funds allocated have been extremely scarce, and there is lack of benchmarks for what is good communication. Coupled with this is the lack of comprehensive evaluation of the communication activities undertaken (Mahanti 2011).

Apart from this, as stated earlier, science communication has so far been top down, with no attempt to take the opinions of common people. One of the major reasons behind this is that the media through which major communication takes place—print and the television—do not have much space for participation. The web is far more interactive now and latest technologies have helped science communication to involve greater participation of people (Alisson 2014).

Website Goals

At the initial stage comes the role of the website as the creator of the first impression, or the image builder. In an age in which a major part of our economic activity takes place through interactions between distantly located individuals and organisations across the globe, this first impression of the organisation that the website creates cannot simply be wished away. It is the website that one visits for primary information about an organisation. The website represents the face of the organisation. This is why the dynamism of an organisation needs to be reflected through its website.

While it is true that a dynamic organisation may have a stale website, it is equally true that a stagnant organisation's website will reflect that stagnancy. This is why it was decided at the DBT that being such a dynamic organisation, it was imperative that its dynamism be reflected 360 degrees, whether through images or approach.

Recent work of the organisation is expected to be highlighted on the home page of its website. Therefore, if an organisation has a website which does not represent any frequent changes distinctly on the *home page*, there is a high probability that it will create an impression of the organisation being stagnant, backward and inactive.

All too often on government websites, changes of personnel are not regularly updated. This creates impediments in communication, when the caller seeks some information from an official whose duties and responsibilities would have changed. This makes the organisation look ossified, and hence dampens the spirit of any new person to communicate with it.

It is this image that influences crucial decisions at critical stages (Barr and Weiss 2012), be it the decision of a bright student to opt for biosciences as a career option, thereby drawing human resource to the stream, or the decision of an international agency to collaborate on a project, thereby drawing funds, infrastructural and intellectual resources to an important area of work.

At the second stage comes the role of the website as a source of accessible information. Therefore, details of the areas of work of the organisation, the designation, areas of work and the contact details of people working need to be represented there.

Beyond this is the role of the website in engaging the public to make science attractive to them, build confidence in the government and remove prevalent myths about the areas of work of the particular organisation—precisely to make up for the false impression that years of lack of communication has resulted in.

Website and Content

Soon after I took over the charge of overhauling the DBT website, the first questions I faced was, "Oh, so you are an IT specialist!", and when I said, "I was not an IT specialist but a communication specialist", I was asked how I could maintain a website.

There is a perception that a website is a technological issue, and it is not seen as a vehicle for communication. The dominant perception as well as the practice in India is that a website is to be made by IT experts and information is to be put into it according to the availability of space. The obverse is correct: a website is only a communication tool enabled, improved, modernised and so on by IT experts. In fact, the IT expert has to tailor the website exactly according to the proprietor's needs. Unless this is kept in mind, a website tends to become technology driven or design driven.

Content-Driven Site

In a website, the latest developments of the organisation are highlighted in very significant manners through the wise use of space, words, fonts, design and colour (Nielsen 2002).

The content on the home page should be timely and fresh. The writer should stay tuned to developments in the industry and the subject matter and write about news and updates that will be relevant to the target audience. The website needs to build its reputation as a source of useful and up-to-date information so that the audience identifies with it as a must-read source (Kramer 2013).

We need to keep the objective of the website and the audience in mind to choose the information that requires priority. Accordingly, there should be a judicious use of words and language.

An editorial calendar can help in organising the content around important dates relevant to the subject (Kramer 2013)—for example, days like the day of DNA discovery, the international cell day and the international health and tuberculosis day are important days for biotechnology organisations anywhere. Special feature stories are to be uploaded on the DBT website on such days.

Keywords are important for organisation's visibility on the information highway, and hence significant keywords that the audience will look for need to be introduced in areas such as the headline, first paragraph and so on (Barr and Weiss 2012). However, one should be careful not to overdo it, for ultimately we write for people, not for search engines.

Language

A large part of the readers on the Internet are teens with very low levels of patience. That is why, on the net we need to write for patience-deficient users. A cluttered screen full of text repels teens more than anything else. With other more attractive options available to distract their attention, the young audience quickly shifts over. They prefer sites that can be easily scanned and illustrate content visually. Appropriate web-writing and formatting techniques are crucial for this high-volume teenage audience. Content should be displayed in small, meaningful chunks with lots of white space, so that students can retain the information even through interrupted reading and obvious distractions of their age like phone calls and text messages.

Writing should be short and snappy, the language free wielding, rather than a stickler to grammar (Nielsen 1997; Loranger 2014).

At the same time, for sites on science communication such as the DBT website, there needs to be another level of information for the interested audience. For example, while a science story can have 'speed reads' that highlight the key points in the story, a 400- to 1000-word write-up, and then there is a link to a deeper level of information—relevant articles on which the story is based. The entire design of the site should be attuned to these requirements.

Design of Site

The ground rule is that it should be simple. Clutter and overload of information should be avoided on the home page even while guiding the user to navigate to the right place for his or her respective information requirement. While this is the thumb rule, an acceptable design is not written in stone.

The massive change in the design of the *New York Times* website is a reflection of the alternation in user demand in websites. Its pages which were earlier full of links to related news articles have now been changed to lots of white space. Few years ago, this design would have been unfathomable to the users. Usually, media sites are built with as many links as possible, so as to give the reader as many entry points as she or he might click. However, with the surge of Internet access through the mobile, the management has projected this design to be more acceptable to the new user (Co-design 2012).

One of the major features introduced is that the comments section is compressed when the article appears, but if one clicks on 'comments', it expands and assumes almost as much importance as the article itself, thus giving the user a choice of higher interactivity and high importance for his or her comments. However, the new design is receiving a lot of criticism (Sullivan 2014). This reflects the need to get users attuned to any drastic change before a formal launch is made.

As far as fonts are concerned, tiny fonts are not preferred on the net. Therefore, options for enlarging fonts should be provided by the site's programming. At the same time, options for changing the colour of the site should also be provided.

Communication Components

Visual communication is one of the significant parts of communication, especially on the web. Websites that can illustrate concepts visually are popular. For example, a visual representation of mitosis and meiosis can explain the concept much better. It is said that a picture speaks a thousand words. Difficult concepts can be made much easier through visual representation. The journey to the discovery of the DNA can also be visually illustrated. Such visuals include timelines, tables, flow charts and many other forms. It also includes flashes, videos and so on.

However, there is a case against overdoing of visuals. Use of multimedia where it is not required also scares the reader away. The best online features are those that teach the young something new or keep people focused on a goal. Some of these are online quizzes, forms for providing feedback or asking questions, online voting, features for sharing pictures or stories, message boards and so on. Depending on its usefulness multimedia can attract or repel the reader.

Easy navigation is a crucial requirement for websites. Readers need to understand the structure of a page. Pages without structural links are like orphans. However, at the same time it is important not to overdo structure. For example, linking all sections of the site from all the pages would confuse the reader more. Rather it would be useful to have relevant links in all pages (Nielsen 2000).

Social Media

The year 2013 was widely regarded as the year social media leapt from dorm room to boardroom. The power of the social media had made itself significantly felt for the first time in 2011. Former President Hosni Mubarak of Egypt shut down the Internet all across his country for five days during demonstration in the country against his regime. This was done to curb the easy communication and airing of public grievances through the social media that was helping the demonstrations to proliferate all over the country (Ali 2012).

Election campaigns have also seen a massive use of social media, be it the 2011 US elections where both the presidential candidates Barack Obama and Mitt Romney placed a huge premium on it, or the major parties in India like Bharatiya Janata Party (BJP), Congress and the Aam Aadmi Party, during the 2014 elections. One of the reasons attributed to the rise in popularity of Barack Obama before the elections was the social media (Davies 2012), while the major reason behind the sweep of the BJP in India was an active social media campaign (Dhapola 2014; Patel 2014).

Twitter, launched in 2006, has more than 500 million followers, making it an important method of communicating. With the energies of a ten year old (2016) it can wreck havoc. Information can be shared in real time and can go viral within minutes (Tumasjan et al. 2010). Continuous tweets during the three-day South Asia Biosafety Workshop helped to popularise the proceedings of the workshop among a worldwide audience who are registered with Twitter, through postings and reposting (Muldoon 2013).

Taking clue from this, the author set up a continuous tweet system from the DBT twitter account during a seminar organised on the occasion of the second foundation day of the Biotechnology Industrial Research Assistance Council (BIRAC). The seminar, which highlighted the importance of partnerships in translating science into research, received a high level of response. Another occasion on which this was used was an inspirational lecture by eminent scientist and Secretary, DBT, Prof. K. VijayRaghavan to students. The tweets were re-tweeted, and the two occasions also increased the number of 'Likes' on the DBT Facebook account significantly.

Social media thrives on timeliness and regularity. Following this principle, a Twitter and a blog schedule has been set up so that the audience has a regular supply of information on which they can depend and look forward to.

The social media has completely changed the manner in which communication takes place. Communication, as mentioned, was earlier visualised as top down—from the authorities to the people. Social media has made communication inclusive of the opinions of the common people. All issues are open to public debates and people's opinions on an issue are out in the open for the public to see and get influenced by.

Approvals

One of the banes of government websites is approval for anything and everything that is uploaded. Initiated as a guard against spreading misinformation, the approval process now restricts information itself. The pace at which the approvals come through acts against the very purpose of news—timely dissemination of information. News becomes out-of-date a day later, whereas approvals for news items from the respective departments take weeks to come.

One of the reasons behind this is the lack of perception about speed in news. Another reason is the completely different manners of functioning of scientists and communicators. While the scientist's world, which consists of peer reviews in journals and publications, the work takes months to appear in print, the journalist's timelines involve days and hours. In order to overcome this challenge there is a need either to imbibe the importance of news among scientists in government firmaments or to find a way to bypass approvals at least for news updates. We need to put our heads together on this.

Speeding up this process has become even more relevant in an age when information dissemination is real time—through the social media. Such approvals would sound the death knell of efforts to engage with the public through social networking.

The web, however, is still not fully effective if it is not used in tandem with other traditional media. For, much as we miss out on reaching out to the youth if we ignore the web and the social media, a large part of our population is still outside this new ecosystem either due to lack of access to the Internet or due to incompatibility with the fast paced world of technological advancements.

However, today reaching out to the gatekeepers of the traditional media too involves communication through social media like Facebook and Twitter. This is because, the journalists also pick up a lot of news from these sources, which are easily accessible and reduce a lot of their legwork.

The website and the social media are powerful tools of communication. They need to be used in the right manner. News is very transient today. Therefore, timeliness, accuracy, simplicity and capacity to draw attention are very important for science communication on the web.

Acknowledgements The development of the website has been a team effort. I use the limited space here to thank all those involved in the entire process: Prof. K. VijayRaghavan the brain behind the concept, Dr Shailja Gupta for her efforts in pushing it, Dr Shandhya Shenoy and other DBT officials for their support, Vigyan Prasar, under the able leadership of Dr Gopichandran, for their support, the web-communication team—Poonam Trikha, Vijay Sonekar, Alpana Saha, Deepika and the e-media solutions team—for all their efforts that went into it, journalist seniors and friends for their crucial suggestions and my entire family for their relentless support during the entire journey, particularly my journalist husband for his critical inputs.

References

Ali, A. H. (2012). The power of social media in developing nations: New tools for closing the global digital divide and beyond. *Harvard Human Rights Journal, 24*, 185–219.

Alisson, E. (2014). Internet interactivity has changed the way that science is communicated. *FAPESP*, http://agencia.fapesp.br/internet_interactivity_has_changed_the_way_that_science_is_communicated/19171/. Accessed 28 May, 2014.

Bajwa, G. S. (2003). ICT policy in India in the era of liberalization: Its impact and consequences. *GBER, 3*, 49–61.

Barr, C. & Weiss, A. (2012). *Million dollar web presence.* Entrepreneur Press.

Bhandari, N., et al. (2014). Efficacy of a monovalent human-bovine (116E) rotavirus vaccine in Indian infants: A randomized, double-blind, placebo-controlled trial. *The Lancet, 383*, 2136–2143.

Co-design (2012). Will the New York Times redesign lead to a new web standard? http://www.fastcodesign.com/1672123/will-the-new-york-times-redesign-lead-to-a-new-web-standard. Accessed 28 May, 2014.

Colecchia, A., & Schreyer, P. (2002). ICT Investment and economic growth in the 1990s: Is the United States a unique case? A comparative study of nine OECD countries. *Review of Economic Dynamics, 5*, 2408–2442.

Davies, R. (2012). Digiactivism, memocracy. Changing face of citizen's action. Digital Natives with a cause, Newsletter Vol 10.3/Dec 12. http://cis-india.org/digital-natives/blog/changing-face-of-citizen-action.pdf. Accessed 16 December, 2014.

Dhapola, S. (2014). Mantris who facebook: How Modi turned BJP into a social media warrior. www.firstpost.com/living/mantris-who-facebook-how-modi-turned-bjp-into-a-social-media-warrior-1552209.html. Accessed 2 June, 2014.

Digital Future in Focus. (2013). https://www.comscore.com/Insights/Blog/2013-Digital-Future-in-Focus-Series. Accessed 2 June, 2014.

EBTC (2013). Indian Biotechnology sector–an overview, European Business and Technology Centre. http://www.ebtc.eu/pdf/130326_PPT_Webinar-on-Biotech-Destination-India_Opportunities-&-Challenges.pdf

India Telecom News. (2003). Teledensity—total, urban and rural. Available from: http://www.indiatelecomnews.com/industrymarket.htm

Kramer, S. (2013). http://www.v3im.com/tag/shelly-kramer/#axzz3EgLJpLeN

Loranger, H. (2014). Break grammar rules on websites for clarity. Nielsen Norman Group, March 23, 2014. http://www.nngroup.com/articles/break-grammar-rules/

Loranger, H. & Nielsen, J. (2013). Teenage Usability: Designing Teen-Targeted Websites, Nielsen Norman Group, February 4, 2013. http://www.nngroup.com/articles/usability-of-websites-for-teenagers/

Madhi, S. A., & Parashar, U. D. (2014). 116E rotavirus vaccine development: A successful alliance. *The Lancet, 383*, 2106–2107.

Mahanti, S. (2011). Developing critical understanding through science popularization in India. In G. Raza, R. E. N. Fujun, H. J. Khan & H. E. Wei (Eds.) *Constructing culture of science:*

Communicating of science in India and China. National Institute of Science Communication and Information Resources (NISCAIR) & China Research Institute for Science Popularization (CRISP).

Muldoon, L. (2013). Using social media to communicate. *South Asia Biosafety Programme Newsletter 10*.

Nielsen, J. (1997). How users read on the web. Niellsen Norman Group, http://www.nngroup.com/articles/how-users-read-on-the-web/. Accessed October 1, 2014.

Nielsen, J. (2000). Is Navigation Useful, Nielsen Norman Group, http://www.nngroup.com/articles/is-navigation-useful/. Accessed January 9, 2014.

Nielsen, J. (2002). Top 10 guidelines for home page usability. Neilsen Norman Group.

Obregón, R., & Waisbord, S. (2010). The complexity of social mobilization in health communication: Top-down and bottom-up experiences in polio eradication. *Journal of Health Communication: International Perspectives, 15*(Supplement), 1.

Patel, A. (2014). India's social media election battle. http://www.bbc.com/news/world-asia-india-26762391. Accessed March 31, 2014.

Quantum Indians (DVD) (2013). Public Service Broadcasting Trust, India.

Sullivan, M. (2014) After the times redesign, complaints and changes, http://publiceditor.blogs.nytimes.com/2014/01/20/after-the-times-redesign-complaints-and-changes/?_php=true&_type=blogs&_r=0. Accessed January 20, 2014.

Tumasjan, A., Sprenger, T. O., Sandner, P.G., & Welpe, I. M. (2010). Predicting elections with twitter: What 140 characters reveal about political sentiment. Proceedings of the Fourth International AAAI Conference on Weblogs and Social Media (pp. 178–185). Association for the Advancement of Artificial Intelligence.

Using Social Media for Research and Reaching Out

Sandhya Sekar and H.S. Sudhira

Abstract Social media is a generic term used for Internet-based applications through which users create, share and exchange information and ideas through virtual communities. The general perception of social media is that it is a way to waste time, meant for young people to share photographs and jokes. However, what needs to be understood is that different social media platforms are merely tools—it is up to the individual to use them in specific ways. With science and scientific knowledge accumulating at a great pace, social media can be a quick, efficient way to exchange information, both between scientists and for scientists to disseminate information to the general public. In this chapter we discuss the concept of social media, introduce the different web-based tools available, explain how scientists can benefit from having a web presence and provide suggestions for getting started.

Keywords Science communication · Social media tools · Facebook · Twitter · Academia.edu · ResearchGate · Mendeley

Introduction

In less than a decade, online tools have exploded in popularity, with social media sites witnessing a meteoric rise in the number of users. Social media is a catch-all term used to refer to a variety of online communication applications, which since early 2000s have created a platform for people to communicate their ideas, like never before.

One fundamental manner in which social media differ from other forms of media is the possibility of immediate reactions and responses to an idea or a piece of

Views expressed in the chapter are the authors' own and not necessarily shared by the editors.

S. Sekar (✉) · H.S. Sudhira
Gubbi Labs, Gubbi, Karnataka, India
e-mail: sandhya@gubbilabs.in

H.S. Sudhira
e-mail: sudhira@gubbilabs.in

P. Bagla and V.V. Binoy (eds.), *Bridging the Communication Gap in Science and Technology*, DOI 10.1007/978-981-10-1025-5_17

information. In traditional forms of media, like the newspaper and television, there is a barrier between the person who disseminates the idea or information and the recipient. If the recipient wishes to respond, she or he needs to employ another communication channel like the telephone or email—or even a letter. In the world of social media, the interaction is immediate, as easy as typing few lines and pressing a key. With the smartphone becoming ubiquitous, responses can also be on the go.

Such interactions on social media, when regarded as just what they are—sharing of information—can go beyond interpersonal relationships, and permeate the professional. In fact, there are social media platforms meant for professional networking. The world of academia is no exception.

Scientists can potentially use the power of social media to speed up the pace at which they develop and share knowledge. This can include both 'inreach', sharing knowledge within the scientific community and 'outreach', sharing with the general public (Darling et al. 2013). There is a growing diversity of 'social ecosystems' that support the scientific and scholarly use of social media (see Box 1). Collaborative project spaces (Google Docs, wikis and blogs), online communities (Mendeley and Zotero) and professional networking sites (Twitter, Facebook, Academia.edu, LinkedIn, ResearchGate) can all be used to develop new ideas and collaborate and share scientific knowledge with others.

Although a number of guides exist online, many researchers still feel overwhelmed and hesitant to be part of the virtual world. There is also a common misconception that social media is meant for leisurely activities that waste precious time; but that is not always true. Different social media platforms are just tools—how they are used is entirely up to the individual concerned. This chapter familiarises the reader with the available tools, discusses the benefits of an online presence and online science conversations, and explains how scientists can efficiently harness online resources.

Box 1: Social Media Tools

Blogs

Traditional, long-form articles. Wordpress and Blogger are two of the most popular blogging platforms, which offer free hosting for your blog and offer easy interfaces for posting text and images.

RSS feeds

RSS is the abbreviation for Really Simple Syndication. It uses web feeds to publish frequently updated information like news headlines and journal content alerts, without having to use a browser constantly. Feedly is one of the best RSS aggregators, with mobile, tablet and computer interfaces.

Twitter (http://twitter.com)

Social networking site and microblogging platform that limits posts to 140 characters (including spaces). With a free account, you can 'follow' other Twitter users, which means that you subscribe to their updates and can see their messages or 'tweets' in your feed. When you tweet, other Twitter users can see and if they want, follow your messages. You can tweet your ideas, links to a great article you just read, images, videos and other multimedia—within the limit of 140 characters. When you want to include a fellow Twitter user in the conversation, you can mention them using the @ symbol. You can 'retweet' tweets you find interesting, too. A hashtag (#) is used to categorise topics. When a lot of people are talking about one topic, that hashtag starts to 'trend' on Twitter.

Facebook (http://www.facebook.com)
The most widely used social media site. In India, Facebook is a lot more popular than Twitter, which is just slowly catching up. A business-like 'Page' can easily be created so that personal and professional matters can be kept separate.

Google+ (https://plus.google.com/)
Similar to Facebook, it is Google's social networking site. It links your Google account to your other activities like Google Maps, Gmail and YouTube. It can be used to share photos and text.

YouTube (https://www.youtube.com/)
YouTube is a video-sharing website where users can upload, view and share videos. It is mostly used by individuals, but media corporations put up selected content on it.

Tumblr (http://www.tumblr.com)
A microblogging site that can publish any type of media very easily and quickly. Users post photos, videos or short quotes as opposed to lengthy written narratives. Tumblr offers automatic forwarding of new posts to Facebook and Twitter accounts.

Pinterest (http://pinterest.com)
A photo-only microblogging site where users define themed 'boards' for posting content (e.g., food, art and marine fish). The 'pins' from Pinterest can also be shared via Facebook and Twitter.

Storify (http://storify.com)
A way to aggregate and organise tweets, videos, blog posts and other media. Storify is especially useful for compiling media on discrete discussions and preserving tweets before they become archived by Twitter. For example, if there is a panel discussion or academic seminar, a Storify can be created that includes live tweets from the audience, videos of the panelists, and links to their publications, websites and social media profiles.

LinkedIn (http://www.linkedin.com/)
A business-oriented social networking site, where you can upload professional details such as curriculum vitae, place of employment and potentially hunt for jobs.

Academia.edu (http://academia.edu/) **and *ResearchGate*** (http://researchgate.net/)
These are social networking sites meant for academics—a place to discuss various questions, share articles and follow research in a particular field.

Mendeley (http://www.mendeley.com/)
This has both web-based and computer-based features. Mendeley Desktop is a PDF and reference management tool and the web version is an online social network meant for discovering new articles in a particular field, sharing articles with others and collaborating.

Why Should Scientists Use Social Media?

Science is poorly understood, almost universally. Citizens of all countries need accurate information to make informed decisions about a variety of issues from climate change to forest policy to medical practices. Education is only part of the answer, because it targets the next generation; to change older people, information needs to reach them in a digestible form. These people depend on the media to learn what they need to know. They are the people who vote and who are taxed to pay for the scientific funding. As a community, scientists need to reach out to them.

As of now, science is almost always a one-way conversation. Publishing an article in a peer-reviewed journal is one step in the conversation—it is just the beginning. These publications are behind paywalls. Even the ones that are open access are too jargon filled for the untrained eye to understand and appreciate. Social media is a cheap, quick and efficient way for scientists to break out of the 'ivory tower' that is academia.

Research Benefits from an Online Presence

If I Google you, by name, what will I find? Potential students, institutes looking to recruit faculty members, journalists on the hunt for an expert source—all turn to Google to look for answers. Lack of an online presence even runs the risk that undesirable search results appear before the desirable ones; one fact about the Internet is that even if you are not speaking about yourself, someone else is speaking about you. One way to keep track of what is being said about you is to join the conversation and step into make corrections where necessary. Even a simple website can be your little corner of the web, your space where you can share journal articles, discuss your thoughts and scientific opinions, attract students and circulate information about professional opportunities and upcoming events (Bik and Goldstein 2013).

Social media can "contribute to the research process as an open access and free service", which can be used to source expert opinions about work in progress (Darling et al. 2013). Instead of an average 30-member academic department, a scientist can have a much larger "virtual department" on Twitter (Darling et al.

2013). When Darling et al. (2013) characterised Twitter followers for the four authors of their article, they saw that "about 55 % of the followers were science students, scientists or science organisations who could be potential collaborators. The remainder comprised nonscientists, media and the general public who may be more likely to be engaged in the dissemination of published scientific findings." With politicians and policy makers increasingly jumping on the social media bandwagon to keep pace with the younger generation, these platforms bring with them the distinct possibility of influencing science policy.

A wider network can result in invitations to give research seminars and talks, chair conference sessions, write articles and book chapters, as well as job offers. Many science grants—at least the ones given out by international agencies—need scientists to disseminate results from their work to a broader audience, and social media can help meet these conditions.

Networking during conferences is also easier with social media. Twitter is especially suited for conferences. If a suitable 'hashtag' is introduced early enough, participants can meet virtually and start discussions before they meet in person; a conference can become even more inclusive and involve scientists from all over the world who could not be physically present. Tweets during a conference can be aggregated using a tool like Storify to archive discussions for thinking about later.

Twitter can be an extremely powerful way for a scientist to promote his or her own work. The trick is conveying complex information in 140 characters or less! If necessary, a link to a longer blog can be used. Social media is also an excellent place to network with distant colleagues without having to spend time and increasing your carbon footprint travelling. Scientists who teach can also use social media, especially Twitter, as an excellent source of teaching material. Bite-sized pieces of information about the latest research can help scientists keep abreast of the flood of information available, when most researchers have the time to go over abstracts or even content alerts.

Social media can also help arrive at the most up-to-date consensus about analytical methods—you are bound to find someone else with the same problems who can help. Software and analysis discussion forums help graduate students and researchers working from anywhere on earth to learn new techniques even without needing to spend money flying down to an expert. Forums for the statistical software 'R' are an excellent example of an extensive, friendly community of researchers who need specific kinds of analyses.

There are some extreme cases of social media directly influencing research. A team of researchers from the Smithsonian's National Museum of Natural History performed a fish survey in Guyana. They had to identify more than 5000 specimens in less than a week to obtain an export permit. Facebook came to their rescue; they put up a catalog of images on Facebook and turned to their colleagues spread around the world for help (Smithsonian Science 2011). Scientific games have been helping researchers solve puzzles impossible to decipher otherwise—Foldit (http://fold.it/portal/) uses players to unravel complex protein structures and Galaxy Zoo gets people to classify galaxies photographed by the Hubble telescope according to shape (http://www.galaxyzoo.org/).

There is some evidence to show that an active online presence and increased social media activity impact a researcher's credentials measured using more traditional metrics like impact factors (Bik and Goldstein 2013). Other studies have shown no correlation between an article's citation count and Twitter mentions; this could mean social media mentions are capturing a previously unquantified impact of a scientist's career. Darling et al. (2013) talk about 'alternative metrics or altmetrics', which try to quantify the broader impact and reach of scientific knowledge beyond traditional journals, an important component of such communication being social media (Darling et al. 2013). These metrics are based on data like: 'the volume of downloads for an article or data repository, the number of mentions on sites like Facebook and Twitter, and bookmarks to online referencing libraries such as Mendeley or CiteULike'. There are also third-party platforms, such as ImpactStory.org and Altmetric.com that help researchers measure the online impact of their science (Darling et al. 2013).

The use of altmetrics is still controversial in the scientific community. "But one must keep in mind that altmetrics are not intended to replace traditional bibliometrics like number of citations. The two approaches are in fact complementary, capturing different types of impact for different audiences" (Darling et al. 2013).

Possible Problems of Using Social Media for Scientific Discussions

This discussion would not be complete if we do not examine some of the problems surrounding discussion on science on social media. The main problem is of intellectual property ownership. Thanks to cleverly written terms of service agreements, most social media platforms retain the ability to republish without permission. Also, one cannot really control who reads what you publish; this can become a problem when discussing data from articles or grants.

An extreme example of this was the 'Climategate' scandal that began in 2009 when internal emails about the International Panel on Climate Change (IPCC) were leaked (Darling et al. 2013). Climate change deniers confused academic debates over data interpretation with a fabrication of results and used the argument to discredit IPCC's findings. Scientists are urged to be extremely careful with their exchanges.

Social Media Tips for Scientists

Adapted from de Lange (2012) and Bik and Goldstein (2013):

How Can I Reflect My True Identity Online?

To establish an online presence and avoid undesirable Google search results, researchers should have a personal website at the very least. It should describe their specific research projects and areas of expertise and be searchable by colleagues, journalists and the public alike. Although professional websites can be established through your university/institute, external hosts (a free site at http://wordpress.com or a custom paid domain) offer more flexibility and are easier to access and maintain.

Where Do I Start?

Start with people you know, people with common interests; follow the social media that they link to and that links to them. It is completely acceptable to 'unfollow' people or groups if their information is not relevant or useful. It can be beneficial to read first without contributing ('lurking') to learn logistics and basic etiquette of different social media platforms.

Which Platform to Use?

When building your online social media profile, try to decide which tools suit your style. For a comparison of the most popular tools see Table 1. For instance, there's no point in setting up a blog if you simply do not have the time and motivation to update it. In which case, perhaps Twitter would be a better option. Twitter is really good for peer support, sharing resources, and building up your networks. Facebook has official page options that you can explore. If your work generates incredible images you are keen to share, try Flickr or Pinterest; if you simply want a more solid and static profile try the academic sites. In all platforms, make sure you complete your profile carefully, and use it to connect to your other online presences. For instance, if you tweet and blog, include a link to your blog on your Twitter profile, and add a Twitter widget to your blog so visitors can easily find and follow you there.

What Type of Information Should I Share?

Share your work, and details of your research to the extent that it is acceptable, but you should certainly check any existing privacy, nondisclosure or social media agreements that you have with your employer or the journals that publish your research. Getting a funny or playful angle around your research also piques interest and draws people in. Try not to post commercially sensitive data and personal information that might impact on

Table 1 A comparison of online tools

Platform	Pros	Cons
Blogs	Longevity; posts are accessible via search engines Robust platform for building an online reputation	Time investment for preparing thoughtful posts Posts should be disseminated and advertised via other platforms
Twitter	Low time investment, short posts Ability to rapidly join in on online conversations The most current source for breaking news and topical conversation	Posts are quickly buried under new content Twitter does not make its archive database accessible to search Gaining followers can be a slow and difficult process
Facebook	Established juggernaut in the social media world Ability to create 'groups' and 'pages' for a person or cause	Privacy concerns
Google+	Integration with Google tools Easily manage privacy/visibility by grouping contacts into 'circles'	User base not unique compared to other sites Users still unsure how to use it
LinkedIn	A renowned, trusted professional network used by many companies Geared towards job searches	It is not very intuitive to use Intellectual property and privacy concerns

Source Authors' work

your professional reputation; watch out for automatic app updates. Be very careful—the Internet never forgets!

How Can I Judge the Right Tone?

Get a colleague to have a look and give you a second perspective when you start blogging to see if the voice is right, says Osborne. When you start, err towards the formal and loosen up later, but, "if your quirky stuff is in good taste, it should go down fairly well."

How Can I Manage My Social Media Activity?

You do not have to spend hours checking and posting furiously on social media. Use an app such as HootSuite or Buffer with which you can schedule posts.

References

Bik, H. M., & Goldstein, M. C. (2013). An introduction to social media for scientists. *PLoS Biology, 11*(4), e1001535.

Darling, E. S., Shiffman, D., Côté, I. M., & Drew J. A. (2013). The role of Twitter in the life cycle of a scientific publication. *PeerJ PrePrints 1,* e16v1.

de Lange, C. (2012). Social media tips for scientists. *Nature Jobs*. Available at: http://blogs.nature.com/naturejobs/2012/09/28/social-media-tips-for-scientists Accessed June 3, 2014.

Smithsonian Science, (2011). Facebook friends help scientists quickly identify nearly 5,000 fish specimens collected in Guyana. Available at: http://smithsonianscience.org/2011/03/facebook-friends-help-scientists-quickly-identify-nearly-500-fish-specimens-collected-in-guyana/ Accessed June 6, 2014.

Part V
Bridging the Gap Between Scientists and the Public

Vigyan Rail: Science Exhibition on Wheels

Vinay B. Kamble

Abstract 'Vigyan Rail—Science Exhibition on Wheels' was a landmark event in the history of science communication of the country at the turn of the 21st century. Vigyan Rail (meaning, 'train of science'), a train carrying scientific exhibits on various themes crisscrossed the length and breadth of the country for eight months during 2003–2004 in an attempt to take science to the people. The project was conceived and implemented by Vigyan Prasar, and executed in collaboration with Ministry of Railways, with active participation of 18 departments/ministries of Government of India engaged in various scientific and technological activities. During its entire journey, Vigyan Rail aimed to motivate the people to learn, understand and appreciate how science and technology (S&T) have helped to improve their quality of life and the new challenges that have emerged through exhibits, working models and slide/multimedia shows and demonstrations. At the same time, the exhibition focused on the arduous path and the determined efforts of our scientists who made this progress possible, immensely contributing to the social and economic growth of the country. Vigyan Rail was a unique experiment that proved how people crave for information and knowledge about S&T that they use in their everyday life, or the environment they live in. This project evinced keen interest in many developing and developed countries around the world as well. Vigyan Rail—its conceptualisation, implementation, impact and the lessons drawn from this landmark project—along with suggestions to make future projects even more effective and the recent efforts are discussed in this chapter.

Keywords Science train · Vigyan prasar · Ministry of railways · Science popularisation · Science exhibition

---

Views expressed in the chapter are the author's own and not necessarily shared by the editors.

V.B. Kamble (✉)
Vigyan Prasar, Noida, India
e-mail: vinaybkamble@gmail.com

V.B. Kamble
Department of Science and Technology, Government of India, New Delhi, India

P. Bagla and V.V. Binoy (eds.), *Bridging the Communication Gap in Science and Technology*, DOI 10.1007/978-981-10-1025-5_18

The Genesis

Railways have played an important role in the social and economic development of the country for over one and a half centuries, besides being a means of communication and transportation. It is the most important network for travel and transportation throughout the country. The railways have brought people living in far-flung areas of the country into the mainstream, and hence have also played a major role in cultural integration of the country. Further, ever since the era of economic liberalisation that began in early 1990s, the pace of growth and development in scientific, technological and industrial spheres have been phenomenal—more so, in fields such as information technology(IT) and biotechnology.

Indeed, modern technology has been connecting the people from far-flung areas of the country together; however, a direct contact with the people with an aim to spread scientific awareness and help them develop a scientific outlook is what will eventually help India to emerge as a technologically powerful and culturally cohesive nation. There is no gainsaying the fact that despite the variety and volume of the programmes on science and technology (S&T) communication today, a need is felt for an efficient means or a network that would allow us to reach the people of the country directly in different geographical regions with relative ease. Perhaps, the vast network of the Indian Railways could be one such means!

Sometime in 2002, thoughts like these prompted M.V.Kamath, the then President, Vigyan Prasar Society, to ask a question in one of its general body meetings: "Can we have a train moving throughout the country carrying an exhibition depicting the achievements of the country in different fields of science and technology?" Well, the idea, strongly supported by Dr. Murli Manohar Joshi (the then Minister of Science and Technology, Human Resource Development and Ocean Development) and Nitish Kumar (the then Minister of Railways) culminated into 'Vigyan Rail—Science Exhibition on Wheels' in 2003, a prestigious project conceptualised, formulated and implemented by Vigyan Prasar jointly with the Ministry of Railways and with active support from the Department of Science and Technology (DST)—in particular, Prof.V.S.Ramamurthy, the then Secretary, DST, and Chairman, Governing Body, Vigyan Prasar.

The project was undertaken with the active participation of various scientific departments/ministries/councils of Government of India. After several rounds of discussions with the officials of the Ministry of Railways, the Vigyan Rail project finally evolved and modalities were worked out. Vigyan Prasar prepared a detailed project proposal with inputs from Ministry of Railways. To discuss the various aspects related to the Vigyan Rail project and to seek the views, suggestions and commitment to the project, a coordination meeting of scientific departments/ministries/councils was held in July 2003. The response was truly overwhelming. DST, Vigyan Prasar and other participating departments/ministries agreed to share the costs for putting together the Vigyan Rail—Science Exhibition on Wheels (Padmanabhan 2006).

The Exhibition Takes Shape

An apex committee under the chairmanship of Secretary, DST and members drawn from DST, Ministry of Railways, National Council of Science Museums and Vigyan Prasar supervised the development of the entire Vigyan Rail project.

The guiding principle followed was that each compartment would be arranged in such a way that it would depict the progress made by India in a particular field—environment, agriculture and so on. In addition, the exhibition would also have exhibits/panels describing scientists and the discoveries they made at appropriate places. The emphasis would be on the arduous path and the approach they followed, illustrating their life as a scientist. This essentially would describe the questioning attitude and a rational approach they followed, or what we call the method of science. This would bring out the excitement and thrill of doing science.

Due to the specific dimensions of the coaches, there was an obvious limitation on the type of exhibits that could be fabricated and displayed. Emphasis was on motivating the people through exhibits, working models and slide/multimedia shows. There would also be experts/volunteers to explain the exhibits to the people in the local language. The experts/volunteers would be chosen and trained from the local areas, as far as possible. Types of exhibits would include interactive exhibits bringing in visitor participation, attractive animations, large screen video/multimedia shows, audio-visual shows, software for sale on various media and so on. Further, the exhibits were to be vetted by the apex organizing committee constituted for the purpose. Texts/panels were handed over to Vigyan Prasar for getting the panels/blow-ups fabricated/assembled. Suggestions by the respective departments/ministries were incorporated.

The Ministry of Railways had previously organised an 'Exhibition on Wheels'—an exhibition on train to celebrate 150 years (1853–2003) of railways in India, which had a run of eight months from 15 August 2002 to 15 April 2003. This made the job relatively easy enabling Vigyan Prasar to build upon the experience and resources of the Railways in running an exhibition train in the country. As in the case of the Railways, Vigyan Rail too would have twelve exhibition coaches and move throughout the length and breadth of the country for about eight months in different parts of India, halting at 56 stations en route. The train utilised by the Ministry of Railways for its exhibition was fully refurbished. A contractor was appointed after the due selection process and the fabrication of the Vigyan Rail began in the right earnest in September 2003 at the Delhi Safdarjung Railway Station, Delhi. Through periodic meetings, supervision and cooperation of the various ministries/departments, the train was ready to be flagged off in early December 2003 (Kamble 2003).

The Exhibits: Salient Features

The National Council of Science Museums showcased Indian heritage in S&T, modern developments and contribution of Indian scientists. Department of Atomic Energy (DAE) put up exhibits that sought to remove the myths about hazards of nuclear power plants and highlighted the beneficial applications of nuclear technologies. The Indian Space Research Organisation (ISRO) highlighted the achievements of our space programme by displaying models of various satellite systems developed for different applications such as communication, meteorology, search and rescue, remote sensing and resource mapping. The exhibits of India Meteorological Department (IMD) provided information on weather forecasting services, seismological studies, Antarctic meteorology and the phenomena such as the super-cyclone.

Department of Biotechnology (DBT) attempted to highlight the progress made in India in this field and its importance in various fields of human activity. One could get to know all about the oceans, its resources and the efforts to tap these resources for national development from the exhibits put up by the Department of Ocean Development (DOD). With a vast coastline of 7500 km India has a long maritime history. The exhibits provided the visitors with information on exciting expeditions to Antarctica, study of ocean dynamics that influences the weather on the mainland, exploration of gas hydrates and so on.

The Department of Information Technology explained its major activities to promote knowledge-based enterprises, development of electronics and promotion of the Internet, e-commerce and IT-based education. The exhibits of the Department of Telecommunications gave a brief overview of the evolution of telecommunication over the years. With a network of 40 laboratories and 100 extension centres, the Council of Scientific and Industrial Research (CSIR) is one of the world's largest R&D institutions engaged in a wide spectrum of activities. These cover new products, processes and technologies in diverse areas ranging from chemicals and drugs to materials and electronics, from safe drinking water to aerospace and from environmental protection to petroleum processing. All these were portrayed in its exhibits.

The Indian Council of Agricultural Research (ICAR) depicted the efforts made through its network of research institutions to ensure food security by developing not only high-yielding varieties of different crops but also technologies to tap their full potential. The Indian Council of Medical Research (ICMR) highlighted the research work being done to control diseases, improve nutritional levels of the population, reduce infant and maternal mortality rates and develop strategies for health-care delivery with a view to achieving the goal of 'Health for All'. National security is no less important, and the Defence Research and Development Organisation (DRDO) put up exhibits that showcased the development of state-of-the-art technologies in the frontier areas of aeronautics, armament, high-energy materials, combat vehicles, combat engineering equipment, missiles, radars, robotics and artificial intelligence.

The exhibits put up by the Ministry of Environment and Forests sought to bring out the relationship between nature and human beings and the need to conserve the environment and ensure sustainable development. The Ministry of Water Resources and the Central Water Commission sought to educate the visitors on the distribution of freshwater in the world and inculcate the attitude towards conservation of water resources. The Ministry of Non-Conventional Energy Sources (MNES) had exhibits displaying a number of initiatives to develop systems based on renewable sources of energy such as sun, wind, water and garbage. Models of these were displayed by MNES with descriptive notes on their cost-effectiveness and benefits.

In addition, Technology Development Board and Technology Information, Forecasting and Assessment Council (TIFAC) also participated with exhibits/information on few specific projects. Vigyan Prasar put up exhibits/information on its major activities, a quiz and short video clips on everyday use of science.

The Journey

Vigyan Rail—Science Exhibition on Wheels was flagged off by Atal Bihari Vajpayee, the then Prime Minister of India, on 15 December 2003 from Delhi Safdarjung Railway Station in Delhi. The Prime Minister said on the occasion, "Ours is an ancient country which is reborn in the modern world. We want to preserve and strengthen our centuries old cultural heritage and traditional values. Also, we want that India should have the top place in the modern world." Dr. Murli Manohar Joshi, the then Minister of Science and Technology, Ocean Development and Human Resource Development, Nitish Kumar, Minister of Railways and Bachi Singh Rawat, Minister of State (Science and Technology) also voiced similar feelings. Indeed, Vigyan Rail—Science Exhibition on Wheels attempted to echo these very thoughts.

The journey began from, and concluded at Delhi Safdarjung Railway Station. The same route, which the Indian Railways followed for its Exhibition on Wheels in 2003, was chosen for this Science Exhibition on Wheels—it was felt that from operational point of view, it would be better to follow the route taken by a similar train earlier. This marked the launch of the project in which for the first time an opportunity was provided to the people in remote corners of the country to have a glimpse of the nation's progress in the field of S&T. It may be noted that the Vigyan Rail halted at 60 stations instead of the 56 originally planned, the unscheduled four halts being the result of excessive demand for accommodating certain areas en route by the local public.

Although the train was flagged off on 15 December 2003, it actually left Delhi Safdarjung Railway Station at Delhi a week later giving an opportunity for the Delhiites to have a glimpse of the exhibits. It returned on 16 August 2004 to the same station after completing a journey of over 15,000 km crisscrossing the country, covering 60 railway stations—including four unscheduled halts! During the process, it had helped rekindle the declining interest in science among the children and the younger generation in the country. At every place it visited, from Rajkot to Tinsukia and Pathankot to Kanyakumari, it received a thunderous welcome. An estimated 5million people visited the Science Exhibition on Wheels during this period. It was winter when the train moved in the northern states, and summer when it moved in the south. And yet, braving the cold and the heat, the visitors came not only from the town where the railway station was located but also from nearby areas.

The Vigyan Rail was designed in such a manner that visitors could enter the first coach and move on to the last coach without having to get down at any stage, as all the coaches were interconnected. Starting with the nation's scientific heritage the exhibits portrayed the progress in diverse areas concerning environment and forests, atomic energy, information technology, communications, water resources, ocean development, scientific and industrial research, defence, non-conventional energy sources, agriculture, health and medicine, space and so on. As mentioned earlier 18 ministries/departments engaged in scientific activities participated in the Vigyan Rail—Science Exhibition on Wheels. The exterior of the coaches was utilised for advertisements and relevant messages for the public by the participating ministries/departments related to their activities. Although the response of other exhibition was overwhelming, there were certain limitations. The period of halt at each station was 2–5 days. At times, this was found to be too short to accommodate a large number of visitors. Although the exhibition time was from 10 a.m. to 7 p.m., often it became necessary to keep the exhibition open till late in the night when the number of visitors was large. The exhibits were in English and Hindi, but in a multilingual country like India, often the visitors expected the exhibits to be in their regional language. This not being feasible, volunteers with science background were arranged to explain the exhibits to the visitors in their local languages.

Vigyan Rail received tremendous support from local authorities, administration and academic institutions wherever it went in terms of publicity, security, volunteers for explaining the exhibits in the local language and so on. Prominent personalities from different spheres were always enthusiastic in making the Science Exhibition on Wheels a meaningful and an enjoyable experience for all the visitors.

The Impact

There is no gainsaying the fact that success of any venture is determined by the public response and media reaction to it. In particular, the impact of a project like Vigyan Rail could be gauged by the number of visitors it had at various places, composition of the visitors at every halt, response of the students and how seriously they took the visit, response of the visitors and, of course, the media reaction. We discuss these aspects briefly in the following paragraphs.

From the viewpoint of impact, Vigyan Rail could be regarded as a resounding success. It helped people become aware of the nation's scientific heritage and achievements, and at the same time helped trigger an interest in science among the youth who visited the exhibition. In a vast country like India, the reach of conventional means of communication can only be limited. True, radio and television have penetrated almost everywhere in the country, but the mode of communication through these media is generally one way, and hence they are effective only in a limited way. However, when one could see and feel the interactive exhibits, and also get information in person from an instructor, the interest in science is immediately aroused. Such interactive exhibits have the potential to communicate the excitement and thrill of science, and of a scientific career. Further, in rural or remote areas, where there are virtually no opportunities to visit a science centre, to obtain information or even discuss the topics of interest with someone, Vigyan Rail—Science Exhibition on Wheels scored a point, and this explains the tremendous response it received wherever it went.

The Numbers

The Vigyan Rail attracted over 25,000 visitors at every halt, and in some places, the figure crossed several lakhs. Bareilly created a history of sorts with a record turnout of 2.5 lakh visitors over a period of 3 days. As stated earlier, despite the fact that the Vigyan Rail travelled the northern region in winter and the southern region in summer, neither the biting cold of the north, nor the scorching heat in the south was a deterrent to the visitors who waited in the serpentine queues for long hours to enter the train and see the exhibits. In all, about 5 million visitors of all ages visited the exhibition and had an opportunity of becoming familiar with the scientific developments in different areas and the scientists who have contributed immensely to the socioeconomic development of the country. A figure of 1 million of a total population of a billion plus may appear insignificant. However, given the vast size of the country and the fact that larger proportion of population live in rural, remote or inaccessible areas, 25,000 visitors or more a day to Science Exhibition on Wheels cannot be considered insignificant.

Composition of the Visitors

Apart from the numbers, the composition of the visitors from different strata of the society brought out the interest and curiosity of the people to know about various aspects of S&T, and the role played by S&T in their lives. We briefly describe the spectrum of the visitors to the Vigyan Rail.

The advance publicity through local newspapers and radio about the scheduled arrival of the Vigyan Rail at the railway station kindled the interest of the people from all walks of life. Rural folks came trekking all the way on foot, bullock carts, bikes, buses or the local trains to see the exhibition. Despite the fact that many of them were illiterate, they evinced a keen interest in the exhibits. Not satisfied with just viewing the exhibits, they asked probing questions to the guides/volunteers about what they and the exhibits aimed to convey. Indeed, some of the exhibits and models held them in awe, say, on defence or space, but they learnt from other exhibits on the rich biodiversity of the country and need to conserve it, need for tapping renewable energy sources, enhancing their farm output and preventing diseases. They returned home with appreciation for S&T and how it has contributed to socioeconomic growth and improving the quality of life.

Students of all age groups were the largest community to have visited the Vigyan Rail. Well, one of the main objectives of the Vigyan Rail was to trigger an interest in science among the youngsters, especially students. Of late, the number of students taking to research as career has been steadily declining, causing a grave concern for the growth and development of S&T in the country. The overwhelming response to Vigyan Rail among the students and the younger generation showed that such an effort could effectively address this issue. True, students far outnumbered the other visitors, and took their visit to the exhibition as if it were the part of their studies! Armed with notebooks and pens, they jotted down important points from the exhibits, put searching questions to the guides and wrote down the replies. They regarded it as a valuable supplement to what they studied in the schools. School authorities arranged buses to transport the students to the railhead. Many schools declared holiday in order that their students could visit the Science Exhibition on Wheels. At some places, schools arranged tests based on the exhibits! At some places, collectors/deputy commissioners even organised essay competitions based on the exhibits! These efforts helped students study the exhibits on different topics and issues in depth rather than having a cursory glance at them.

Without any shadow of doubt, Vigyan Rail offered a unique opportunity to both students and teachers, especially in rural and remote areas, to keep abreast of the advances that had taken place, or were taking place, in the field of S&T affecting their lives. At the same time, they learnt about the lives of scientists, the arduous path and the rational approach they followed, and what is known as the method of science. All this would go a long way in triggering an interest in science among them and also help them in their higher studies.

Parents of the students also evinced keen interest in exhibits. This is quite important, since if the parents are convinced about the importance of S&T, there are good chances that they will encourage their wards to opt for science stream and later take to research as a career. For most of the parents, the exhibition was an eye-opener. In fact, many of them got a feel of S&T for the first time at one place.

Besides the groups described in the preceding paragraphs, there were industrial workers, farmers and professionals—people from almost every walk of life visited the Science Exhibition on Wheels in the entire country with great enthusiasm.

Response of the Visitors

The exhibits generated varied but positive responses from the visitors. "A thrilling experience", "educative" and "wonderful" were some of the epithets used by the visitors. After visiting the exhibition, a schoolgirl in Chandigarh wrote, "Science is so wonderful! Now I want to be a scientist!" "*Yeh pradarshani to gagar mein sagar hai*" (This exhibition says so much in just few words—an ocean in a pitcher!), commented a visitor. Another commented, "*Kuan khud pyase ke paas aya hai*" (The well has walked to the thirsty!). The former comment brought out the depth and the breadth of information provided by the exhibits, while the latter highlighted the way the thirst for knowledge had been quenched by the exhibition. Surely, Vigyan Rail had helped them know many things, which they had not known earlier. This was a major objective of the Vigyan Rail, and the comments of the visitors showed that the exhibition had largely achieved its objective.

Media Reaction

The media is always regarded as the mirror of the public opinion and the public aspiration. Judged from this point of view, Vigyan Rail could be considered a great success. The eight-month journey of Vigyan Rail received a wide coverage both in the print and in the electronic media at all the places where it halted. The common theme running through the media reports was that Vigyan Rail was a repository of information, and it provided a rare and unique opportunity to people in far-flung areas to know about advances made by the country in the field of S&T. The coverage was quite exhaustive and kindled the interest among the general public to view the exhibits. *The Hindu* described it as "a unique mission" and noted that its success was due to its mobility, which provided connectivity to people even in remote areas.

Vigyan Rail Creates Ripples in Several Countries

Vigyan Rail created ripples in several countries. There were requests from France for information on Vigyan Rail. A scientific delegation from Argentina visited Vigyan Rail in Delhi. A member of the delegation expressed, "If we are carrying something back to our country, it is the concept of Vigyan Rail!" A diplomat from the American Embassy who visited the train exclaimed, "Now I want to have a science train in California, my home state!" There were numerous requests on Vigyan Rail from Asian countries as well. Vigyan Rail proved to be a veritable access model to communicate S&T to the people using traditional communication technology blending with modern learning technologies. The author received a special invitation to share his experiences on Vigyan Rail at the International Conference on Distance Learning Technologies for the countries of the South and South-East Asia held at Siem Reap, Cambodia, in September 2004.

The Lessons Learnt

The overwhelming response to Vigyan Rail showed that there is thirst for knowledge at all levels and in all strata of the society, and that Vigyan Rail proved to be an extraordinarily efficient method to quench this thirst. And yet, given the vastness of the country with a population of a billion plus and with diversity of the languages spoken in different parts of the country, Vigyan Rail was only a drop in the ocean. Although Vigyan Rail was an efficient means, there is still a long way to go to achieve the goal of inculcating scientific temper and promoting greater awareness about the value of S&T among the general public. The Vigyan Rail could touch only 60 stations, many places were left out; to cover most of the country by the Science Exhibition on Wheels, it is imperative that it should be continued with exhibits updated periodically.

The experience in running Vigyan Rail has brought certain lessons, which should be addressed in future endeavours. One of the lessons relates to the journey itself. In view of the fact that the weather conditions in our country vary significantly from one region to another, adequate thought needs to be given while drawing up the journey schedules so that the visitors are not put to inconvenience due to severe cold or scorching heat.

The period of the halt ranged between two and five days. This was found inadequate especially in view of the overwhelming public response that the exhibition generated, and the resultant rush. As a result, the visitors had to wait for long hours in the queue before gaining entry into the train. The period of halt therefore needs to be increased.

In view of the fact that the space available in a train is limited, too many working models could not be put up. The number of working models was rather small, and those too were put up by only few organisations/departments. The ways and means need to be explored to put up more working models in the future exhibitions.

One important aspect that needs attention relates to the number and level of guides/volunteers who could explain the exhibits to the visitors in the local languages. Often the visitors were not able to understand the exhibits despite the explanation given (in English or Hindi) since explanation tended to be a bit too technical. Recognising this, the participating departments/ministries had deployed experts to explain the exhibits to the visitors. However, in the rural areas, where people conversed only in the local language, neither the experts nor the guides could explain the exhibits adequately to the visitors. To address this problem, the local coordinators recruited guides/volunteers from the faculties of the local educational institutions who were briefed and trained in the art of explaining the exhibits to the visitors in the local language. But as it turned out, the number of such guides/volunteers was not adequate. This aspect would need to be kept in mind for similar endeavours in future. Otherwise, the basic objective of the exhibition, that is, to reach out to the common people, may not be achieved fully. Perhaps, the advances in ITcould come to our rescue.

Moving Ahead: A Few Suggestions

Despite the limitations, the epic journey of Vigyan Rail was indeed a very fruitful and an educative experience for Vigyan Prasar. Truly, this was due to the cooperative efforts of the different participating agencies and effective coordination by the officials of the Ministry of Railways and Vigyan Prasar. Since the value and importance of the exhibition has been established beyond doubt, it may be worthwhile to set up a permanent organisation under DST to conduct such science exhibitions on a continuing basis.

Vigyan Rail, for obvious reasons, could halt only at limited number of railway stations. True, many people from the surrounding areas visited the exhibition through different means, but a large number of them still could not. Could there be a 'feeder service'—an integral part of the exhibition, so that selected scaled-down version of the exhibition could be taken to the nearby towns/villages for the benefit of the people who could not come to the railhead? This was an idea mooted by Prof. V.S. Ramamurthy after the conclusion of the first phase of the Vigyan Rail*, and is certainly worth pursuing.

Vigyan Rail could not travel to hilly terrains in the country having no railway connectivity, for example, in the northern and the north-eastern states. A scaled-down version of science exhibition, developed specifically for these regions could be organised to travel by road on two or three vans to be specially fabricated for the purpose. This would help people in the far-flung areas of the hill states to get familiar with some of the advances and how applications of S&T are transforming their lives.

The impressive show made by Vigyan Rail in its first cross-country run made the then Vigyan Prasar President M.V.Kamath prescribe a much larger role for Vigyan Rail. In an interview to *Frontline* Kamath mooted the idea of sending Vigyan Rail to Pakistan, Bangladesh and eventually to all SAARC countries playing the role of a 'peace train' (published on 7 May 2004). Although this idea fascinated many, both at Vigyan Prasar and at the government, it could not be implemented. Yet, such an endeavour has immense possibilities for the exchange of information among the SAARC countries, and maybe, collaboration on projects of mutual interest. It could surely act as a 'peace train'.

Considering the need and utility of the project and the vastness of the country, it is only appropriate that we have more of such trains regularly running through the country on short intervals. Projects of this type are definitely good investment for the future, since besides giving useful information they also help generate an interest for science among the youth who may take up a career in S&T in future.

Well, this would truly be one of the best investments in terms of country's future scientific progress and social and economic growth.

Bridging the Gap

Despite the fact that Vigyan Rail has proved to be an important means of S&T communication that takes science to the people, it is necessary to realise that development of scientific outlook requires continuous exposure to science in various forms and through different means. Organisation of events like Bharat Jan Gyan Vigyan Jatha, national campaigns built around natural celestial phenomena (such as solar and lunar eclipses, and planetary transits), or concerted countrywide programmes built around international events (e.g., World Year of Physics 2005, International Year of Planet Earth 2008 and International Year of Astronomy 2009) would go a long way in triggering an interest in science among the people in general and students in particular. Uses of media, both traditional and modern, have proved to be extremely effective in spreading scientific information. Programmes such as countrywide classroom using two-way audio and two-way video satellite channels have proved to be extremely effective in dissemination of scientific information, as demonstrated by Vigyan Prasar using Edusat and Insat channels. Since 2000s, Internet has added a new dimension along with several social media, such as Facebook and Twitter to science communication. However, we must realise that encouraging children to take up investigative projects with scientists significantly contributes towards developing a scientific outlook among them. It is heartening to note that governmental organisations such as National Council for Science and Technology Communication (NCSTC), Vigyan Prasar and many nongovernment organisations are trying to organise and spread such activities in a concerted manner.

Vigyan Rail is a unique concept that has proved to be extremely successful in bridging the gap between science and the people. It has helped in promoting awareness among the people about S&T in various fields of immediate relevance to them and encouraging the younger generation to evince interest in science as a career. In 2005, NCSTC organised yet another run of Vigyan Rail through different areas of the country where it had not reached earlier. Later, NCSTC organised an event called Science Express that carried German exhibits on fundamental science in 2007 and had four runs in the country. Since 2012, it runs a Science Express—Biodiversity Special, and had two runs in the country. Well, the idea has caught on!

References

Kamble, V. B. (2003). The journey begins. *Dream, 2047*(6), 35.

Padmanabhan, B. S. (2006). *Vigyan rail: science exhibition on wheels*. Noida, UP: Vigyan Prasar.

Organising Children's Science Congress: Challenges and Opportunities

Anuj Sinha

Abstract Since 1990s, a major platform, 'Children's Science Congress' (CSC) has been provided annually for young researchers under 18 years of age. They work in teams on open-ended projects around a preannounced theme and present their findings at school and district events. Theme development is a consultative process among subject experts, state academic coordinators and the members of National Council of Science and Technology Communication Network. Mentors are sensitised to the philosophy of the open-ended research projects at state and regional workshops. They are familiarised with the theme of the congress and exposed to the range of projects in the various subthemes. Competitions are organised during October and November at the state level to shortlist entries to the National Children's Science Congress (NCSC) and Rashtriya Bal Vigyan Sammelan (RBVS, meaning national children's science conference) organised concurrently with the annual session of the Indian Science Congress since 1999. Presentation in regional languages or English is another characteristic that encourages participation from state-run schools. Outstanding projects are selected both in the junior (age under 14 years) and senior (age 14 to under 18 years) categories. On the flipside, while projects with potential have been picked for internship a good number of ideas have been abandoned after the national event. There has been little effort to develop long-term association with delegates of the earlier congresses despite the advancements in information technologies.

Keywords National Children's Science Congress · Rashtriya Bal Vigyan Sammelan · Science fests · Science exhibitions · Student scientists

---

Views expressed in the chapter are the author's own and not necessarily shared by the editors.

A. Sinha (✉)
Network of Science Communicator Organisations, Delhi 110 091, India
e-mail: cpranuj@yahoo.com

P. Bagla and V.V. Binoy (eds.), *Bridging the Communication Gap in Science and Technology*, DOI 10.1007/978-981-10-1025-5_19

Introduction

Since 1990s there has been a colossal experiment of encouraging investigatory science projects by teams of young researchers. Every year thousands of talented youth join the movement and experience the joy and tribulations of open-ended experimental projects. They are involved in generating new knowledge and communicating the methodology and results to their peers and experts.

A major platform called National Science Congress (NSC) is being provided to children (under 18 years of age) annually, on which they present their work on open-ended experimental science projects. From this event, the innovative projects and pathbreaking results of the researchers are selected by judges who evaluate the presentations and examine the logbooks. Only the abstract is necessarily in English with the choice of language of the presentation being left to the researchers. Screened projects are presented at the next (inter-district) stage, and the ones further shortlisted participate in the National Children's Science Congress (NCSC).

After providing a brief background to these efforts, this chapter offers insights into the process, the impact of participation on the researchers, increasing familiarity with project methodology and applied statistics and some missed opportunities.

School education aims at developing necessary learning skills, uncovering creativity and nurturing talent among the learners. Memory, hand–eye coordination, observation and analysis, critical thinking and improved self-esteem are nurtured at every level in a conducive environment. Teachers have a significant mentoring role and contribute to a new generation of decision makers who are environmentally conscious and responsible citizens.

High school students and teachers, through intense interaction and within the scope of school curriculum, can emerge as advocates and practitioners of 'the method of science'. The years in high school will have to transform into an open laboratory adopting and adapting every opportunity into a structured step towards the larger objective.

Students would emerge as creative youth with a thirst for new knowledge and high curiosity to understand the processes of nature. It would discourage forces of fundamentalism and encourage citizens to be open minded (Tyagi 2012; Jaideep 2006).

This was certainly not the scenario in late 1980s in India. In 1991, a science-based organisation in Gwalior, the Gwalior Science Centre, pioneered the Children's Science Congress (CSC) as a small but significant step in the right direction (http://www.ncsc.co.in/). The next year it registered a larger participation, and Department of Science and Technology (DST), Government of India realised the potential of this platform.

Network for Indian Science Communicator Organisations (NISCO) had been registered as National Council of Science and Technology Communication Network (NCSTC Network) around this time with representatives in each state and union territory. NISCO began a consultative process and launched the NCSC in 1993 to take the process forward.

Table 1 Themes of annual Children's Science Congress (http://www.ncsc.co.in/themes)

Year	Venue for national congress	Theme
1993	New Delhi	Know your environment
1994	New Delhi	Clean up India
1995	Guwahati	Clean up India
1996	Hyderabad	India of our dreams: let us go for it
1997	Bhopal	India of our dreams: let us go for it
1998	Chennai	Nature: let us share, care and conserve
1999	Goa	Nature: let us share, care and conserve
2000	Kolkata	Indigenous scientific knowledge for a better tomorrow
2001	Pune	Indigenous scientific knowledge for a better tomorrow
2002	Mysore	Food systems: towards nutrition for all
2003	Lucknow	Food systems: towards nutrition for all
2004	Guwahati	Harness water resources for a better future
2005	Bhubaneswar	Harness water resources for a better future
2006	Gangtok	Biodiversity: nurture nature for our future
2007	Baramati	Biodiversity: nurture nature for our future
2008	Dimapur	Planet earth: explore, share and conserve
2009	Ahmedabad	Planet earth: explore, share and conserve
2010	Chennai	Land resources: use for prosperity—save for posterity
2011	Jaipur	Land resources: use for prosperity—save for posterity
2012	Varanasi	Energy: explore, harness and conserve
2013	Bhopal	Energy: explore, harness and conserve

Source Author's work

Analysis of the Project

Since 1990s, CSC as a major platform for young researchers under 18 years of age has been functioning annually. Participants work in teams on open-ended projects around a preannounced theme and present their findings at school and district events. Outstanding projects are selected both in the junior (age under 14 years) and in the senior (age from 14 to under 18 years) categories. Each theme is retained for 2 years to allow more depth in the projects. A new theme is introduced to provide new opportunities for the young researchers. It is also a disincentive for old research being recycled with cosmetic alterations (Table 1).

Distribution of CSC Projects According to Their Type

About 45 % of the projects selected at the state level are field-survey-based projects, followed by scientific-experiment-based projects (30%) and technology-focused model-making (13%) projects. The projects dedicated to testing of a scientific idea and developing technology-transfer prototypes constituted, respectively, 7% and 5% of the total projects in 2005.

The large number of survey type of projects indicates preference for one type of knowledge generation and the flexibility of completing the study in a short time. Often questionnaire design, the sample selection, size and analysis of the data leave much to be desired.

Theme development is a consultative process among subject experts, state academic coordinators and network members. This process is necessarily prolonged and results in a guidebook for use both by mentors and by young researchers. How distinct are the activities suggested in every edition and do these result in encouraging open-ended precollege research? We examine these questions in the subsequent paragraphs.

Mentors are sensitised to the philosophy of the open-ended research projects at state and regional workshops. They are familiarised with the theme of the congress and exposed to the range of projects in the various subthemes.

Competitions are organised during October and November at the state level to shortlist entries to the NCSC and RBVS, organised concurrently with the annual session of the Indian Science Congress since 1999. Most delegates and guide teachers who attend the NCSC use the opportunity well and the feedback is largely positive. Bonhomie develops between the young researchers and by valedictory time many links are forged.

Time Spent on Various Project Activities

The researchers report spending 43% of the total project working time (excluding report writing) on collecting data in the field and taking photographs of the locations. They spent 21% of this time on fabrication work in school workshop and shopping for the necessary materials, 20% on computer work in the school and 18% on conducting experiments either in the school or college laboratory or in an institute. As regards to collection of information on the project topic, the respondents dedicated 25% of the total project working time on studying either in school library or in other libraries in the area, and 10% on Internet surfing.

Methodology for Investigation

For 64% of the child scientists, the guide teacher and their project team developed the methodology for project investigation by consulting the 'Activity Guide'. Significantly, 28% of them state that they developed this methodology all by themselves by consulting the relevant scientific literature; 6% received the help from other science teachers and laboratory staff of their schools; 4% from the district coordinators; and 5% of the respondents reported using the expertise of local scientific/engineering organisations. Some respondents took the help of their parents and former child scientists.

In the case of survey type of projects, amazingly 37% of the child scientists reported developing the questionnaire by themselves; 35% of them were helped by their guide teachers, 7% by other science teachers and 6% were assisted by the scientists at local science and technology (S&T) institutions; 10% of the respondents say they designed the questionnaire using inputs provided either by their class fellows, former child scientists, district coordinator or their parents.

Access to Internet

The Internet support was best provided in Andaman and Nicobar Islands and Mizoram (by all the teachers), followed by Odisha (83%), Kerala and Punjab (each 75%) and Karnataka (60%). While a large number of teachers from some of the backward/remote states and union territories provided this access to their students, those from relatively more developed states and union territories were far less in number.

Correction for rural researchers and quota for each state—a factor of the district competitions—results in equitable representation of participants at the national congress. The option of presentation in regional languages or English is another characteristic that encourages participation from state-run schools. Merit of the project alone therefore does not ensure selection for the next stage; 80% of the child scientists, who participated in the CSC 2005, were from towns and rural areas of the country, and the rest from the metropolis/big cities (categorised as A-class cities).

The three main criteria used by the evaluators (in decreasing order of importance) for evaluating the projects were: (i) quality of data and its analysis, (ii) validation of selected problem by experimentation and (iii) overall scientific knowledge of the students.

About 45% of the evaluators reported making additional efforts in assessing the reports of the child scientists who wrote project reports and made presentation in their vernacular language, whereas 22% among them either took the help of interpreters, or made consultations with the judges familiar with the language in question. About 16% of the judges assessed how the presenter correlated the data with the research outcomes; 13% of them quizzed the child to assess his or her clarity of concept and another 13% requested for a summary in English. Significantly 11% of the judges reported familiarising themselves with the vernacular terms in English before the presentation.

Kendriya Vidyalaya Sangthan (an organisation of central schools primarily catering to the educational needs of children of transferable government employees) has adopted the programme and encourages students to conduct research and present findings across its branches in all regions. The national-level selection is organised to dovetail into the NCSC event in December.

Typically, nearly two-thirds of the child scientists are from government schools, 30 % from public schools (elite fee-charging schools in India) including missionary schools and the rest 4% from the Kendriya Vidyalayas. This shows that the CSC organisers have evolved very wide networking among a large number of schools located in the backward and remote areas of the country.

Academic Performance

The CSC attracts bright students who benefit from the opportunity and exposure. An analysis of NCSC 2005 revealed that 38 % of the child scientists scored more than 80% marks in the last board examination, 42% obtained 60–80 % marks and 20% below obtained 60 % marks. This shows that the CSC system tends to promote not only the children who are academically the best, but also those on the lower rungs of academic achievements.

The academic performance of the child scientists can also be assessed by the rank they achieved in their class and in the board examination. One-third of the child scientists ranked among the top 5 in their class, although as mentioned earlier, only 38% of them performed very well in the last board examination, scoring more than 80% marks.

Source of Project Idea, Project Working Time and Time Available for Project Supervision

According to 25 % of the child scientists, the time available to them for carrying out the project work was less than one month; 27% of them reported a duration of 1–2 months and 39% reported 3–4 months. For 9 % of the respondents, the project completion time was fairly long—more than 4 months.

However, the guide teachers indicated a different time span available to the child scientists for carrying out the project work. According to 57% of them, this time was less than one month; 8% of them reported a duration of 1–2 months and 11% reported 3–4 months. For24 % of the teachers, the project completion time was fairly long—more than 4 months.

Significantly, as many as 48% of the child scientists reported that they themselves chose the project topic (even without the help of their guide teachers). For 42% of them, the project idea was worked out by the resource persons and 12% got it through science-related coverage in the media and Internet. Among those who were helped by resource persons, 33% worked on the idea suggested by their guide teacher, 8% by their parents/relatives and 2% by a scientist/professor known to them.

For 61% of the child scientists, the schools provided access to laboratory equipment to carry out the experiments pertaining to the project work and even the equipment for carrying out investigations at the project site. About 20% of them reported purchasing the necessary tools and devices (such as water pitchers, ropes and filters) by themselves. For 19% of the respondents, the research facilities were provided by the linked organisations. For some of the children, simple tools and measuring instruments were made available by family members and schoolteachers.

Assistance in Project Report Writing

According to 60 % of the child scientists, they were helped by their guide teacher in writing the project report. While 20 % of them reported that they wrote it themselves with the help of team members by consulting the 'Activity Guide', for 9 %, this assistance was provided by science teachers, drawing teachers and computer instructors of their schools. Further, 5% of them got the friendly help from their class fellows and former child scientists and 4% got assistance from family members (mother or sister in many cases). Only 2% of the researchers were able to mobilise this help from a scientist, college professor, hospital staff or nongovernmental organisation (NGO) specialist known to them.

For the design of the presentation, about 62% of the child scientists received the help from their guide teacher, 8% from other science teachers, 5% from the principal/head master of their school and for 2% of them it was the state/district coordinator, who offered this assistance. About 11% of the researchers reported that they themselves prepared the presentation all by themselves, while 13% of them were assisted either by their class fellows/friends, former child scientists, family members, a scientist or a college professor.

Of the project reports prepared by the child scientists 41% were in English, 36% in Hindi and 23 % in regional language. This shows that about 60% of the young researchers tend to do science in their regional language including Hindi.

In 37% of the project reports, the number of illustrations (graphs, sketches and pictures) was 5 or less. The pictorial content in 28% of the reports was 6–10 illustrations and in 20%, this number was 10–20. The remaining 15% of the works of the child scientists were very well illustrated with more the 20 pictorial presentations.

According to 49% of the children, they themselves drew the graphs and sketches and clicked photographs at the project location, and some produced the illustrations on computer. However, 19% of them borrowed this material from published sources and 11% downloaded the pictures from the Internet/compact discs (CDs). Other sources were science teachers, local NGOs (for their illustrated reports and other material used in campaigns) and 8% of the young scientists developed the illustrations by obtaining inputs from the community, panchayats (village-level government bodies), municipalities and water-works departments in the areas of their investigation.

International Presence

Representation from Bangladesh, Sri Lanka and Nepal has been an irregular feature. Nor has there been any effort at encouraging their participation as it tends to divert attention from the main objective of providing enriching opportunities to over 600 delegates.

Association of South-East Asian Nations (ASEAN) Secretariat has ensured regular participation of young researchers from member countries. About 10 entries, one from each member, are deputed along with guide/mentor. The projects and selection criteria adopted by different countries are neither revealed nor discussed. The benefit of such participation needs analysis.

Project Follow-Up

According to the stakeholders, 53% of the projects undertaken by the child scientists were accepted by their schools for further study/application, 52% were reported in a science magazine/newspaper/school and 47% of the projects were taken up by local community/government department for application; 10 % of the research outcomes got the acceptance of local industry for studying their application potential (for instance, developing water-purification systems). Only 5% of the works of the child scientists were considered for further support under the Technical Entrepreneurship Promotion Council. Their participation at the Intel International Science and Engineering Fairs (Intel ISEF) was also low—only 4 %.

High-Quality Projects

More than half of the state coordinators (60 %) considered quality of CSC projects either very good or good and 33 % felt it was satisfactory; 54 % of the district coordinators considered the quality of projects generated in their respective districts satisfactory; 46 % of the network members found the quality of the CSC projects presented by the child scientists at various CSC meets very good or good and 50 % considered the quality of projects satisfactory. About half of the evaluators found the projects presented for rated evaluation were of low quality.

Projects with potential are now being picked up for internship. Good ideas have, until 2013, been abandoned after the national event. There has been little effort to develop long-term association with delegates of earlier congresses despite the advancements in information technologies.

IIT Kharagpur hosted about 80 bright scholars (investigators of about 20 projects) and for three weeks exposed them to frontier research through lectures, discussions, visits to laboratories and some hands-on experience with sophisticated instruments. They also participated in some friendly competitions and honed their communication skills. Such internship is a regular feature now.

Concurrent Developments

Based on the impact of NCSC, National Curriculum Framework has recommended introduction of elements of open-ended research projects in formal schooling both at high school and at higher secondary stages. While the report has been accepted by the government, specific strategies for its introduction have not been given shape yet.

Continuous Comprehensive Evaluation (CCE) scheme introduced by the Central Board of Secondary Education (CBSE) has found a positive reflection with CSC researchers by earning marks for their projects and participation in many schools.

Many opportunities are available for nurturing interest in science among school students, and major ones are listed in Table 2. However, NCSC has evolved with its own characteristics and is not a duplication of any other platform. According to the stakeholders, the CSC projects differ significantly from those carried out by children under other project-based programmes both in content as well as their local relevance/applicability. The distinguished features of the CSC projects cited by them are:

(i) They do not pertain to model making but promote creative and innovative ideas on a big scale highlighting the applicability of science in solving local problems.
(ii) They offer opportunity for research guidance to teachers.
(iii) They can be easily undertaken by rural children.

Many children, however, do participate in more than one competition/activity and take advantage of the opportunities available. Projects from NCSC are screened and selected for National Initiative for Research and Innovation in Science Fair from the past few years.

Table 2 Major platforms for promoting research interest in precollege students

Platform name	Organisation	Year of start
Pt Nehru Science Model Competition	NCERT, New Delhi	1958
National Children's Science Congress	NCSTC, DST	1993
National Science Seminar	NSCM	1999
Initiative for Research and Innovation in Science	NCSTC, DST, CII and Intel	2006
INSPIRE Science Model Exhibition	DST	2011

Source Author's work

Personality Development

The improvement in self-confidence and self-esteem of young researchers who participate in such activities is remarkable (Tyagi 2007). According to child scientists, the CSC project work has led to the following benefits in learning:

(i) Improvement in their understanding of several topics in botany, physics, environmental sciences and social sciences as contained in the Secondary School Board (SSB) syllabus.
(ii) Thorough understanding of the method of science to solve a problem in a systematic way; knowledge gain in the specific fields of water-related issues, health, atmospheric science and geography.
(iii) Increased curiosity to gain knowledge from published scientific literature.
(iv) Awareness about the grassroots situations where they carried out the investigations.
(v) Scoring better marks in practical examination.
(vi) Development of abilities in writing and presenting reports leading to better language skills.
(vii) Development of attitude towards working in team mode.

The CSC project work has also led to other benefits such as creation of water resource facilities, for example, installation of filtered drinking water and rainwater-harvesting systems in their schools.

According to 50% of the parents, the main benefit accrued from the project-based study for their wards was a wide exposure to learning science by doing and acquiring skills in conducting scientific investigation and field surveys, leading to new learnings and ability to work out new ideas. Other significant benefits reported included first-hand exposure to local problems, building of self-confidence and leadership qualities, and exposure to scientific literature.

Policy Initiatives

The main thrust areas of the CSC project systems should be as follows:

- Use of scientific method in solving local development problems along with feasibility of transfer of research outcomes to local conditions.
- Generation of S&T inputs necessary for eco-friendly sustainable development all around.
- Promotion of the spirit of scientific research among children in the context of the changes taking place in today's science.
- To become a viable channel for promoting the spirit of scientific enquiry among people at large with their participation in the investigations and with the child scientists as the interface between science and people.

The suggested steps forward are, therefore:

(i) Link CSC with mainstream science by focusing on definite scientific fields of study to generate productive research outcomes.
(ii) Discard the current state-wise quota system with an open-ended system so that the states could present a large number of high-quality projects at the national level.
(iii) Provide children incentives like annual scholarships/advantages in admission to higher classes in order to attract the best brains .
(iv) Set up expert groups at local level who would visit schools and hold discussions with children and teachers to promote scientific thinking on tackling local problems.
(v) Bring in place a long-term policy of providing off-school facilities to children for experimental work and consultation.
(vi) Incorporate the CSC in the country's educational system.
(vii) Adopt a clear-cut child-centric policy in order to encourage child scientists to take up research topics of their interest.

The secondary issues associated with CSC are as follows:

(i) Provision of adequate funds to children for carrying out the project work.
(ii) Bringing in experts to participate in projects/workshops and in assessing quality of projects through random checks.
(iii) Introduction of combined teacher–student projects as joint ventures.

The guide teachers have made a series of suggestions for improving the content and utility of the 'Activity Guide'. These are as follows:

- Include a section on how to conduct field survey, select samples and carry out statistical analysis.
- Include a section on how to carry out a laboratory-based project.
- Include a section on how to write abstract and its format with an example.
- Provide as an example one complete project report.
- Select subtopics that are related to the theme.
- Include bibliography on the focal theme and addresses of libraries and websites .
- Include addresses of resource persons with e-mail IDs.
- Include schedule of district/state/national-level meets.
- Issue adequate number of copies in all languages to schools directly.
- Hold teachers workshop on how to use the guide.
- Attach a soft copy on CD with the guidebook.
- Hold orientation programme on focal theme for education department officials.

Teacher Orientation Programme

The measures suggested by the guide teachers for making CSC orientation programmes more productive are as follows:

- Programme should impart training including short project design workshops.
- Programme should involve competent resource persons, grassroots experts from local NGOs with experience in field surveys and former child scientists.
- At least five to six teachers from each district should be invited.
- Selection of guide teachers must be according to the theme.
- Booklet highlighing the rules and regulations and the research protocol should be followed by guide teachers and references for research should be distributed.
- Certificate of participation should be issued to teachers as an incentive.

The following set of measures may be considered for revamping the CSC evaluation system.

Revision of Evaluation Guidelines

1. Originality of the research idea in terms of new concepts and methodology, and its local relevance in solving a problem.
2. Quality of data collected and quality of the experimental validation.
3. Overall scientific knowledge of the student in the context of research problem.
4. Quality of the presentation of research and the final report.
5. Mechanism to assess whether work was done by students themselves.

Mode of Evaluation

1. Not summary and presentation, but full project report should be examined.
2. Evaluation should be in two parts, namely project work and presentation.
3. Evaluation sheet currently being used should be redesigned to make it short and less cumbersome.
4. Separate presentations for junior and senior projects should be scheduled.

Quality-Control Measures

Projects selected should be reexamined by a panel of experts, and systematic random field checks should be carried out by external experts for ascertaining the validity of data.

Selection of Judges

1. Only senior experts should be involved in project assessment work.
2. Evaluations at district/state level should include one judge from national level.
3. Evaluators should be changed every year.
4. A comprehensive training programme should be organised for evaluators.
5. Judges should have adequate competence to effectively communicate with children.

The following suggestions regarding recognition of the CSC projects need to be considered:

I. Declare at the national event the outstanding projects in various categories: locally relevant projects, survey-based investigations, experiment-based and technology-transfer projects.
II. Declare the best 50 projects.
III. Change the notion of competition so that all the participatory projects are treated equally and issued certificate of participation.
IV. Adopt a ranking scale (+A, A, +B, B and so on) in the citation certificates given at the state level (for this purpose, software should be developed and provided to the state coordinators).
V. Hold competitive CSC at international level.

Even after 22 editions of NCSC there is inadequate awareness of the platform and the benefits that participants derive. Specific measures are needed to improve the situation. Some suggestions are:

1. Appoint persons involved in science popularisation activities and national-level participants as science counsellors/CSC campaigners at district level and conduct every year CSC campaigns from school to school in every district holding exhibitions and lectures and using banners/booklets/CDs and other means indicating the benefits of participating in the CSC.
2. Set up CSC clubs at district level to hold training for children and teachers in conducting research.
3. Popularise project-based research among local people and therefore among children about the social benefits of research done under the CSC.
4. Build CSC awareness among parents.

The print, electronic and digital media need to be exploited and utilised. Some suggestions are:

1. Supplement the actions suggested in the preceding paragraphs with annual media campaigns including advance announcement of the year's focal theme of the CSC in newspapers, children's magazines and science magazines, as well as on radio and television channels covering all regional languages and giving information on the landmark projects with names and photographs of the winners.

2. Launch CSC project website/newsletter spelling out new project areas and topics and guide/facilitate children to submit their projects directly on website.
3. Discuss in media about the job prospects in science.
4. Publish articles based on CSC projects in other countries in order to motivate the young participants.
5. Distribute NCSTC Newsletter among school libraries free of cost.

Acknowledgements The author acknowledges the support of Ashok Raj, Centre for Science Technology and Environment Policy Studies (CSTEPS), New Delhi, who presented a study titled 'National Children's Science Congress: Strengthening Its Space in Indian Science by Improving the Quality of Projects and Outreach among School Children' in November 2007, the data from which have been used for analysis in this chapter.

References

Jaideep, B. (2006). *Children's science congress. An introspection*. The Orchid, Souvenir, 14th National Children's Science Congress, NCSTC Network.

Tyagi, B. K. (2007). National children science congress-2005: A survey. *VIPNET News, 5*.

Tyagi, B. K. (2012). Communicating science through children's science congress: The biggest experiment on informal science education in India. *Sociology of Science and Technology, 3*, 130–144.

Bridging Educational Institutions for a Citizen Science Project: A Case Study from Malappuram District, Kerala, India

V.V. Binoy, S. Radhakrishna and A. Kurup

When spider webs unite, they can tie up a lion

—Ethiopian proverb

Abstract Strategic bridging of various institutions into a functional network and enhancing communication between the nodes is a low-cost and effective approach for gathering grassroots level information and dissemination of scientific knowledge to various actors of a multi-stakeholder society. Such a cross-talking network of institutions, for example, schools, colleges, nongovernmental organisations, youth clubs and interested individuals, is the heart of participatory research programmes, such as citizen science projects. Although citizen science projects have gained popularity since mid-1990s, very few studies have critically analysed the dynamics and efficacy of interinstitutional networks in India. Here we share our experiences of connecting two major compartments of the educational system, school and college, in the Valanchery region of Malappuram District, Kerala, India, via a 'student network' for a citizen science programme to monitor the local mammalian diversity.

Keywords Citizen science · Student network · Student–scientist partnership · Bridging of the institutions · Science education · Science popularisation

Citizen science projects that operate through the services of laypeople who may not have a research background to conduct investigations, interventions, advocacy and education in various sectors under the guidance of experts is gaining popularity in

Views expressed in the chapter are the authors' own and not necessarily shared by the editors.

V.V. Binoy (✉) · S. Radhakrishna
School of Natural Sciences and Engineering, National Institute of Advanced Studies, Indian Institute of Science Campus, Bangalore 560 012, India
e-mail: vvbinoy@gmail.com

A. Kurup
School of Social Sciences, National Institute of Advanced Studies, Indian Institute of Science Campus, Bangalore 560 012, India

P. Bagla and V.V. Binoy (eds.), *Bridging the Communication Gap in Science and Technology*, DOI 10.1007/978-981-10-1025-5_20

the Indian subcontinent. Public participation in research enhances the capability of the scientist community to conduct labour-intensive and field-based research over large spatial and temporal scales (Gommerman and Monroe 2012), which would otherwise be very difficult due to the huge amount of financial and human resources required (Lepage and Francis 2002). Furthermore, scientists interacting with lay citizens obtain the opportunity to understand people's awareness concerning their areas of study, while interaction with researchers can deepen people's understanding of natural phenomena and expose them to the methodologies and rigour invested by professional scientists to produce meaningful knowledge (Nerbonne and Nelson 2008). Being the perfect blend of principles from 'public understanding of science' and 'public engagement in science' (Haywood and Besley 2014), citizen science initiatives can work as powerful tools for communicating scientific information to common and arm them to understand, refine and 'scientise' the policies (Crona and Parker 2012). Based on the degree of involvement of laypersons, there are three categories of citizen science projects: *contributory projects*, in which volunteers only do the data collection for projects designed by scientists; *collaborative projects* that give the freedom to refine study design, analyse data or communicate findings to participants and the third category of fully *democratised'co-created projects*, in which public may actively engage with researchers through all steps of the scientific process (Bonney et al. 2009).

Citizen science programmes broadcast scientific information and cultivate scientific temper in a society through social learning (Trumbull et al. 2000; Krasny and Tidball 2009). However, in a multi-stakeholder and multicultural social system, effective social learning (that is, learning from another member of the society; Bandura 1977), which requires a long-term working relationship between stakeholders, can be achieved by bridging institutions formally or unofficially through strategic agents and enhancing interaction between them (Olsson et al. 2007). According to Brown (1991) such a network of cross-talking institutions can function as "a conduit of ideas and innovations, a source of information, a broker of resources, a negotiator of deals, a conceptualiser of strategies, (and) a mediator of conflicts", along with disseminating scientific information. Hence, a network of institutions not only promotes exchange of ideas between the nodes but also works as a point of convergence for views of people from different backgrounds, which could be exploited by researchers and policy makers to enhance the quality of the decision made for a particular society (Muller et al. 2012; Aslaksen and Myhr 2007; Code 2006). Hence, citizen science programmes always try to include members from different stakeholder institutions such as schools, colleges, youth clubs, faith-based communities and so on along with scientists and academicians.

One common attribute of the citizen science projects working all over the world is the active inclusion of students as volunteers (Weinstein 2012). Researchers usually show very little inhibition to share their work and experience with students, the scientists of the future. Additionally, there is a general consensus among scientists promoting citizen science that young individuals and students "can be entrusted for the quality of their data because so often they approach issues without the longer lived experiences of adults who may be jaded or biased, tainted or torn" (Kozol 2005). If trained properly, the quality of data collected by the students are comparable to the data generated by professional scientists (Delaney et al. 2008), and by being the part of such initiatives they could acquire knowledge and skills required to address local problems (Mueller and Tippins 2012). Unfortunately, in many citizen science programmes the activity of a student volunteer ends with the collection of data suggested by the scientists (Muller et al. 2012), and their familiarity with other members of the society—an advantageous element in effectively communicating information and promoting social learning via active and passive modes (Lamberson 2008)—is often underutilised. Moreover, many of the citizen science projects are temporary in nature and rarely make any backup plans to maintain the network of institutions formed. However, if schools of an area are linked with colleges and students from these two important but rarely interacting compartments of education systems are allowed to communicate with each other under the auspices of research institutions, virtually or in person, such networks may be developed into a permanent citizen science project framework for collecting diverse kind of data and communicating science to various strata of the society. Additionally, selected college students trained by and actively interacting with professional researchers can coach schoolchildren to conduct scientific research. Similarly, the continuous interaction between teachers and scientists could enhance the knowledge level of the former, which in turn would enhance the quality of science education and bridge the gap present between different segments of the academics (Brossard et al. 2005; Braschler et al. 2009).

Being a geographically large nation whose population speaks different languages and follows different cultures, effective communication of science and technology and collection of vital information from the grassroots levels of the society are not possible in India without the active involvement of local people (Rist et al. 2007). Although citizen science is gaining popularity in India and the student community is a major segment of the volunteer population (Balakrishnan 2010; Quader and Raza 2008), few studies have attempted to analyse the dynamics of bridging educational institutions and its impact on the production and dissemination of knowledge in the Indian context (Rist et al. 2007; Balakrishnan 2010; Danielsen et al. 2011). Such studies are essential for successful implementation of citizen science programmes in India due to the rigid power relation existing in every sector including education, "family and caste-based use of natural resources, top-down approaches in the planning and implementation of developmental programmes, relatively strong interference of government institutions and NGOs, lower degrees of autonomy, problems with collective resource management and presence of stark digital divide" (Rist et al. 2007). In this context, our team explored the possibility of linking schools with nearby colleges and research institutions in order to initiate a citizen science programme to monitor wild mammal species in a human-dominated area located near the Western Ghats biodiversity hot spot.

Building a Network

We selected as our area of study, the district of Malappuram in the state of Kerala in southern India. Relative to other states of India, Kerala has several unusual markers; the literacy rate is the highest in the nation; per capita spending on health care and education is the highest among all states; finally, the standard of infrastructural and basic services are remarkably high despite low per capita income (Franke and Chasin 1990; Ramachandra 1995, 1997; Parayil 1996). Most significantly, the state has a history of promoting environment sustainability and an enviable record in citizen movements successfully influencing development policies to move in pro-environment directions (Parayil 1989, 1996). Malappuram in north-central Kerala is the most underdeveloped district in the state—a report on the human development status in Kerala ranks Malappuram's human development index as the lowest in the state (Centre for Developmental Studies 2006). Physiographically, Kerala forms a large part of the Western Ghats biodiversity hot spot, and Malappuram district enjoys a forest cover of 758.86 square kilometres (about 22% of the total geographical area) and possibly due to the slow rate of development, large swathes of prime wildlife habitat are relatively undisturbed.

In order to build a student network, schools representing different economic strata (government schools, private schools and residential schools owned by charitable trusts for poor children) and different mediums of instruction (English and Malayalam medium schools) located in the Valanchery area of Malappuram district were visited by the research team to discuss the objectives of the project with interested teachers. Similar meetings were conducted with the staff and students of three arts and science colleges and a teachers training college located in this region. A one-day workshop was organised at a college in Valanchery for teachers and student representatives from 14 schools, the three arts and science colleges and the teachers training college. The focal theme of the workshop was the protection of biodiversity at local level and the necessity and potential of involving the student and teacher community as active participants in the conservation programmes. The need for connecting schools, colleges and research institutions and working together as a single unit for understanding and guarding the diversity of life in the surroundings were also discussed. Students and teachers participated actively in the discussion and most of them expressed their interest to become the part of any movement for the protection of biodiversity. During the workshop a team of experts from the National Institute of Advanced Studies (NIAS), Bengaluru, conducted focal group discussions on biodiversity issues and assured training and technical support if members could build a student network.

The workshop provided the opportunity for school students and teachers to interact informally with college students, college teachers and the research team. The participants of the workshop reached a consensus to focus on and monitor wild mammalian species present in their surroundings. Many of the houses in this region have large backyards that abut forest areas and form extended habitats for small wildlife species; hence, living in this region provides scope for close interactions with different mammalian species. Citizen science programmes have been criticised because they rarely permit student volunteers to study and incorporate their own interests into the project; instead their skills are utilised solely for data collection (Moore 2011; Muller et al. 2012). Hence, we did not 'hand out' specific problems for the students to work on. Students were requested to observe mammals in their surroundings and choose species and topics of research according to their interests and to contact the college teachers or the research team from NIAS either directly or with the help of college students if they required any assistance. Schoolteachers offered to start off a unit to monitor mammalian diversity by gathering interested students in their school, whereas college students, college teachers and the NIAS team took on the job of providing necessary materials and help to schoolchildren for conducting the study. Additionally, 40 undergraduate students were selected and trained to guide schoolchildren to conduct research. A team of four college students, comprising first, second and final year degree students, was linked with each school participating in the project. Care was taken to add at least one member who had completed his or her school education from the focal school in the team of 'student mentors' so that the sociocultural dynamics of the schools were understood by the 'mentors' team. Inclusion of junior and senior college students in the team ensured the maintenance of the link with school units even after final year students passed out from the college. The role of student mentors included monitoring the activities of the school units and finding solutions (with the help of college teachers and the NIAS team) for any problems faced by school students in their research efforts. College students visited schools and discussed with schoolchildren and teachers the issues regarding their research. Members of the NIAS team also informally discussed with teachers the status of the studies being carried out by the students as well as their attitude towards such programmes.

The Constraints!

Although the mood at the workshop was one of enthusiasm and energy, student mentors from the colleges subsequently received very feeble responses from schools. Students and teachers from schools showed little enthusiasm for the project and almost all of them failed to start a unit in their school. When the college students visited the schools to discuss the matter, many of the teachers indicated that they were uninterested to continue the activities of student networking in their school. Further conversation revealed that student and teacher communities were not interested in studying mammalian species present in their surroundings either because they were too familiar with them or because they did not know much about them. They felt that studying rats, squirrels and so on was a useless exercise because "everybody knows everything about them". They were not enthused about studying endangered species such as the Malabar civet, because they were completely unfamiliar with the animal. Many of the students complained that as most of the animals were not 'attractive' they were teased by their friends for being interested in their study. Interestingly, many of the students showed an aversion for canid species (jackal, fox and so on) due to a religious taboo against dogs prevalent among the believers of Islam in this region. Many students complained that they were not able to formulate a question to continue their study; an observation that underlines the incompetency of the examination-oriented science education system prevalent in India in developing a scientific temper and critical minds in the classroom. Many students expressed the opinion that involving in activities such as citizen science did not provide them any school benefits, such as grace marks, hence they preferred to spend their free time in other entertainment activities, namely watching television, engaging in sports and so on. Classrooms in most of the schools and especially in government schools have little access to non-textbook information or media sources of knowledge acquisition, such as Internet. This also limited teachers' and students' ability to engage in dialogues about scientific issues (Driver et al. 1996) as well as the successful implementation of the activities of the student network.

Many schoolteachers held attitude similar to that of students towards the mammals. Teachers confessed that they were not very familiar with mammalian diversity because although many of them were science graduates, they were untrained in biodiversity issues, and the nonavailability of information sources about local species often caused further problems. Interestingly, many teachers expressed an interest to start a citizen science programme for studying plants, in particular medicinal plants, as they were familiar with plants and believed it to be a venture useful to the society. They also added that unlike animals, observing and studying sedentary life forms like plants is easier with their busy schedules and that the school management, state medicinal plant and biodiversity board, as well as the general public would appreciate such activities. The majority of colleges in this region offer masters in zoology with specialisation in entomology or fisheries and aquatic biology. Therefore, the availability of college teachers who are familiar with mammalian fauna is also rare in this region. Many teachers were also of the opinion that any activity connected with ecology is the duty of the teacher in charge of National Green Corps (NGC), a programme launched by the Government of India to enhance ecoliteracy in schools. Another notion prevalent among the teachers in the region was that ecological activities are against development; one teacher complained that he was socially excluded by his colleagues as he was involved in environmental awareness movements. Even though teachers who participated in our study had opportunities to interact with teachers from other schools during 'cluster training programmes', they never tried to work in association with their peers from other institutions.

According to Balakrishnan (2010) attempts to enhance awareness and thus reducing snakebite deaths in another area of the Malappuram district via a citizen science project was a success, and students actively participated in the venture. Galloway et al. (2006) suggest that participants' preconceptions may lead to biases in data acquisition and misinterpretation of results and that the volunteers will be motivated to continue their work, only if they believed that they were making valuable contributions to the scientific study (Phillips 2008). Participants in our study did not consider studying small mammals valuable or important. This observation underscores need to understand the mindset of volunteers and choose a topic that is appreciated by the participants for successful bridging of institutions and implementation of projects involving student's participation.

According to Hogan (2002) and Muller et al. (2012) existing power dynamics can also affect success of citizen science programmes at any level. Although students interacted with the research team and college students during the workshop, they were reluctant to contact either the members of the research team or students and teachers from colleges, after the meeting. Both schoolchildren and teachers believed that college teachers and researchers were beyond their reach in terms of scientific knowledge and that their questions and doubts would be of no value to them and would be ignored. Another factor that blocked the communication between school students and college teachers was the medium of communication. School students and teachers believed that they would have to converse in English while communicating with college teachers or the researchers, and as they possessed insufficient command over spoken English they avoided chances of interaction. In contrast, college students were very enthusiastic by the activities of the project and successfully completed the duties assigned to them proving that they can be an excellent group of volunteers who can catalyse the success of any citizen science programmes.

It can be concluded that a citizen science project that bridges academic institutions is an excellent tool for data collection, science communication and science education, since it occupies a position in between the scientist-driven top-down approach and democratised methodologies coupled with local concerns, prevalent in a bottom-up approach utilised for capacity building and development (Ely 2008; Engel and Voshell 2002). However, the success of bridging academic institutions and developing such a network into a permanent and successful citizen science project would be possible in a multi-stakeholder and multicultural social system like India only if the basic elements of science such as systematic recording, precision, logic, criticism and professionalism are blended in right proportion with the components of social science research, namely beliefs, culture, attitude, traditions, freedom, history and so on (Trumbull et al. 2005; Muller et al. 2012). Only then will volunteering students become capable of noticing local problems, asking their own questions, obtain guidance from experts, develop protocol, collect data and build their own models and grow into successful communicators who can bring science to common.

Acknowledgements We express our gratitude to the principal, staff and students of MES Keveeyam College, Valanchery for their support during this study.

References

Aslaksen, J., & Myhr, A. I. (2007). "The worth of a wildflower": Precautionary perspectives on the environmental risk or GMOs. *Ecological Economics, 60*, 489–497.

Balakrishnan, P. (2010). An education programme and establishment of a citizen scientist network to reduce killing of non-venomous snakes in Malappuram district, Kerala, India. *Conservation Evidence, 7*, 9–15.

Bandura, A. (1977). *Social learning theory*. Prentice-Hall, New Jersey, USA: Englewood Cliffs.

Bonney R., Ballard, H., Jordan, R., McCallie, E., Phillips, T., Shirk, J., & Wilderman C. C. (2009). *Public participation in scientific research: Defining the field and assessing its potential for informal science education*. Washington, DC: Center for Advancement of Informal Science Education (CAISE)

Braschler, B. (2009). Successfully implementing a citizen scientist approach to insect monitoring in a resource-poor country. *BioScience, 59*, 103–104.

Brossard, D., Lewenstein, B., & Bonney, R. (2005). Scientific knowledge and attitude change: The impact of a citizen science project. *International Journal of Science Education, 27*, 1099–1121.

Brown, L. D. (1991). Bridging organizations and sustainable development. *Human Relations*, 44, 807-831.

Centre for Development Studies. (2006). *Human development report*-2005. Thiruvananthapuram: State Planning Board, Government of Kerala.

Code, L. (2006). *Ecological thinking: The politics of epistemic location*. Oxford: Oxford University Press.

Crona, B. I., & Parker, J. N. (2012). Learning in support of governance: theories, methods, and a framework to assess how bridging organizations contribute to adaptive resource governance. *Ecology and Society, 17*, 32.

Danielsen, F., Skutsch, M., Burgess, N. D., Jensen, P. M., Andrianandrasana, H., Karky, B., et al. (2011). At the heart of REDD+: A role for local people in monitoring forests? *Conservation Letters, 4*, 158–167.

Delaney, D. G., Sperling, C. D., Adams, C. S., & Leung, B. (2008). Marine invasive species: Validation of citizen science and implications for national monitoring networks. *Biological Invasions, 10*, 117–128.

Driver, R., Leach, J., Millar, R., & Scott, P. (1996). *Young peoples's images of science*. Buckingham, United Kingdom: Open University Press.

Ely, E. (2008) Volunteer monitoring and the democratization of science. *The Volunteer Monitor, 19*, 1–5.

Engel, S. R., & Voshell, J. R. (2002). Volunteer biological monitoring: Can it accurately assess the ecological condition of streams? *American Entomologist*, 48, 164–177.

Franke, R. W., & Chasin, B. H. (1990). Development without growth: The Kerala experience. *Technology Review, 93*, 42–51.

Galloway, A. W. E., Tudor, M. T. & Vander Haegen W. M. (2006). The reliability of citizen science: A case study of Oregon White Oak stand surveys. *Wildlife Society Bulletin*, 34, 1425–1429.

Gommerman, L., & Monroe, M. C. (2012). Lessons learned from evaluations of citizen science programs, School of Forest Resources and Conservation, Florida Cooperative Extension Service, Institute of Food and Agricultural Sciences, University of Florida. Original publication date May 2012. Visit the EDIS website at http://edis.ifas.ufl.edu.

Haywood, B. K., & Besley, J. C. (2014). Education, outreach, and inclusive engagement: Towards integrated indicators of successful program outcomes in participatory science. *Public Understanding Science, 23*, 92–106.

Hogan, K. (2002). Pitfalls of community-based learning: How power dynamics limit adolescents' trajectories of growth and participation. *Teachers College Record, 104*, 586–624.

Kozol, J. (2005). *The shame of the nation: The restoration of apartheid schooling in America*. New York: Three Rivers Press.

Krasny, M. E., Tidball, K. G., & Sriskandarajah, N. (2009). Education and resilience: Social and situated learning among university and secondary students. *Ecology and Society, 14*, 38.
Lamberson, P. J. (2009). Social Learning in Social Networks. MIT Sloan School Working Paper 4763-09, MIT Sloan School of Management.
Lepage, D., & Francis, C. M. (2002). Do feeder counts reliably indicate bird population changes? 21 years of winter bird counts in Ontario, Canada. *The Condor, 104*, 255–270.
Moore, K. (2011). Survey: How much would you pay to save the bay? *Asbury Park Press*. Retrieved from http://bbp.ocean.edu/pages/109.asp?item=584
Mueller, M. P. & Tippins, D. J. (2012). Citizen science, ecojustice, and science education. In B. J. Fraser, K. Tobin, & C. McRobbie (Eds.), *Second international handbook of science education* (pp. 865–882). New York: Springer.
Mueller, M. P., Tippins, D., & Bryan, L. A. (2012). The future of citizen science. *Democracy and Education, 20*. Article 2.
Nerbonne, J. F., & Nelson, K. C. (2008). Volunteer macroinvertebrate monitoring: Tensions among group goals, data quality, and outcomes. *Environmental Management, 42*, 470–479.
Olsson, P., Folke, C., Galaz, V., Hahn, T. & Schultz. L. (2007). Enhancing the fit through adaptive co-management: creating and maintaining bridging functions for matching scales in the Kristianstads Vattenrike Biosphere Reserve, Sweden. *Ecology and Society,* 12, 28.
Parayil, G. (1989). "Science for social revolution": Science and culture in Kerala. *Impact of Science on Society, 155*, 233–240.
Parayil, G. (1996). The 'Kerala Model' of development: Development and sustainability in the third world. *Third World Quarterly, 5*, 941–957.
Phillips, A. L. (2008). Of sunflowers and citizens. *American Scientist, 96*, 375.
Quader, S., & Raza, R. H. (2008). Migrant watch: A citizen science programme for the study of bird migration. *Indian Birds, 3*, 202–209.
Ramachandran, V. K. (1995). A note on Kerala's development achievements. *Monthly Review, 47*, 19–24.
Ramachandran, V. K. (1997). On Kerala's development achievements. In J. Dreze & A. Sen (Eds.), *Indian development: Selected regional perspectives* (pp. 205–356). Delhi: OUP.
Rist, S., Chidambaranathanb, M., Escobarc, C., Wiesmannd, U., & Zimmermanne, A. (2007). Moving from sustainable management to sustainable governance of natural resources: The role of social learning processes in rural India, Bolivia and Mali. *Journal of Rural Studies, 23*, 23–37.
Trumbull, D. J., Bonney, R., Bascom, D., & Cabral, A. (2000). Thinking scientifically during participation in a citizen-science project. *Science Education, 84*, 265–275.
Trumbull, D. J., Bonney, R., & Grudens-Schuck, N. (2005). Developing materials to promote inquiry: Lessons learned. *Science Education, 89*, 879–900.
Weinstein, M. (2012). Schools/citizen science. *Democracy and Education, 20*, 1–3.

Communicating Science to the Common: Perspectives of the Science Communicators from India

D. Balasubramanian, Milind Watve, S. Ramadorai, K. Kasturirangan, R.S. Sundar, Prabir Purkayastha, T.V. Jayan, Rajendra Singh and Arvind Gupta

Abstract Although science has not left any aspect of the life in the modern society untouched, communicating science with general public effectively is not a trivial task. One of the major barriers for effortless communication between science and society is the complexity of the language of science, which is difficult for the society to comprehend. In India, the zenith of cultural and environmental diversity, where people hold different values for the same resources living in the same community with complicated power structure, reaching the developments of sci-

---

Views expressed in the chapter are the authors' own and not necessarily shared by the editors.

D. Balasubramanian (✉)
L. V. Prasad Eye Research Foundation, Hyderabad, India
e-mail: dbala@lvpei.org

M. Watve
Department of Biology, Indian Institute of Science Education and Research (IISER), Sai Trinity Building, 3rd Floor, Chemistry Wing, Room No. 305, Sutarwadi, Pashan, Pune 411 021, India
e-mail: milind_watve@yahoo.com; milind@iiserpune.ac.in

S. Ramadorai
Management Council, National Institute of Advanced Studies, Bengaluru, India
e-mail: ramadorai.office@tcs.com

K. Kasturirangan
Indian Space Research Organisation (ISRO), Government of India, Bengaluru, India
e-mail: k.rangan@gov.in

K. Kasturirangan
Planning Commission, Government of India, New Delhi, India

R.S. Sundar
Kudankulam Nuclear Power Plant (KKNPP), Kudankulam PO, Radhapuram Taluk, Tirunelveli 627 106, Tamil Nadu, India
e-mail: rssundar@kknpp.com

P. Purkayastha
Delhi Science Forum, D-158 Lower Ground Floor, Saket, New Delhi 110 017, India
e-mail: prabirp@gmail.com

P. Bagla and V.V. Binoy (eds.), *Bridging the Communication Gap in Science and Technology*, DOI 10.1007/978-981-10-1025-5_21

ence and technology is highly depended on people who can translate the language of science into a format that is appreciable to the layperson without losing the essence. This chapter is a collection of the perspectives and experiences of the luminaries of science communication in India—D. Balasubramanian, Milind Watve, S. Ramadorai, K. Kasturirangan, R.S. Sundar, Prabir Purkayastha, T.V. Jayan, Rajendra Singh and Arvind Gupta.

Keywords History of science · *newsclick.in* · *Chandrayaan-1* · Public–scientist interaction · Science toys · Educational resources · Levitating pencil · Touching slate · Science project · School laboratory · Science journalism · Technology communication · Science popularisation · Student–scientist interaction · Kudankulam Nuclear Power Project (KKNPP) · Nuclear energy · Risk communication · Tarun Bharat Sangh · Jal Purush · Water harvesting · Natural resource management

Communicating Science with the Public and Policy Makers

D. Balasubramanian
L. V. Prasad Eye Research Foundation, Hyderabad, India

The interaction between scientists and society at large has gone through distinct stages of evolution and engagement. Very early on, science was not even called science but part of natural philosophy. It was practised by groups of people who retreated from everyday life into monasteries for academies, discussing among themselves about such recondite themes as the nature of things, reason, truth, the meaning of life, heaven and earth. These were the scholars of Upanishadic discourses (relating to *Upanishads*, the philosophical treatises of ancient Hinduism) in India and similar groups in ancient Greece, China, Latin America and elsewhere. They were the philosophers (literally meaning lovers of argument), who mattered little to the commoners, but respected and held in awe by them, and indulged by the

T.V. Jayan
The Telegraphh, Kolkata, India
e-mail: tvjayan@gmail.com

R. Singh
Tarun Bharat Sangh, Tarun Ashram, Bhikampura, Alwar District, Rajasthan, India
e-mail: jalpurushtbs@gmail.com; watermantbs@yahoo.com

A. Gupta
Inter-University Centre for Astronomy and Astrophysics (IUCAA), Pune University Campus, Pune 411 007, India
e-mail: arvindtoys@gmail.com
URL: http://arvindguptatoys.com

kings. These practices of pursuit of knowledge in isolation from the everyday life continued from antiquity until well after the first millennium.

All this while, the commoner, in his own way was developing tools and technology that helped in daily living—be it taming fire, making tools or starting agriculture—all heuristic in nature. Technology thus precedes science by centuries and millennia. It is when we turn to half way past the first millennium around AD 1500, or the Age of Reformation in Europe, that science and the art of logical reasoning take on as an independent sharp discipline—as a systematic knowledge of the physical or material world—from the recondite to the rational. People keep talking of Francis Bacon or Rene Descartes as key individuals in this enterprise, but there must have been other Bacons elsewhere too—logicians who applied their craft of reasoning to understand phenomena, and to build theories to explain how these extant technologies worked. They engaged with everyday phenomena and everyday people, communicating with, and interesting them with their explanations and analyses.

Science as Educator and Entertainer

Already, the conflict starts between scientific deductions and traditional thought. When Galileo explained the law of gravity through his experiment at the Tower of Pisa, it was acceptable as an advance in knowledge. On the other hand, when he propounded the heliocentric idea, it went against the church and he paid for it with his life. This starts the first debate and intolerance of new ideas that go against 'God given' knowledge. Galileo paid for his challenge, but even at death said, "e pur si muove" ("and yet it moves").

Indeed, science did move on after Galileo, despite the church, as we progress beyond the 16th and 17th centuries. This unease apart, popularisation of science and its concepts and applications begin by the 17th century. In England, the famous Royal Society gets started and begins flourishing. Elsewhere in Europe, science academies get initiated by royalties, and science starts capturing the imagination of at least the landed gentry. By the 18th century, the Age of Reformation turns into the Age of Enlightenment, and science becomes increasingly the talk of the town. When the Royal Institution at Albemarle Street in London organised its periodic lectures by the great scientists, such as Humphrey Davy, there was often a traffic jam in the street, caused by the coaches in which the landed gentry arrived. This then is the age where science moves on from being simply celebratory into one that educates, entertains and enhances.

This tradition of expanding knowledge has been carried on to the present day. There is more to the world than we knew, there is more within the world than we knew, and there is more to the universe than we knew. Science communicators such as Jacob Bronowski, George Gamow, Arthur Clarke, Lewis Thomas and Stephen Jay Gould wrote in simple language, using analogies and no graphs, pie charts or mathematics, with humour and in conversational style. They shared their joy and excitement with us. When Lewis Thomas says we should announce our presence on earth to extraterrestrial and extragalactic life forms simply by streaming the music of Johann Sebastian Bach, he makes us think why. When Clarke writes of a lift or elevator to space, his science fiction soon becomes a reality, and when Gould wonders about the smile of the flamingo, we appreciate the play of evolution. These authors, like artists, are the antennae to the future.

These communicators are my heroes, and theirs are the ways to enthuse and to educate on the sly. Such writings build on existing knowledge base and expand it. They made us realise that there is only one earth and that natural resources are limited and need to be conserved and recycled.

Science and the Social Belief Systems

There is, however, a worry, a tension and a challenge. Many scientists themselves fall short while communicating their ideas. They talk science, they attempt to 'sell science' and not attempt to think of the context. Society and communities take time to internalise their novel scientific ideas. We should not forget, discard or decry the belief systems, social order and practices, however 'wrong' they may appear to the 'science-first approach' that scientists often take. Darwin's 'theory' of evolution is still not accepted by some groups, and on several counts. One is, it goes against the biblical ideas of creation. The other is the puerile argument: "it is only a theory and we know how theories come and go." Yet another is the attempt to include new developments into the existing belief system; examples are inclusion of the uncertainly principle and quantum physics as but a reflection of yin–yang, or the Tao of Physics.

Belief systems are tough to crack and unyielding, and a science communicator cannot dismiss them as superstition, illogical and out of date. He cannot 'preach' or take the 'high ground'. Despite all the explanations, the Ganesha statue is still believed by many to drink milk when offered by a devotee (I am referring to the reports that had appeared in 2010s in the Indian media). Handle the issue deftly, do not make fun of superstitions; attempt to dialogue, not diatribe nor deride. It does take time for society to absorb and include new scientific thoughts and developments. About 50 years ago, when the first test tube baby was born, there was unease in the civil society, but today, a three-parent baby does not shock people. Dolly the sheep was cloned, and yet why is human cloning a no-no? When a scientist claims that it is no different than Dolly, why does society not accept it? The context and the canvas are important. Note that developments in science and technology move at the speed of a gazelle, while societal and legal systems move at glacial speed.

Science Bred Technology, Used to Colonise the World

Historically speaking, science began not only to explain but to apply its principles to invent materials, develop new technologies and applications to daily life and to help society through such inventions and technologies. From explaining how technical materials worked, science now develops technology, based on its laws and principles. One major consequence of this shift, particularly the technology of seafaring combined with those of rudimentary global positioning system (GPS) using the longitude, the technology of swords, gunpowder and guns, was the conquest and colonisation of a significant part of the world by the technologically adept people, and the technologies they developed and used spread across the world. The Industrial Revolution was born and bred. Technology has helped people live better lives, with less drudgery. Be it the steam engine, electricity, appliances, chemicals, pesticides, fertilisers and detergents, they were all adopted with gay abandon. Manufacture changed the whole picture. Before the Industrial Revolution, manufacture was a collective affair, involving rural households, skilled craftsmen and artisans, for whom it was livelihood. With the advent of private companies, household manufacturing gave way to machinery and assembly-line automation, which replaced manpower remarkably. Here starts another conflict—this time not between science and society, but the effects of commercial interests and capitalism on the lives of people, animals and plants. Look back for a moment to think about Mahatma Gandhi and his spinning wheel to make *Khadi* (the fabric weaved from home-spun cotton).

Against Science or Technological Tyranny?

It is important to point out here that much of present-day controversies across the world are not always against science, but against the effects of technology on the lives of people and other living beings. Objection to Darwin was (and is) at the belief level, while objection to cotton mills came from Gandhiji and his spinning wheel at the level of livelihood of people. As manufacture and large-scale production of labour-saving and comfort materials started in Europe and the USA, degradation of the environment also started. Recall Rachel Carson and her book 'Silent Spring' in 1962. She not only noted that the birds sang less and less, but also the Ohio River full of detergent froth. The chemical industry lashed out against her with fury. At long last, good sense prevailed and controls were set in place, even if half-heartedly, and the Environment Protection Agency (EPA) was born to restrict such runoffs and attempt to establish a sort of compromise between economic development and environmental protection.

The Scale of Things

It is also the scale of the thing. Science deals with astronomical dimensions on one side and subatomic on the other. What Newton explained to the world was easier to grasp by the average citizen, but what Higgs explains about the boson is harder to grasp. Science communicators need to keep that in mind. Then again is the scale of time; while we, in daily life, are familiar with time in terms of seconds on one end and years, centuries (even millennia) on the other end, concepts concerning evolution involve billions of years. Darwin and natural selection are harder to grasp, while Lamarck is easier to grasp, since it appears to fit with our own time scales of generations. A similar situation arises with the ideas of climate change. The scale here is in terms of centuries. When we are told that glaciers will melt and sea levels will rise in few decades, because of changes brought about over centuries, it is hard to appreciate. We would like to have proof or events that can be appreciated in months or years, else it becomes too abstract and long term and falls off the list of our concerns. And commercial interests exploit this. Also, a government that has a lifetime of few years tends to pass the buck. The scales of the physical dimension and time dimension that science deals with are different than what daily life deals with, and are thus rather 'unnatural'. This is an issue to be kept in mind when communicating with the public. Lewis Wolpert describes this as the unnatural nature of science.

Then, there is the scale of numbers right here in our planet itself. It is the very nature of technology to reach to larger and larger numbers. Companies make their profits only if the scale of sales is high. While flooding everyday life with products like television, cellphones and social media products, and with the reach of the worldwide web, they have also brought discomfort to time-honoured family practices and social mores, as well as interactions and conventions. Virtual bonding is not the same as warm body bonding.

Look at the genetically modified organism (GMO) tangle. The issue is far more to do with patenting and cartel-based monopoly than safety itself. 'Activists' claim damages to the environment and biodiversity and claim that even diseases are caused by some GMOs. My claim here is that they cherry-pick (better: dirt-pick) some literature; I support discarding Bt and the promotion of chosen GMOs and allowing no patenting at all. Let them be public property just as the Borlaug wheat, or the Sona Masoori rice have been, based on governmental policies. In addition, the Nobelist Richard Roberts goes all the way for Golden rice (which contains the vision-protecting vitamin A in it) and says countries that ban it should be sued for manslaughter. Extreme arguments have held forth so far, and a dialogue, not diatribe, is vital for the food safety of billions with an ever-shrinking arable land across the globe.

What Are the Issues in Communication?

How then do scientists communicate today with the public and policy makers against this backdrop? It is a tough task that needs an understanding of the state of mind, the nuances in argument and the plus and minus of any new development. Society has awakened, perhaps more than society, the civic activists, who occasionally throw the baby with the bath water. But then we need to look at what baby we are talking about and how good or bad is the bathwater.

How Then Do We Communicate?

Scientists cannot communicate in an adiabatic, science-first fashion. A communicator should advocate, educate, entertain, share experience, and criticise when needed. Above all, he should not preach or take the high ground, but treat the other party as an equal. They should study the context, realise whether antiscience ideas and policies come due to threatening long-held beliefs and religious ideas, political interests or commercial pressures. These three sit on the opposite side of a table. Science communication is an advocacy. It pitches science and technology as a desirable thing for our lives, health, environment and long-term advantage. Yet, extolling or recommending it for human welfare (or otherwise) has to take into account the following: language simplicity, audience and its background, and the counterpoints, namely vested religious, political and economic interests. Any recommendation that ignores these will fail. Science-first-based advice is inadequate, just as vested-interest-based ones are.

What then would I advocate? More science in the use of our daily lives; that is the rational thing to do. I advocate GM food, vaccines and health practices, cutting greenhouse emissions and attempts to strike a mutual balance between environment and development. Madhav Gadgil is correct on environment and K. Kasturirangan is well meaning on space research programmes. It is the job of the advocates to bring forth a mutually acceptable compromise. The main reason why India progressed from a 'basket case' to an almost-developed nation in two generations is the use of science just as much as the attempt to reconcile with developmental norms. It is a tussle, a compromise. Nevertheless, with this balance in mind, we can attempt to have a set of people who should eat their stomachs full, enjoy their lives, sing their songs, dance their steps, live long and healthy lives and leave the place safe and healthy for their children, grandchildren, great-grandchildren and further on. Just prayer and philosophy will not do. Science and rationalism are needed. And for this we need to engage the community and their rulers on mutually understandable and acceptable terms. Not an easy job, but not one that we can shirk.

A Two-Way Traffic: How Scientists Are Benefitted by Communicating with People

Milind Watve
Department of Biology, Indian Institute of Science Education and Research (IISER), Sai Trinity Building, 3rd Floor, Chemistry Wing, Room No. 305, Sutarwadi, Pashan, Pune 411 021, India

In the popular perception, the communication between scientists and people is a one-way transmission of knowledge and information. The scientists enlighten people about some interesting frontiers of research and their relevance to people. This is no doubt an essential part of the interaction. What I want to point out here is a lesser known side of the interaction that can tremendously benefit science if we make proper use of it.

Ignorance is a highly valuable contributor to science, but unfortunately we mostly forget to make the right use of it. Useful ignorance comes from people most naive to a subject. An ignorant individual is likely to ask most fundamental questions. This becomes more and more difficult with education, training and experience over the years in a subject. Education generally begins with textbooks. The way textbooks are written and taught is such that everything written there is believed to be true. This 'truth' makes the foundation on which a student progresses in any subject. Some of the students take up research as a career and start addressing questions on the cutting edge of the field. The cutting-edge psychology is so strong that they are not interested in returning to issues that were addressed and answered several decades ago. This is what they learnt in textbooks. This is what was established a long time ago. This is old stuff, and there is no point in spending valuable time on worrying about it because one is most unlikely to get novel and high impact papers from the old stuff.

Reluctance to return to old stuff is extremely common in the psychology of active and competent cutting-edge researchers who are aggressively pushing the frontiers of knowledge. This is rewarding quite often, but there is a serious risk in it. It is quite likely, and has happened many times in the history of science, that all the active and high-profile research is being pursued on a conceptual foundation, which is not so sound, has some fundamental flaws or perhaps is absolutely wrong. The reason why such wrong foundations persist in science for quite some time lies in the process of science education and the course of training of a researcher. Theories and hypotheses are built from time to time as a part of the scientific process. These are based on the standing state of evidence and contemporary thinking. Proposers of these hypotheses are well aware of their limitations and are generally open to alternative ways of thinking. However, if these hypotheses are not falsified quickly enough, they enter textbooks. The way textbooks are written and taught is substantially different than the process of scientific enquiry. A new generation of researchers are born with the foundation of this textbook knowledge. They simply assume the textbook knowledge to be 'true' and want to do frontier science based on this foundation. As this generation builds up, it becomes more and more unlikely that the fundamentals would be challenged or reexamined. As a result, research might continue on a completely misguided line.

A very promising solution to this problem is interaction with naive audience and undergraduate students. It is necessary to return to the fundamentals in order to communicate with people. One possibility is that just because of the need to return to basics, scientists rethink about the conceptual foundation and find flaws, if any, in it. The other possibility is that naive people ask naive questions, and naive questions are more likely to expose fundamental flaws in current thinking.

I have recently come across a case where the fundamental thinking in the field is seriously flawed. The gaps, traps and logical flaws are glaring enough to be noticed by undergraduates. There is already enough evidence to falsify the current theory. And still the cutting-edge research community continues to work assuming the textbook theory to be valid without realising that they are working on a wrong foundation. The case is that of etiology and pathogenesis of type-2 diabetes that is currently believed to be as follows:

1. Obesity is a consequence of positive energy balance.
2. Obesity causes insulin resistance (IR).
3. The body tries to compensate IR by making the pancreatic beta cells produce more insulin.
4. There is a progressive degeneration of beta cells resulting in relative insulin deficiency/insufficiency (RII); that is, insulin levels being short of compensating for IR. A combination of IR and RII results in hyperglycemia.
5. The increased blood sugar levels lead to all other pathophysiological consequences of diabetes.

The sequence appears very logical and convincing at a glance. That is what I thought as I began. I made a mistake at some stage of using this theory as an example to discuss in an undergraduate gathering how to ask questions, how to design experiments and how to test and support or falsify hypotheses. We started wondering on how to test one or more of the set of hypotheses mentioned by designing specific experiments. We ended up designing experiments although we could not do most of them ourselves. Then we started looking at literature to see whether someone had already done some of these experiments. Some of them had actually been done. And then came the surprise. Experiments that could have potentially falsified one or more of the preceding hypotheses, whenever done, had clear-cut results that would reject the hypotheses. This was true for all of the five hypotheses (Watve 2013). I briefly relate only few of them.

We first examine the hypothesis that beta cells produce more insulin in order to compensate for IR. In the insulin-resistant state there is a high level of IR as well as high levels of insulin in blood. Association is not a proof of causation. Alternative possibilities also need to be examined, namely (i) insulin overproduction is primary and since high levels of insulin can lead to hypoglycemia, IR develops as a compensating mechanism and (ii) both are secondary to an unknown primary change that stimulates both the processes simultaneously.

One suggestion came from the undergraduate group that whatever comes first could be the cause and whatever follows is the effect. Fewer studies have followed the time course of development in sufficient details. In all these experiments the first detectable change was either high levels of insulin and/or hypoglycemia. A better way to resolve between the alternatives would be to selectively induce one of the processes and see the response of the other. Such situations do exist naturally; experimenters have achieved such feats. In certain types of pancreatic tumours insulin is overproduced. This overproduction is due to the tumour, and therefore can be safely assumed to be primary. An insulin-producing tumour is always accompanied by IR, and if this tumour is removed, insulin sensitivity increases. A reverse experiment in which IR was artificially induced is also available in the literature. The muscle tissue is responsible for majority of insulin-dependent glucose uptake in the body. If insulin receptors in the muscle are specifically knocked out, insulin levels remain surprisingly normal. Let us also look at the mechanisms available that bring about the respective compensatory responses in either case. If pancreas has to compensate insulin resistance, it needs to know the level of IR of the body and accordingly increase the insulin-production level. The predominant mechanism that, according to the classical theory, stimulates insulin production is raised blood glucose. But ironically the blood sugar is not raised in the early insulin-resistant state. In fact, in the early hyperinsulinemic phase there is slight hypoglycemia. Therefore, insulin production does not appear to be stimulated by glucose. No other mechanism has been suggested even speculatively that can measure the level of IR and instruct the beta cells to produce appropriate amounts of insulin. On the other hand, we found more than half a dozen pathways by which high insulin levels could induce resistance.

All the evidence together raises doubts about whether hyperinsulinemia is an attempt to compensate IR or vice versa. Evidence is clearly in favour of the later; that is, hyperinsulinemia is a primary response and IR arises as a compensatory reaction to hyperinsulinemia. If hyperinsulinemia comes first, the logical links between postulates 1 to 2 and 2 to 3 are in trouble, and so is postulate 4. If hyperinsulinemia comes first, beta cell degeneration cannot be said to be the cause of failure of compensation, and if that is so, what makes the blood sugar go up? This could be a fatal blow to the classical thinking. If we accept that hyperinsulinemia comes first, we need to worry about what causes hyperinsulinemia.

The classical reasoning of hyperglycemia is that IR remains high, but insulin levels are insufficient to compensate. How can we test this postulate? A simple way to test would be to mimic the hypothetical process and see whether the result is the same. For example, let us take a normoglycemic insulin-resistant and hyperinsulinemic individual. Now if insulin production in this individual is artificially and specifically suppressed, sugar levels should go up. If we can demonstrate this, we will be able to demonstrate that IR and insulin insufficiency are 'sufficient' to explain hyperglycemia. Such experiments have actually been carried out repeatedly. The big surprise of these experiments is that when insulin production is suppressed, insulin sensitivity increases rapidly such that the sugar level remains more or less unaffected. The result is consistent across at least five different agents for suppressing insulin, and these experiments have been performed by different laboratories independent of each other and involving both rodent and human models. The most logical interpretation of these 'inconvenient' results is that when insulin production is suppressed, IR actually seems to adjust itself to keep the sugar levels normal.

Another experiment that raises doubt about IR–RII being sufficient cause of hyperglycemia is again that of MIRKO (muscle insulin receptor knockout) mice. In these mice, insulin signalling in muscle is reduced by over 95%. This is an extreme possible case of muscle IR. By classical thinking we would expect insulin levels to increase in compensation, but they do not, as seen in the preceding paragraphs. Now if there is extreme muscle IR and no compensatory insulin production, it must be an ideal situation for blood glucose to go up. However, most surprisingly, that does not happen. Sugar levels remain normal in MIRKO mice. This means that extreme IR and lack of compensatory insulin response are not sufficient to increase blood sugar.

What I have quoted here are few representative examples of evidence against the textbook theory of type-2 diabetes. In reality there is much more evidence against all the five postulates. On the other hand, looking back, the evidence based on which they were suggested looks very flimsy. The surprise is that in spite of the compelling evidence against the textbook theory, diabetes researchers continue to believe in it and also continue their research on the foundation of a wrong theory. Interestingly, when I tried to talk to leading diabetes researchers they were least interested in talking about these experiments. They mostly thought it was something that they learnt during their first year of medicine, and therefore, not open to debate. 'This is too preliminary. Now we are at the cutting edge, why should we worry about some old stuff?' The reason I came to realise the paradoxes and the falsifying evidence that have destroyed the theory is that I tried to talk to undergraduates who did not have a prior 'knowledge' and were therefore open to think and question from the very beginning.

I suspect that the case of type-2 diabetes is not unique. There may be more areas in science that have similar stories. If researchers in all areas try to talk to naive but curious people, they would realise flaws and paradoxes more easily and keep on correcting their paths if wrong. This is potentially the great role naivety and ignorance can play in the progress of science. Therefore, I think the interaction between scientists and people is not a one-way flow of information and knowledge. Questions from ignorant people can keep scientists on their toes all the time. One should not forget what we all learnt as a child from the famous story: if the emperor is naked, only a child would say so, all the grown-ups and knowledgeable ones would not. Therefore, the knowledgeable ones have much to learn from kids (read *the novices in science*).

Reaching Science and Technology to the Common

S. Ramadorai
Management Council, National Institute of Advanced Studies, Bengaluru, India

We are living in an era not only dominated but also often driven by technologies, and that too, fast-changing technologies. Today, there is no aspect of human life that is untouched by technologies in one form or the other. The way we communicate, do business and plan our workdays and holidays are dominated by information and communication technologies. In fact, it is all of us who make up the marketplace and often dictate the choice of technologies.

In addition, while we use these technologies they are concealed behind simple, easy to use, gadgets; most of us could not be bothered about the science behind them at all. Normally this would be fine, but consider the fact that most social media users are unaware of the fact that every time they are on the Internet they leave behind bits and bytes of data about themselves—the simple click of the 'like' on Facebook makes them vulnerable. Millions of youth in India, even in remote villages, are on Facebook. Therefore, should there be more awareness among the new generation on what constitutes sensible Internet behaviour, should there be more public engagement around cyber-security and should this be a part of our general education curriculum? The answer to all of these questions is a resounding *yes*. Clearly there is a deep connection among science, policy and society.

In a knowledge society, we all need to work to promote scientific literacy so that everyone is up to speed, empowered and ready to contribute to the great debates about science, technology and the future. This is possible through public engagement; which is nothing but different ways in which the activity and benefits of science, technology and research can be shared with the public, with a view to interact and exchange ideas with the goal of generating mutual benefit.

Today chasm between scientists and institutes of higher learning on one part and society on the other is very wide in India; they both talk different languages, and therefore the communication of science in a manner that is appealing and simple to people, the key to bridge the gap, is very rare. On one hand, scientists need resources and forums to have meaningful conversations with the public and on the other, we need forums that bring together citizens and scientists.

Consider this simple fact. In Mumbai at the onset of the monsoon, the seafront is a popular place for people to gather to see the waves in all their glory. In the recent past, we had particularly high tides and furiously high 5-metre waves lashing the coast drowned several young people. People are assigning such unusual phenomena to the climate-change effect. Now climate change is one topic that is absolutely amenable to public engagement; people are affected, and they would benefit with scientifically presented knowledge on climate change. On the other hand, the government can leverage such forums to drive behavioural changes in people leading to conservation of resources, such as consuming less water, electricity and so on. We have to find ways to create forums to reach the average citizen on the streets where talking about scientific issues could become as natural as talking about football or cricket.

Science festivals are one way of public engagement. As an example, the British Science Festival is one of Europe's largest celebrations of science, engineering and technology with over 250 events, activities, exhibitions and trips taking place over a week in September. This event offers something for everyone with activities for families and schools groups, adults looking for entertainment and stimulating debate as well as events aimed at professionals interested in the latest research. Similar events could bring people face to face with leading scientists with a gift for storytelling and the knack of explaining complex research.

Museums are another opportunity for public engagement. Museums must compete with cinema for weekend family outings. Research shows that parents believe that museums are 'good places to take children to learn,' and several studies support the idea that families use museums as socially 'mediated' learning environments. In one study, children indicated that they often prefer to visit museums with their families, rather than school groups, because they get to look at more things of interest to them personally and talk to their families about what they are seeing. Similarly, there are over 2500 botanical gardens around the world with facilities that schools can use. Innovative outreach programmes that engage families with science activities through workshops, family science day and so on must be launched in India. In Michigan Technological University, National Science Foundation has sponsored a programme where Ph.D. students are placed in middle-school classrooms to teach students about watershed science. In the classroom, the graduate students hone their communication skills, and help cultivate the next generation of scientists. However, the most important objective of this programme is to increase each Ph.D. student's ability to communicate the content of their research in a way that middle-school students can understand.

Many of us are familiar with the 'TED talks'. TED (technology, entertainment, design) is a global set of conferences under the catchy slogan: "Ideas Worth Spreading". TED used to be an exclusive conference in Monterey, California when it first began in 1984. It was only in 2006, when TED talks were launched using YouTube as a platform, that its popularity really soared. TED talks have been a great medium for scientists to bridge the gap with the public. There were 411 TED talks on science alone.

Clearly, there are sociocultural impacts of the new technologies; there are known and unknown future risks. In the early years of the technology era, it was adequate for the governments to take the views of few specialists to arrive at decisions on technology options and enabling policies. Today's world is indeed very different. On one hand, countries across the world are moving towards democratic forms of government where all policy options including technology options have to be taken with direct public involvement and approval. On the other hand, the technologies are getting more complex and informed decision making by the public at large is becoming more and more difficult. It is important for all the stakeholders including the governments to understand the public perceptions. It clearly calls for a two-way communication between the stakeholders and the public at large.

Many countries are seized with the need for wider public participation in scientific discussion and have put in place a national strategy for the same. Institutions such as National Institute of Advanced Studies (NIAS), Bengaluru and others must take the lead in preparing a white paper to be presented to the Ministry of Science and Technology to initiate such a policy for India as well. In 2010, the Australian government released a report titled, *Inspiring Australia: A National Strategy for Engagement with the Sciences*. The purpose was to establish "a national approach for the communication of science and its benefits to a broad range of stakeholders, including policy makers and the public, to create 'a scientifically engaged Australia., " Earlier, in 2009, UK had released a report prepared by the Science for

All Expert Group, on *The Challenges for Twenty First Century Science: A Review of the Evidence Base Surrounding the Value of Public Engagement by Scientists*. It aimed to assess "whether the increasing amount of activity makes a difference to improving the environment for science, and what certainty we might have whether further increasing engagement would further improve the UK's scientific environment." This report reviews the evidence underlying the idea of public engagement, to better model the relationships between scientists and public shaping science's special societal function. The British governments' Department for Business, Innovation and Skills has also commissioned a series of studies looking at public attitude towards science in the UK.

In the USA, Center for Advancement of Informal Science Education (CAISE) released a report titled, *Many Experts, Many Audiences: Public Engagement with Science and Informal Science Education. A CAISE Inquiry Group Report* in 2009. The report seeks to serve as a prompt for discussion and exploration of public engagement in science. Following up on work done in connection with this CAISE report, the Museum of Science in Boston, with funding from the National Science Foundation, collected information on over 200 projects that involved public engagement and convened a workshop of practitioners of public engagement to identify strategic directions for future work within the field.

In a country as diverse as India, a large rural–urban divide, multiple languages and a political system that diffuses the borders between public good and private good, all render policy making a challenge. We have seen this during 2010s in the cases of siting of nuclear power stations, genetically modified food, land acquisition for large infrastructure projects, water sharing between states and so on. While public involvement is intense in all the challenges, it cannot be said that public understanding of the underlying issues is as healthy as it should be. This calls for more public dissemination of knowledge through the right forums. I believe that to achieve this goal, our institutions of higher learning need to play a lead role.

Effective Communication of Science and Technology: The Story of *Chandrayaan-1*

K. Kasturirangan
Indian Space Research Organisation (ISRO), Government of India, Bengaluru, India
Planning Commission, Government of India, New Delhi, India

On the first Atoms for Peace Award, presented to Niels Bohr, an outstanding physicist of the 20th century, a principal figure in elucidating the revisions of the philosophical foundations of physics for comprehension of quantum phenomena, James Killian, had the following to say about Bohr:

> In your profession, in your teaching, in your public life, you have shown that the domain of science and the domain of the humanities are in reality of single realm. In all your career, you have exemplified the wisdom, the humaneness, the intellectual splendor which the Atoms for Peace Award would recognize.

In the context of increasing role of science and technology to improve the quality of life of the humankind, the preceding statement about Bohr is not only pertinent for the present-day scientific community, but probably even more relevant.

'Science Meets the Public: Bridging the Gap' is a theme which is timely in the context of India's major ambitions to use the instrument of science and technology for the socioeconomic development in the coming years, on a scale, which has been unprecedented. Innovative use of advanced technologies is central to the process of accelerating development. A wide range of priorities drive the need for advanced technologies—from energy to water management, to farm production, medical research, waste disposal, health care, communications and transportation. Advances taking place in frontier areas of research such as micro-miniaturised electronics, advanced sensors, alternate energy sources, synthetic biology, materials and nanotechnologies are being integrated rapidly into the application systems. As a driver for future, strategic fundamental research is also undertaken through fields such as nuclear energy, aeronautics, space and life sciences. While such new technologies are expected to bring revolutionary changes to our lives and environment, there is a flip side too. There are often unintended consequences. While the usefulness of nuclear power, biotechnology or digital technologies are evident, there are also perceived and real negative consequences that trigger conflicts between government policy and some segments of the society.

The reasons are not difficult to imagine. Along with marvelous benefits of high technology, world has seen deadly weapons of mass destruction, unabated pollution of environment and increasing health risks to large sections of populations. Public, not only in our country but elsewhere also have often come to take a sceptical view of government's communications and its positions. Scientific community also suffers partly as has been shown by a recent survey abroad on genetic technologies, which revealed the lack of trust of public in scientists in placing society's interests above their own goals. We have also witnessed much public outcry about the ill effects from the use of many pesticides and the ethical issues connected with stem cell research and applications. All over the globe, public engagement is increasingly being considered as central to the development of science policies. Let us examine few cases in this connection, particularly, to show the fact that there are no tailor-made solutions to deal with these issues. The strategy needs to be made with respect to specific theme or subject under consideration. Let us examine this with reference to biotechnology advances, particularly because of the public interest in the introduction of genetically modified (GM) organisms into the food chain.

Biotechnology advances have been making revolutionary impact in the fields related to food, nutrition, health care and environmental sustainability. Their applications enabled crops resistant to pests/diseases and abiotic stress, and crops of high-yielding varieties, all of which assume importance in the context of addressing the food security. Much debate and controversy had arisen recently; however, when

GM pest-resistant version of brinjal, known as Bt Brinjal (created by inserting a gene from the soil bacterium *Bacillus thuringeinsis* Brinjal into brinjal), was proposed for approvals for commercial cultivation, the opposition came from a section of scientific community itself, besides some environmental group pointing risks in terms of the presence of an unwanted transgene in this GM food crop, which encodes resistance to certain antibiotics. Transfer of such resistance to humans could be a cause for concern. Several lacunae were pointed out with the evaluation process, for instance, that it was inadequately focused and ignored realistic environmental and other vital anxieties of the public. The proponents for this GM crop on the other hand argued that this transgenic crop would reduce the use of insecticide for the most prevalent pest, namely BFSB (brinjal shoot and fruit borer) by 70% and of pesticides for other insects by 42%. Since brinjal is cultivated over a large area in India with a production of 8.2 million tons and the pest BSFB is causing losses of 60–70% in commercial plantings, this development should be welcomed. The controversy, however, culminated in an indefinite moratorium, which has been recently reviewed.

The case of Bt Brinjal is typical of the strategy normally adopted by scientific establishments with respect to the public engagement, which is limited to a top-down process of communicating and obtaining endorsement of science, which is already planned for implementation by them. Especially, when an interested party like developing organization performs evaluation, there arise questions on credibility, genuineness or otherwise. An important lesson from Bt Brinjal case is the need for an independent agency to evaluate and engage with public groups to assure full transparency of relevant information and a comprehensive evaluation and compliance. In general, it is also important to have a systematic and constant monitoring of GM crops in scientific terms and the policy making should be guided by such feedback for corrective measures. Proper evaluation of the nature of public concerns, objective responses and ability to create necessary level of confidence in public mind need expertise beyond pure scientific domain. The need for resultant flexibility in policy measures is central to many areas such as environment, health and safety where complex interplay of several disciplines is inevitable, and the very evolutionary process of knowledge has to be factored in.

The then minister of environment and forest and the author addressed the presidents of all the six academies concerned, in our attempt to find out whether they could come up with a comprehensive report on the Bt Brinjal that could become a basis to defend the case in public. The academies together did an excellent job of creating a document which is available on the Indian National Science Academy (INSA) website. It was the first time the presidents of the six academies came together to look into a single issue of this nature. However, even this, unfortunately, failed to convince the public! Academy report can be authentic scientifically, but when it comes to communication certainly there are additional dimensions that we need to recognise, which probably was missing.

Another interesting area relates to the development of nanotechnology. Nanotechnology springing from the multidisciplinary convergence of advances in condensed matter physics, engineering, microbiology and chemistry deals with materials and systems with at least one dimension of about 1–100 nanometres and involves processes that modify features of molecular structure in terms of physical and chemical attributes. Rapid advances in this new discipline have been triggered by the development of instruments capable of imaging and manipulating single molecules or atoms. The National Science Foundation of the USA had earlier estimated the total global market for the nanotechnology products and services reaching a trillion dollar value by the middle of this decade. While the technology is still evolving, its applications are already found in several products such as sunscreens, antibacterial surfaces, automobile tyres and some anticancer drugs, and there is promise for more number of benefits. Nanotechnology's potential use can extend to every field related to science and technology.

Use of nanotechnology is also fraught with risks. Looking at the health hazards, it has been demonstrated that the inhalation of buckyballs (buckminsterfullerene) and nanotubes can affect the lungs. The deleterious effects become apparent after years. Scientific evidence also indicates serious toxic nature of nanoparticles of gold and silver, even though in the bulk level they do not exhibit the same. Hence, scientific certainty about toxicity of materials in 'nano' form requires detailed investigations to evaluate and map the nature, level and remedies of risks. Apart from technology issues, there are ethical issues of privacy protection that come up when the revolution in nanotechnology allows vast amounts of information to be stored and transmitted via tiny devices.

Nature of a proactive process adopted by the Dutch government for assessing public attitudes and level of risk acceptability for evolving policy on nanotechnology merits a mention in this context. The government of the Netherlands set up an independent committee to organise a dialogue—Dutch Societal Dialogue on Nanotechnology—thus sending out a strong message that it was not intending to merely push its own agenda on the policy for these technologies, but significantly enhance the credibility of the process. The committee devised a three-step process which included (1) providing information to public, (2) increasing awareness and (3) two-way interactions with public. The initial steps of communicating to public and enhancing their awareness were considered necessary in view of the novelty of technology and dynamically progressing knowledge. This made the dialogue more effective. The overall dialogue activity was divided into some 40 projects and a number of scientists, nongovernmental organisations (NGOs), companies and individuals were engaged through open competitive process to carry out the same. The public agenda was made representative with wide range of people, and the process was accessible through a variety of media including science cafes, theatre plays and teaching materials apart from television and Internet. End result of this multi-pronged process has been positive, resulting in rising awareness among the public of the risks of nanotechnologies. The most striking result of the Dutch Societal Dialogue on Nanotechnology is that, after the dialogue, the general public in the Netherlands displayed *more* awareness of the risks of nanotechnologies, and at the same time expressed *more* support for further development of nanotechnology.

This example provides a lesson in the governance of risks felt by populations in introducing new science. Empowered public participation should be enabled by a process that is credible and insulated from overplay of interest of any one community among diverse actors such as government, scientific institutions, business organisations and media. This process expands horizons, even beyond regulations, to influence society to accept risks associated with modern scientific endeavours and to mandate their progress. In the democratic traditions of India, relevance of such mechanisms needs no emphasis.

Chandrayaan-1 is a fine example of public engagement implemented by the Indian space programme. This is another kind of model that we adopted. What distinguishes India's space programme from many others is the process it has been adopting in bringing in stakeholders early in its planning. Soon after the learning phase during 1970s, the programme developed interface with users and made it driven by the users. Regular two-way communication process was set up to derive inputs for definition of programme as well as for implementation and validation of applications. Being user driven also meant being accountable and creating necessary transparency. Such transparency and accountability were consistently practised, and these translated into an organisational culture that fostered participative role for all stakeholders. This generated a strong *trust surplus* in contrast to the trust deficit, which often results from the one-way flow of information. Trust surplus (generated through user-driven programmes) had opened the door of big public support for Indian Space Research Organisation (ISRO).

Chandrayaan-1 is an offshoot of the organisational culture discussed in the preceding paragraph. It represented a new dimension of the country's space endeavour, calling for not-so-insignificant financial and other resources and initially lacking visibility of direct benefit to majority of general public. Embarking on such a programme called for eliciting public opinion with reference to aspects of enrichment of knowledge, enhancement of national prestige, technological leapfrogging, igniting the minds of young generation and, equally important, placing India in an exclusive club and thereby creating the necessary credentials for international participation in collaborative framework.

The concept for *Chandrayaan* mission—as an *avant-garde* interplanetary mission—was proposed during a forum organised to celebrate the National Technology Day in May 1999 with participation of several parliamentarians, policy makers, scientists and technologists, as well as the media. Subsequently, the scientific goals and challenges of the mission were presented in the Indian Academy of Sciences which was followed up by presentations to the Astronautical Society of India. Both these preeminent professional bodies strongly endorsed the proposed plans, considering the scientific and technological outputs of such an endeavour and the future opportunities it can open up. The third element was the approval of the standing committee of the Parliament, which also strongly endorsed the plans in this regard, sounding the positive political will, cutting across different party lines. The contribution made by Pallava Bagla, who was the first to report about the possibility, in popularising India's moon mission through the media also should be remembered with respect (Bagla and Menon 2008).

The proposal itself was further developed in full detail in scientific and technical terms by a task force and was peer reviewed by a national committee of experts. Space Commission, the apex policy body, gave its approval for the project recognising its role in advancing the scientific frontiers, expanding the technological horizons, opening up new international cooperative opportunities and fostering synergy among multitude of institutions that could galvanise a new generation of scientific talents. It is only in this backdrop that on the Independence Day in 2003, the Prime Minister Atal Bihari Vajpayee announced the nation's decision to enter the era of planetary exploration by undertaking the mission *Chandrayaan-1*. It is not out of place to mention that although the entire process of preparatory work took four years, due to diligent preparation, subsequently, the actual development and launch of the spacecraft took just four more years.

It is to be noted that when the structured interaction process was undertaken, there were few individuals who expressed reservations for undertaking such a mission, considering its cost. It was assuring to see that the concerned members of public were actively willing to look at both pros and cons, which is also very important. Indeed, total absence of any negative view would have given us a little bit of discomfort! This is just one example of the way we went through with respect to a major programme, which this country undertook, where we got the support from several segments of the society in different ways before we decided upon embarking on the same.

Another interesting dimension related to India's space programme from public standpoint is the initiative we took to subject the entire activity to a cost–benefit analysis. This analysis was conducted by an eminent economist who went into great depths about the different facets of the programme: the projects undertaken, their application aspects, issues of research and development, establishment of infrastructure within the organisation and outside, optimisation of the human resources deployment, alternate approaches to realising the same objectives and several other similar considerations. The assessment called for making available the entire database related to the elements mentioned earlier in a transparent manner. I should say that this report submitted to ISRO clearly brought out the unique role of India's space programme in the context of socioeconomic development, creation of industrial capability and potential commercial benefits. This report was later published as a book by Oxford University Press (Sankar 2007).

Kudankulam Nuclear Power Project: How Effective Communication Conquered the Fear of Nuclear Energy

R.S. Sundar
Kudankulam Nuclear Power Plant (KKNPP), Kudankulam PO, Radhapuram Taluk, Tirunelveli 627 106, Tamil Nadu, India

I am a slow walker, but definitely never walk backwards
Abraham Lincoln

Kudankulam Nuclear Power Project (KKNPP) was established at the southern part of mainland India in Tirunelveli district of Tamil Nadu state as a part of the inter-government agreement signed between India and Russia for technical collaboration for building nuclear power reactors in 1988. Implementation of this project, with two 1000 megawatts electric (MWe), 3+ generation water–water energetic reactor (VVER) type nuclear power reactors, had begun in 2001. As the project activities started, we were all thoroughly involved in the progress of the project, but our communication with the local population was quite limited. Basically, Nuclear Power Corporation of India Limited (NPCIL) personnel who are engaged in building the plants were all technical specialists understanding the man–machine interface much more than any other issues. Our public relation activities were limited to corporate social responsibility (CSR) activities—health camps, relief camps during tsunami and so on. A workshop for the journalists, organised at Tirunelveli and Nagercoil and a programme conducted in the nearby area of Valliyoor along with Rotary Club were some of the few events conducted to sensitise media and people from the geographical location of the power plant. Additionally, the local MLA (Member of the Legislative Assembly) and Panchayat (village-level governmental body) members visited KKNPP site in March 2002, and a visit was organised for a team of Panchayat members to one of the NPCIL's nuclear power plants (NPPs) to make them aware of the process of power production in an NPP.

The process of building reactor was moving smoothly, but the disaster that happened at Fukushima reactors in Japan on 11 March 2011 totally changed the scenario. Activists from various corners of the nation began to suspect and talk about the safety of the KKNPP. After Fukushima nuclear disaster, our Chairman cum Managing Director instructed all site offices to interact immediately with the public about the accident and how Fukushima reactor compared with the Indian NPPs. As the then station director of Units 1 and 2 along with the site director and a team of other members, the author organised two separate interactive sessions. First one was at Chettikulam village, where our residential apartments are also located and the second at Kudankulam village. The participants included Panchayat presidents, leaders of the villages, youngsters and ladies. However, the people who attended the meeting had no concerns whatsoever as to what had happened at Fukushima. At that point of time they were all worried about how many more jobs they could get directly or indirectly, how many more CSR activities would be carried out that would benefit the society and when KKNPP Units 3 and 4 will become functional so that more employment opportunities are generated. These meetings were conducted in a cordial and peaceful atmosphere. We had given a PowerPoint presentation explaining the events that had led to Fukushima reactor, comparing with the design-safety features of KKNPP. We were satisfied completely and returned to our work.

Trust Dies and Mistrust Blossoms

The work of commissioning of Unit 1 was at its peak in the month of June 2011. Hot-run activities, namely testing of steam-relief devices and steam-line flushing, a process mandatory to make the system ready for further commissioning, had started. Around 1000 employees and a contract workforce of about 2000 were engaged in the work at that time. We would monitor the flushing operations standing in the nearby areas from where the steam was getting released. Full of enthusiasm to complete the activities, being conducted in full swing, we continued our work 24 × 7. This was the first mistake we committed; we were oblivious of the fact that the noise emanating during nighttime was disturbing the local villagers.

Off-site emergency exercise is a mandatory requirement prior to initial fuel loading. In order to conduct this exercise, we launched an awareness programme involving various district officials and the local population. Activists who were far more powerful speakers than us actively worked at this juncture and the rumour-mongers grabbed the opportunity. Rumours were spread that villagers in the vicinity of the plant will be evacuated forever once it becomes operational. Land brokers stepped in, offering land at a nearby place. People living in the vicinity of the reactor panicked and launched protests adversely affecting the law and order situation that threatened the safety of workforce involved in the work of building the reactor.

Agitations took place both at the plant site and at the township; Marine and Tamil Nadu Police forces had to be called in to control the situation. From September 2011 we plied to and from the work in convoys with police escorts. In October 2011 the protesters blockaded the way and our operation crew was held up inside the plant premises for more than 90 hours. This was one of the toughest situations we faced. I personally spoke to family members of the crew assuring that their husband/brother/son was safe. We urged all our employees not to get provoked at any cost. Every day the bus entering the office premises would be protested by the agitators who would hurl abuses on the team. Restraint was our best option. At no point in time were we critical of the protestors, who were our neighbours, and sincerely hoped that their fears would be allayed. From 13 October 2011 until 19 March 2012 only a minimum number of the employees (little more than 100) would come to work at the plant site to maintain the systems operation.

Winning the Hearts of People

We had established offices at the township and were making preparations for further activities. A Public Awareness Committee (PAC) was constituted with author's chairmanship. We brought out small pamphlets on fishing, earthquake, tsunami and radiation in three languages, namely Tamil, the local language, English and Malayalam. These were widely distributed in and around Tamil Nadu and Kerala. It was a humble beginning to engage the people about the truth of nuclear energy. One of the major fears of the people was that seawater may get severely contaminated due to the activities of NPPs. Therefore, small skits were prepared on consumption of seawater at the plant. We told them that the seawater is the life source for us as well, that we used desalinated water for domestic and the industrial use in our township and no water was sourced from any of the local water bodies.

Advertisements depicting the safety of nuclear energy were telecast for quite some time during the peak viewing hours on many television channels. We participated in debates on various television channels and explained in simple terms the mechanism and safety of the power generation in the NPP. Several television channels showed that the fishing activity had not been affected at Kalpakkam, which had been in service for almost 25 years, and at Kaiga in Karnataka. We put up banners depicting the facts about nuclear energy on the public transport buses and aired awareness programme on FM radio channels. Informative ringtones were created and installed in all our telephones. Measures were also taken to inform the people that many processes involved in producing steam for moving turbines of the generator in an NPP were similar to those in a thermal power plant. Additionally, the role the Tamil Nadu Pollution Control Board in monitoring temperature rise caused by KKNPP and the mechanisms installed in the plant to ensure the stipulations of the norms laid down by Ministry of Environment and Forest (MoEF) were also projected. Colleges, industries, Rotary Club, Lions Club and so on invited us to deliver talks, seminars and exhibitions on the various aspects of nuclear energy and also the safety features of KKNPP.

While the efforts to create awareness on nuclear energy and to dispel the unfounded apprehensions of people about KKNPP continued, the Government of India formed an expert committee to assess the safety aspects of KKNPP in October 2011. Dr Muthunayagam led this committee comprising nuclear scientists, Indian Institutes of Technology professors, oncologists, doctors from Tata Memorial Hospital and many other specialists. Tamil Nadu state government also constituted a separate committee to study the safety features of KKNPP and to understand the concerns of local people. After thorough analysis, both the committees submitted their reports to the concerned authorities. The reports proclaimed that the KKNPP processes had adequate safety features in line with the requirements of Indian and international standards, KKNPP reactors can withstand earthquake and tsunami and that there were no radiation hazards to the public and the environment. This report helped us in reopening the power project and resuming the activities at the site in March 2012.

Seeing Is Believing

After reopening of the plant the PAC continued its efforts with more vigour and our activities assumed new dimensions. A structured public outreach programme was put in place by launching a dedicated committee on public awareness which was chaired by the station director. A multifaceted approach including lectures, exhibitions, distribution of booklets, radio and television programmes, open discussions and so on was adopted to reach out to different layers of the public. With the support of district officials, 22 Panchayats covering more than 40 hamlets were given briefings at their doorsteps on the importance of nuclear energy and safety features of the KKNPP.

In addition to the other ongoing public awareness programmes, to allay the unfounded fears of the public on nuclear power and KKNPP, a strategy was adopted to bring the public to the plant site and provide them right perspective on nuclear energy and the safety features. Any Indian citizen with a valid identity and above 14 years of age was allowed to visit and observe the equipment available at KKNPP with prior appointment. Large number of people from various sections of the society, such as schoolchildren, college students, school and college teachers and even priests of various faiths from Tamil Nadu and Kerala visited the KKNPP plant site. Groups of people were taken into the plant area and were shown reactor, turbine building, desalination plant and other facilities for them to get first hand information on nuclear power reactor. The first visit of villagers to KKNPP was organised on 2 October 2012 and, these visits continue as of 2016. Visitors are given a presentation in the language they understand, namely Tamil, Malayalam, Hindi or English which is followed by lunch and a question–answer session. Their queries, mostly relating to protection of marine organisms and handling of waste, are answered to their satisfaction by the experts. The moment they see the greener environment and a normal life being led by us on the NPP premises, they became our brand ambassadors in spreading the positive messages on nuclear power (Table 1).

We have been participating in and organising exhibitions in schools and colleges and during the state-sponsored programmes, which are attended by people in huge numbers. Our volunteers distribute pamphlets, booklets and cartoon books produced by our corporate communication group to those who visit our stalls. We have found that the response is overwhelming and have conducted similar programmes in many locations in Tamil Nadu and Kerala.

In my opinion, there is no standard operating procedure that can be followed to raise awareness about a technology like NPP. Every situation is unique, but the

Table 1 A brief account of public awareness activities conducted during the impasse period (November 2011 to March 2012) to raise awareness of the public about the safety of the NPP

1	80 lecture programmes were conducted for schools, colleges, industrial associations, clubs and so on across Tamil Nadu and the nearby districts of Kerala
2	14 exhibitions were organised on nuclear energy and safety aspects of KKNPP in major cities of Tamil Nadu
3	An informative advertisement on nuclear energy was published in the Tamil dailies
4	Two one-minute radio jingles (Tamil) were broadcasted through FM channels in nearby districts for one month
5	18,00,000 pamphlets on nuclear energy, livelihood, radiation and environment and earthquake, in three languages (Tamil, English and Malayalam), were distributed
6	75,000 copies of informative publications, including an article by late Dr. A.P.J. Abdul Kalam, Honourable former President of India, were printed and distributed
7	A month-long television campaign was launched in association with the Tamil television channels

Table 2 Outreach activities organised during the period 2011 to June 2015

1	1300 programmes, exclusively on public awareness, including lectures, seminars, exhibitions and rallies were organised
2	24,00,000 copies of informative publications on nuclear power were distributed
3	64,000 villagers from about 24 districts of Tamil Nadu visited KKNPP
4	1,30,000 students from several educational institutions across Tamil Nadu and Kerala got opportunity to visit the NPP
5	Organised and participated in over 125 exhibitions/seminars to showcase the benefits of nuclear power and safety aspects of KKNPP, and reached out to about 4,00,000 people
6	Delivered about 135 lectures to reiterate that nuclear power is the need of the hour and on the safety features of Indian NPPs to about 40,000 people
7	14 rallies were organised to promote nuclear energy and environment safety with a participation of about 1,20,000

common factor is that we definitely have to respect the sentiments of the local villagers. Never underestimate their capacity; understanding and 'perception' matter a lot. Introducing activities to upgrade the standard of living of the neighbourhood can work better than compensation. Skill development is one such activity, which could be beneficial to both the industry and society. Providing hospitals, primary health centres and school buildings, and improving the hygiene and so on are also found effective in building a cordial relationship with the people living in the vicinity of the NPP. Key lessons and our experiences can be summarised as follows:

- Maintain direct communication with the local population and take them into confidence.
- Maintain direct communication with the electronic and print media to prevent rumours. Silence leads to more rumours.
- I sincerely believe that the public awareness team at KKNPP is doing its best (Table 2). An information centre at the premises is already constructed and is being equipped.

The Clash of Public and Private Interests: Communicating Using NewsClick Platform

Prabir Purkayastha
Delhi Science Forum, D-158 Lower Ground Floor, Saket, New Delhi 110 017, India

When we talk about science communication, we are really talking of both science and technology communications. When you talk about science, it is understanding how nature works. Science is something that can be at least argued is objective, is dispassionate and is rational. One may question all these words, but at least one can argue that. When it comes to technology, let us be very clear, technology is always for the interest of people, and people are not homogeneous. There are people who benefit out of a certain project, and others who lose out due to that project. There is an obvious conflict that takes place when technology comes into play. Some people lose, some people gain. We may believe that in the overall, society as a whole has gained, but that is no consolation to the person who lost his or her livelihood.

Therefore, it is not simply a matter of communication, but it is also a matter of what kind of roles and interests different technologies play. What technology we choose has also interests of different sections in it. That means people may choose a certain kind of technology, but in the process of making this technological choice, they are also building a certain kind of society that may be against the interest of others. Issues, which come up like this, are really not as they are being presented by at least some scientists and policy makers—that it is a clear issue where there are only objective rational decisions to be made; that there are some wise men or wise women now, who can scientifically tell how society should develop; and that it is relatively poor and uneducated people, who are afraid to lose their livelihoods, who are holding society from going forward. We need to get out of this mode of thinking.

I would perhaps not agree wholeheartedly to the existence of different knowledge systems, and to the need for, democratically, to look at all knowledge systems. What I would hold is that there are clear class interests involved. Technology, unfortunately, is in the business of production, and production means ultimately the political economy of the country. And political economy, whether you like it or not, is a matter of politics as well. When people talk of ideologies—for instance, somebody saying that those opposing a project or a certain kind of development are ideological—that itself is an ideology. Therefore, it is important for us to understand that there is no ideology-free state of the world. Only scientific statements such as 'sun rises in the east' can be made nonideological. However, when we want a certain kind of technology, it is implicit in it that we are proposing a certain kind of technology for the benefit of a certain section of the people; sometimes, majority of the people, sometimes the minority. Therefore, these are issues that are social. Technology ultimately is decided by the society at large, and there is no way of deciding societal issues without politics.

When we talk about science and technology communications, we must understand that it is not just an issue of communication, but it is a fundamental issue of what kind of society we want. Who is losing? Who is gaining? And the struggle between them is also reflected in the struggle within science and technology. Hence, one should be clear that when we are talking about science and technology, we are talking as if it is just science. But in reality it is not. Therefore, this context must be considered—when we talk about either nuclear energy, or genetically modified (GM) rice, or GM Brinjal and so on—that these issues are unfortunately political. Let us take a very quick example of nuclear energy. It is not a question of whether or not nuclear energy is harmful. The question is, what is the cost of making nuclear energy safe? And, if the cost of making nuclear energy safe rises, every time an accident takes place, we finally may come to the conclusion that maybe it is much too expensive for the society to produce energy that way, right now. I am leaving it open at the moment, although my beliefs are different. Well, 50 years later, we will see. But right now, solar energy maybe a cheaper alternative—even though it is much more expensive than coal or nuclear energy from imported Areva reactors. This is something that the nuclear energy establishment has to come clean on; what is the cost of Areva energy? We do not know. The chairman of Areva says that it has been kept secret at the insistence of the Government of India. Nuclear Power Corporation has asked Areva to keep it a secret; therefore, they are also not able to tell us what the agreement is. The author believes that the agreement is still not settled. Therefore, if authorities do want to keep certain things secret, then do not comment on people who may not accept your statements, because ultimately, it is you who are creating the distrust.

Let us move to GM, a very important area because transgenic, GM organisms and marker-assisted breeding, all these issues need to be talked about. But the question, why the resistance is coming up, is not really addressed properly as yet. The consumers in India and elsewhere, who were middle class, believed "if GM technology is used, it may benefit the farmer, but I am being asked to eat something which I don't know whether it is safe or not. Why should I take the hit for the farmers' benefit?" This is also the underlying issue that comes up in GM, and unless we address that, we are not going to perhaps get any major traction on this issue. Perhaps the reason also why it has taken this form, is that it is the first bunch of GM as food crops—the Bt Brinjal and so on. Bt Cotton had much less resistance, because we do not eat it, but wear it. As you can see, we are quite happy to ingest various biotechnological products as medicine. However, we are unhappy if it is in our food. Therefore, there is a certain issue that is involved here that goes beyond science and technology; it really goes back to what relationship GM crops have to the people. Which people? The people who are producing it or the people who are consuming it?

Communication is not just how effective you are—that is not the only issue. For instance, there is a study conducted by Nyhan et al. in 2014 to know the peoples perception about the vaccination in USA. There are people who oppose vaccination for their children in the USA—the antivaccination people. They might be crazy, but they are there', and these people think that vaccination creates autism. Hence, the authorities used a bunch of four approaches based on very scientific material to change these people's beliefs. And all the approaches had taken pain to explaining, well, in brilliantly different ways, talking about the perceived threat from the harm of these diseases, the benefits of vaccination and how this autism connection is completely bogus. All of them failed. What does it indicate? Their belief system somewhere is anchored in such things. For instance, if I accept climate change, I will have to cut down my energy consumption; if I cut down my energy consumption, my American lifestyle is going to change; so instead of changing my lifestyle, I would rather not believe in climate change. And this is exactly what the American public seems to be doing. Hence, the issues of larger societal change—how we approach it—is not just a science and communication issue, it is connected very closely with what is really driving it. Also, unless we are able to hit those elements of it, we are not going to make a success of converting people. Let us forget, for the time being, the people who have nonscience, nonsense, outlook—forget them. There are still huge numbers of people who are in the middle ground. That is where our science communication has to be really oriented to or targeted.

There is a responsibility for the media as an institution to bring science to the people, the way they bring politics to the people. It is really, communicating the excitement of science; it is communicating what our people are doing in different laboratories and other places and what are the advances being made. This excitement of science needs to be communicated. Unfortunately, here we come to the reality, the part that is painful; the institution of the media today, just as institution of science, is in the thrall of private capital, and it therefore sees how to maximise its profit. What is a newspaper? One of the most powerful medium of communication in India. Newspapers are essentially a means of selling advertisements that, incidentally, use news; for the eyeballs it is 'a saccharine- or sugar-coated bitter stuff; they coat advertisements with news so that they can sell advertisement.' If that is the business model, then what are the consequences? Science does not buy anything. It really does not sell newspapers, and only the newspapers receiving more eyeballs get the advertisements. Therefore, page 3 is now in page 1; well, what is really on page 1 today in *Times of India? Preity Zinta versus Ness Wadia*! Now, that is what sells! And if that is what sells, science—unfortunately till the time we have got celebrities—is not going to sell newspapers! Hence, this whole model of the media, the private profit- and advertisement-driven media, essentially means that science communication does not have takers within the media. How many science editors are there? We do not even know who are the editors of the major newspapers; forget about the science editors! But science editor is a virtually extinct tribe within the media as a fraternity.

Now it is a problem that the media does not find any interest in promoting science. In fact, it would prefer to promote the babas (high-profile gurus with large number of followers)—which they do—and promote things that are illegal under the Drugs and Magic Remedies (Objectionable Advertisements) Act. These are being promoted widely on television! So today, the media does not have any stake in science communication. It does not believe that it is its task to bring news and knowledge to the people. On the contrary, it believes that it is there to sell goods and get eyeballs, and whatever gets eyeballs is what it should sell. In this context, serious news, anything serious, are at a disadvantage. And that is why the author turned in desperation to create an organisation, and an entity on the Internet called *newsclick.in*. This platform has been running since 2011 and is trying to bring out news from areas that Indian media does not cover generally.

Take for instance, news gathering. How many reporters of Indian newspapers and Indian media you know are placed abroad? Probably, all news agencies put together, it is not more than 10 or 20. Now, even when Indian media was far weaker than now, it used to have journalists in all the important parts of the world. It had them in different parts of Asia and Africa, and it had them in the Middle East and many other parts of the world. Now I think all the newspapers put together have just one reporter in the Middle East, and as it happens, functioning from *Dubai*, a place that really could not be called Middle East, but a little extension of Kerala perhaps!

We get international news from the Western eyes because we get it from the AFP (Agence France-Presse), Reuters and so on. This is what I wanted to change.

Therefore, http://newsclick.in/ was formed by taking the advantage of very low-cost opportunity, wider reach and the potential of a large audience provided by the Internet. This platform chose two things: covering West Asia, which is of vital importance to India and which we get only through the Western lens today, and science and technology, which are not covered seriously by the Indian media. However, it is also a fact that Internet has seen a much bigger monopoly than we have seen earlier in the media, and this is not the sole solution for reaching science to the general public.

If the interest in the Indian view of the world and interest in science and technology is to be revived again, we may have to use various available options through regulation, laws, public media and public broadcasters. It is a fact that scientific inflows to our schools and colleges are dwindling, and consequently our pipeline of scientific man/woman power is drying out. And we are going to face a crisis of science, because we cannot even teach school-level science, forget about higher-level sciences. This is the misfortune that is before us. If science and technology are to play their emancipatory role, which they need to do, it is imperative in India's current stage of development to have science and technology come in various ways—not for private profit but for larger social good. For all that, we need to look at our scientific institutions; and certainly the media today is not playing the role it should. How do we either compensate that or regulate that needs further discussion.

Challenges of Science Journalism in India

T.V. Jayan
TheTelegraph, Kolkata, India

Introduction

Science and technology (S&T) have immensely transformed the way we live, work, interact, inform and entertain ourselves. The technological advancement of late has been so rapid that many in our younger generation may find it difficult to believe that their parents and grandparents lived in a world—at least for a good part—where there were no computers, cell phones or the Internet. Today, every other household in India boasts of having a colour television set, but not so long ago having even a black-and-white television was considered luxury. Advances on the medical front have helped us find cure for many diseases once considered incurable.

By definition, science is the human pursuit to understand physical and natural environment around us. This leads to the accumulation of knowledge. Technology, on the contrary, is application of that knowledge. S&T share a symbiotic relationship and feed on each other. Breakthroughs in science lead to technological advancements, and tools of technology are in turn harnessed to create new science. And this cycle goes on and on.

More often than not, the first step towards any new technology is taken in a scientific laboratory—far away from limelight—invariably decades before a potential application can be ascribed to it. History of science is replete with many examples. Take for instance, the charge-coupled devices (CCDs). The principle behind CCD was discovered in 1969 by two American scientists, Boyle and Smith, working for AT&T Bell Labs (Janesick 2001). They won a Nobel prize for this discovery 40 years later in 2009. The only application that technologists could immediately think of then was in satellites or in telescopes that peer into the night sky to discover new cosmic objects. Most of us who happily go around clicking with our smart mobile phones or digital cameras would be unlikely to know that it was this same CCD technology that made digital photography a reality.

This does not mean that all the things that scientists do in a laboratory are of material benefit to us. In fact, most of it helps us to understand nature better, appreciate its wonders, know how we have evolved as a species or how the earth or for that matter the entire solar system has come into being in the first place and continues to function.

Those involved in science communication—or its subset, science journalism—have an enviable job of announcing such scientific and technological breakthroughs to the world. They get an opportunity to keep abreast with the developments in the field of S&T and to report on exciting findings that emerge from scientific laboratories. This, however, is just one of the many things that they get to do.

In her famous book *Selling Science: How the Press Covers Science and Technology* (Nelkin 1987), American sociologist of science Dorothy Nelkin says science journalism, among other things, should do three things; it should keep people apprised of scientific advancements, assess the appropriateness of scientific research and help the laypeople make choices related to perceived personal risks.

Nelkin in a way succinctly sums up the role the media has to play in taking science to people. She wants the media to 'cheerlead' good science and at the same time critically analyse research pursued by scientists and be a conveyor of risk issues associated with different technologies so that laypeople can individually or together make informed decisions.

The Practice of Science Journalism in India

According to a report released in 2014, in India there are 94,087 registered newspapers, of which 12,511 are dailies. In addition, there are nearly 400 news and current affairs television channels as of 2016. No country can arguably claim to have more newspapers or television channels than India. The fact that only a dozen or so media organisations among these have dedicated science reporters only illustrates how little importance is given to science coverage in India.

However, does that mean there is no coverage of science in Indian media? Certainly not. You may be surprised to see that newspapers published even from smallest of towns in the country carry science news. On the other hand, a daily published from Belgaum will not have much coverage about research being carried in Bengaluru, which is 500 kilometres away; but its readers will certainly get to read about a breakthrough achieved by scientists working in a laboratory in California, which is not less than 10,000 kilometres away. An avid reader of an Indian newspaper will know more about things being done in a US or European laboratory rather than feats achieved by fellow Indian scientists.

What is wrong with that, you may ask, considering the universal nature of science? Many years ago,during mid-1990s, I went to a reputed newspaper office in the capital looking for a science reporter's job. I was very confident that I might land one as I was freshly out of what is said to be the first ever batch of 'science communicators' ever passed out from an Indian university. The senior editorial person who interviewed me for the job was rather blunt. I can still vividly recollect the smugness on his face when I told him that I could cover science for his newspaper, which was then one of the leading newspapers published from the capital city. "What is there to cover in Indian science other than an occasional satellite launch or one too many disease outbreaks that are reported from different parts of the country year after year", he asked me. My feeble attempts to explain that there were over 100 national institutes in the country and that scientists there are doing some research worth writing about fell on his deaf ears as he made up his mind that no good science is being done in India since the days of Sir C.V. Raman, Jagadish Chandra Bose and Meghnad Saha. Needless to say, I was not selected.

In those days too, Indian newspapers used to carry articles on S&T—mostly of foreign origin. It was so naive of me to think that the newspapers were not publishing science stories from India because not many made an effort to report them. But it became clear to me why there was (and still continues to be) such a profusion of foreign science stories in Indian media a year later when I joined India's premier news agency, The Press Trust of India (PTI), as a trainee subeditor on its general desk. PTI had—and still has—a tie-up with the international news agencies like Reuters and AFP to release their news articles to the Indian press. Generally, probably because of embargo set by different research journals regarding the release of stories based on research articles appearing in them, science stories invariably get released earlier in the day than hard news stories, which are still developing through the day. Many newspapers, which go to bed early, found it convenient to use these science news articles to fill the pages—a practice, I believe, is followed even today. A paper on S&T coverage in English-language Indian dailies bears out this fact. The paper that reports study of S&T coverage in 37 Indian English newspapers found that a whopping 5385 science articles were carried by these newspapers, covering a total news space of 9,86,534 square centimetres during a 3-month period in 2008—that is, 1.62 articles per day per newspaper. Do we still complain that science is not covered in Indian media?

Of course, this is not the entire picture and is too good to be completely true; there definitely is something amiss here. A further analysis of the data presented by the researchers let the cat out. *The Times of India,* which can hardly claim the numero uno position when it comes to reporting on scientific research that happens in India, topped the list by cornering 27% of total articles and 19.2% of column space devoted to science. This was followed by *Asian Age* and *Free Press Journal,* both of which could not have claimed having dedicated reporters to cover science on their payroll. Curiously, *The Hindu,* the newspaper that could boast of having the higher number of dedicated science reporters, was at a poor 10th position, *Deccan Herald* at 13th and *The Telegraph,* the daily I work for, at 20th position with a negligible share of 0.8%!

Seriousness of newspapers is to be judged not only by science coverage, but also by dedicated staff for covering science vis-a-vis the staff on roll for covering other beats such as politics, law, finance and business, sports, films and so on.

A critical examination of science content in the Indian news media, thus, points to the fact that there has not been much coverage of science that is done within the country. Isn't it important to encourage the media organisations in the country to write more about scientific research that goes in the country—that too at a time when things are looking up for the S&T enterprise in India? As a person who has a ringside view of S&T developments in India since mid-1990s, the author can say with pride that the quality of S&T research pursued in India has improved by leaps and bounds during 2000s or so thanks to the creation of many world-class institutions as well as innovative fellowship schemes that have been successful in attracting many bright young scientists working abroad to come back home. Undoubtedly, this has led to an upswing in the quality of research being done in India, and this is quite visible as the number of publications by Indian authors in top-level research journals has steadily increased over the years.

Despite this, there has not been much improvement in the coverage of Indian science in the media. And that is something that needs to be pondered over, considering that our leaders—both political and scientific—have never passed up an opportunity to harp on the role of scientific temper in building a modern India. Scientific literacy is vitally important in fostering scientific temper, and media does play an important role in creating science literacy.

Who Should Be Blamed?

The Indian media is often blamed for not encouraging coverage of science, particularly of the local flavour. That leads to the question whether anybody has dispassionately analysed the reasons for such a shortcoming. Probably, not. Scientists (or those who analyse the science content in newspapers) often complain that sports get so much coverage, while coverage of science is almost nonexistent. Such a comparison, however, is a nonstarter, considering the humongous interest a game like football or cricket generates. A careful analysis will show that not all areas of sports receive equal prominence on the sports pages, and the column space often dedicated to a sport is directly proportional to the interest that readers have for that particular field.

A closer look at science coverage, however abysmal it may be, also shows a similar trend. Space and military science account for much of the coverage in the media. This is followed by exotic discoveries that capture the curiosity of the reader. Examples are the possibility of finding water on Mars, possible existence of many exoplanets, wonders of microscopic world and so on. Many editors would like to play up scare stories for the simple reason that it would attract readership.

There is a question, which is often asked, but mostly has gone unanswered: who is interested in getting science covered in the media, particularly in a developing country like India?

In many Western countries, one can understand, informed public opinion influences budgetary allocations for scientific research. Similarly, corporations dabbling in applied research know that a positive press can help tilt their fortunes in the market. For universities, public funded or otherwise, media provides to showcase the fruits of their research, and this can help them in many ways. It helps attract quality talent and also research funding from government, industry and charity organisations. More importantly, research grants sanctioned in many of the developed countries have an additional built-in component, which emphasises on public understanding of science. In other words, a minute portion of the money is set aside for publicising the research outcome. Experts have often stated that this is mutually beneficial; it makes researchers accountable for what they do and, on the other hand, forces them to share the research outcome with a laypublic in a language they can understand.

The situation in India, however, is just the opposite. It is not surprising considering that the government accounts for more than 80 % of the research money spent in the country. Although we pretend otherwise, what appears in the media does really influence what is funded and what is not. The government rather discourages public discussions on pros and cons of a research project, and public-sector scientists are often threatened with disciplinary action if he or she interacts with the press. Besides, India lacks a vibrant university system when it comes to S&T, and it is rather sad to say that the quality of scientific research is often so pathetic that there is not much to write home about, even though university researchers enjoy the freedom to interact with the press.

Why Indian Science Journalists Suffer the Most

Searching for a potential science story in India is a monumental task in itself. Barring few, none of the scientific institutions has a press officer who is proactive enough to alert science journalists. These institutions never believe in preparing a press release, unless, of course, to mark a milestone like a golden or platinum jubilee or a visit of a dignitary.

Although the number of research papers from India appearing in peer-reviewed journals has improved tremendously both in quality and quantity, science reporters never get to know about a new work unless they stumble on it while scanning research journals that they can access. Many Indian scientists actually consider taking time out to talk to a journalist about their or others' work as a waste of time. They are also worried that reporters would trivialise their work and misrepresent facts.

As a practitioner of the trade, one often gets a feeling that it is only science journalists who are interested in science news in this country because it is their bread and butter.

Does the scientific community have any right to complain about misreporting by the media or about its role in fanning controversies, be it about nuclear energy, transgenic crops or promoting unsubstantiated treatment modalities? Probably not, considering that there has been no or very little effort on its own part to engage with the media and educate the media practitioners about the ways of science or how scientific research is carried out.

To get an idea about how nonserious we have been about science communication/journalism, one has to just have a peep into the syllabus of science communication courses offered by a handful of universities in different parts of the country. The curriculum followed by these finishing schools can hardly prepare the students for a professional break. Science journalists needs to be nurtured if India has to become a flourishing democracy.

Community-Driven Approach for Artificial Recharge: Tarun Bharat Sangh Experience

Rajendra Singh
Tarun Bharat Sangh, Tarun Ashram, Bhikampura, Alwar DistrictRajasthanIndia

This is a saga of complete ecological change through community action in the desert and semidesert regions of Rajasthan, brought about by water harvesting.

Back in 1985 four young men resolved to take up the challenge of stalling the ecological disaster of desertification. However, before I talk about this, let me tell what was happening prior to the community action on water harvesting invoked in 1985, to see the impact of changes on the ecological front today. Also let me relate clearly the role of water in climate change.

Environmental Scene in Rajasthan in 1985

In the semiarid region of Rajasthan, especially in Alwar district, excessive drawing of groundwater for agriculture resulted in its shallow water-table aquifers dwindling to very low levels, as low as 100–120 metres.

Deep-confined aquifers, sometimes as deep as 300 metres, are either salty or too deep to pump. This left only about 11% of the lands under single cropping of which only 3% of the land of Alwar district remained under double cropping. The forest cover in the area, which includes the famous Sariska Tiger Reserve, got reduced to about 7%, that is, about 6500 square kilometres. The people were deserting their villages to earn a meagre living elsewhere as very little farming was possible; precious little was left as pasture lands for grazing the cattle and no new tree growth or forest regeneration was taking place. Thus, the farmers of this region could grow neither much food, nor fodder or pasture, nor any firewood for their rural energy needs. It was a harsh desert climate becoming unbearable due to complete exhaustion of groundwater. When the rainy season arrived (average rainfall of about 16–50 centimetres per year), it would rain only in three to four high-intensity events during 7–10 days of the rainy season, resulting into its loss as runoff, leaving behind thirsty lands.

Water Harvesting for Carbon Fixation and Sequestration to Reverse Climate Change

In the absence of water there is very little growth. When there is no vegetation, there is no carbon fixation into the soil and no carbon sequestration. Thus, the temperature increases. I am aware that various models of climate change predict an increase in temperatures of 2–4°C in the coming 100 years, and a disaster is predicted. However, with correct community efforts, if encouraged by all, especially if facilitated by governments in power, it can be reversed.

The carbon build-up in the atmosphere is to be arrested on all fronts where carbon is pumped into the atmosphere. However, it is only vegetation (crops, pastures, trees and forests) that sequesters it, and turns it into oxygen, which makes the atmosphere livable. Organic agriculture fixes crop residue by its incorporation into the soil or by recycling the organic residue as composts, taking carbon out of the atmosphere. However, for vegetation to flourish, it needs water.

Following are the stories of community efforts in water harvesting carried out to stop environmental degradation and climate change.

An Old Village Lady Provides Vision and Becomes My Water Harvesting Guru

The four youths, including myself and Nathi Bhalai, who went to live in Mangu Meena's village in 1985 for rural upliftment, were inspired by Gandhian model of rural development. Our idea was to bring prosperity to this semiarid village, but seeing no solution around soon we were very disappointed. Two among us left, convinced that nothing could be done. However, even though disappointed, I was not willing to give up. In this mood of desperation, Mangu Meena advised me to build *talabs* (ponds; small water dams to stop, store and recharge the groundwater, also called *johads*). Nathi Bhalai (a destitute woman, low in the Indian caste system) consoled me saying, "you fool, you have not understood what is needed." You need to build *talabs* so that the water does not run off but is held back to percolate to the ground. With Nathi Bhalai on my side—whom I call my guru, who showed me the way—I decided to dig for building a dyke in the village to create a *talab*. Seeing two of us digging dykes, some of the village youth laughed while others started joining us in digging and making the dyke. Rest is history now. Once the first small *talab* was ready, the villages saw water. The ecology of the surroundings changed in the very next season. This became an example of how to start restoring the ecology for better food production, vegetation rejuvenation and river restoration.

The idea spread like fire in the parched villages. Villagers started approaching me for helping them to do the same in their areas.

India has a tradition of water harvesting which, in the modern times, seems to have been lost. Hence, let us look at some of the traditional water-harvesting systems in India as reported by Sheena (1997). Carrying a spade and basket and accompanied by Nathi Bhalai of Gopalpura village, we had started building a dyke to hold water in the hot sun in village Gopalpura. That was in November 1985, more than 30 years ago. By 2008, the community had built over 8600 such *talabs* in over 1068 villages in the district of Alwar alone over an area of 6500 square kilometres under the leadership of Tarun Bharat Sangh (TBS, meaning, *Indian Youth Association*). There are many such efforts now going on in India.

Revival of Ecology by Talaabs, Johads *and* Chaals *in India*

Traditional methods of water harvesting are flourishing today through the efforts of noble Gandhian souls. Anupam Mishra who is recognised as an authority on traditional water-harvesting systems in India, rightly says, "hundreds and thousands of traditional, small reservoirs (tanks) did not appear all of a sudden from the blue in India in older times. For each promoter of tanks, there were tens of people who actually worked on these tanks. These 'ones' and 'tens' combined to make hundreds and thousands. But a society brought up on superficial modern education (system) that it has acquired in the past 200 years, has reduced these tens, hundreds and thousands into nonentities."

The stethoscope of modern thinking has pronounced many traditional rural cultures and self-sufficient societies as backward, obsolete and incapable of sustaining themselves without comprehending or admitting to the causes of disruption and destabilisation of the independent self-sustaining rural societies. Yet, in many parts of the world examples of sustainable management systems can be found. In India, small water reservoirs in the desert areas were part of a complex interlinking natural resource management system. It is based on this knowledge that all the civil society movements for water harvesting in India are being built up. There are examples aplenty.

Johads in Alwar

In the district of Alwar, the TBS, the organisation that I spearhead, is today regarded with reverence, pride and sometimes awe by many living in villages close to the Sariska Wildlife Sanctuary in Rajasthan. Referred to as the 'Ashram' (a religious retreat) by the local people, it has become synonymous with the *talabs* or *johads* today, rippling on the once eroded landscape of the Aravalli Hills. As mentioned earlier, today at least 1068 villages have altogether more than 8600 small and big johads, built with the active and increasing local participation in an area of 6500 square kilometres. In the Aravalli Hills most *johads* are built along contours of the mountain slopes for arresting and storing rainwater. During a heavy monsoon downpour in 1988, the people's *johads* stood firm while the government-built dam at Jaitpura, costing more than INR 1,00,000, was washed away.

i. Bhaonta: An Alwar Village Empowered, Recharged and Re-greened

The Babajiwala Johad, as it was named, the 160-metre-long structure, has an earthen embankment with a masonry spillway and was built by the community. The embankment is 13-metre wide at the base and 1.3-metre wide at the top. It is 4.5-metre high and has a catchment area of 10.25 square kilometres. The details that have gone into the making and maintenance of the dam are etched on a 1.3–1.6-metre-high stone wall meant to keep cattle from getting on to the embankment.

In the valley below, the Sankhara ka Bandh (a *bandh* means a dam and *sankhara* is 'narrow' in English), which owes its name to the site of construction that was a narrow passage between the steep slopes of the adjoining hillsides, was renovated by the community. The cemented, stone-lined embankment has been built to withstand the great force and swift flow of water. With a catchment of 9 square kilometres, the crescent-shaped, convex embankment has been raised across a length of 260 metre and is 7 metre high. Its strong base is 7-metre wide, narrowing to 2.6 metre at the top.

These small dams have not only recharged all the wells in the village downstream but also contributed to the revival of the river Arvari, one of the main tributaries of which originates in Bhanonta–Koylala. The forest in the catchment area of the *johad* does not look very green due to the semiarid nature of the ecology of the region, but its colour brightens up during the rainy seasons. On the embankment of the Sankhra ka Bandh, a painted sign says 'Bhairondev People's Wildlife Sanctuary', a mark of the respect the village has towards the wildlife, which is an inherent part of the forest ecosystem. The village elders have put numerous self-imposed restraints, which are fully observed by the village in obtaining forest produce. These have regenerated the forest in the valley.

The village leaders decided to select this site to build these small dams for following reasons:

1. The site was the narrowest in the valley.
2. The rock formation at that particular site was such that the dam just merged into it.
3. The site ensured that the embankment did not bear the brunt of the high-speed runoff.
4. From past experiences, they knew that groundwater recharge would be the maximum at that site.

ii. A River Revived: How Farmers Revive Rivers and Turn the Sand Dunes Back

The beauty of the regeneration efforts facilitated by TBS in the villages of Alwar (as discussed in the preceding paragraphs) is that the fruits of toil are ubiquitous although they may not seem very obvious. More to the point, they are permanent and their impact is more visible downstream, as in the case of the river Arvari. Nowhere more so than at Hamirpur, a village downstream of Bhaonta–Kolyala, where the statue of river Arvari blesses all at the entry point itself. The perennial flow in the river is the result of the 95 *johads* and dams, such as the Babajiwala Johad and Sankhra ka Bandh, built by villagers upstream.

The village elders would tell you that this river was a mere *barsati nullah* (a monsoon drain or seasonal rivulet) that used to flow for less than a week during the rainy season. As a direct result of harvesting the runoff upstream, the river became perennial in 1996, and has not ceased to flow since then. Even in the hottest months of summer, there is plenty of water in the river. This is attributed to the 'base flow', caused by the upstream small dams built by the farmers, making the river perennial.

In fact, the government had issued fishing licence to contractors from outside the region soon after fish appeared in the river. When the residents of Hamirpur protested, the contractors poisoned the river with insecticide aldrin to kill the fish. However, the community did not give up the fight against the abuse of a river that belonged to them. After initial indifference the police department appreciated the gravity of the situation and assisted the villagers in preventing any further unfortunate incidents. Thanks to the fishes, despite new surface water resources, no incidence of malaria is reported.

On way to Bhaonta from Thana Gazi, a narrow road divides the landscape ahead into two parts. The botched up efforts of government agencies can be seen alongside the success story of people's efforts based on traditional water-harvesting models. The systematic process of conservation beginning with recharging of groundwater by percolation tanks, protecting the scant vegetation and allowing it to regenerate shows people's very own model of development. On the contrary, a top-heavy approach of the government departments building boulder check dams along eroded mountain slopes and contour bunds have only escalated soil erosion and gullies.

iii. **Extracting Prosperity from Sand: Nimbi's Green Revolution**

In contrast to Alwar villages, the area in the adjoining Jaipur district is sandy. A dam built about 400 years ago had been rendered useless by shifting sand dunes. The village had fallen on bad days—agriculture was negligible, lands were barren and sand was all about as a result of the strong gusts of wind in the Bhanpur Kalan gap between two parallel ranges of the Aravallis. Around the year 2000, Nathu Lal Gujjar from village Nimbi, the representative of the village in the local Panchayat (village council) came in touch with TBS and met Nanak Ram of TBS at village Agar. By then, all his pleas for help to government agencies had failed. The village was about to be abandoned by the residents.

The first thing that TBS assessed was the capacity of the community as well as the organisation's capacity to help the village. It then got an idea of the kind of intervention needed to improve the environmental conditions in the village. At Nimbi, the people could have tried to restore the old dam straight away. But the village and TBS jointly decided that it was important to first rehabilitate the smaller dam. The dam was completed much before the schedule, which showed how hardworking the people of this village were. And the result was for all to see.

On the other side of the village, an earthen embankment built to connect the breach in the larger old dam to the far end of the valley was constructed. This earthen structure was built at a cost of about INR 5,00,000. And the result was that crop production in the area downstream of the embankment increased eight times.

iv. **Impact of Water Harvesting on Environmental Regeneration, Agriculture and Climate Change**

I am neither a scientist, nor a professional water engineer or a climate-change expert. I am a small constructive worker of Gandhi and I mobilise the civil society and the community for action on natural resource management and conservation for rural upliftment in India. Here I am recording the impact of our work on the ecology of 6500 square kilometre area in Alwar district during 1985–2007. Since 1985, 8600 small water-harvesting *talabs* in 1068 villages of Alwar district covering 6500 square kilometre area have been built. This has resulted in the shallow aquifer recharge in groundwater bringing up the water table from about 100–120 metre depth to 3–13 metre. The area under agriculture increased from 11% to 70%, of which the area under double cropping increased from 3% to 50% bringing prosperity to the farmers. The forest cover, which used to be about 7%, increased to

40% through agroforestry and social forestry, providing sufficient firewood and sequestering carbon from atmosphere.

The 1068 villages formed an *Arvari Sansad* (Arvari Parliament) to frame rules of water use and restricting areas from growing high water-consumption crops including paddy. It fixed the rules for pasture use on the basis of rotation. The forest use was limited to lopping of tree branches for fuel or cutting poles for domestic use, but never any tree felling.

I believe that this is the way to save the environment and bring prosperity to farmers in Indian villages. Based on this experience, many civil society organisations formed the *National Jal Biradari* (National Water Community) to save the Yamuna river. Alwar is a part of the catchment of Yamuna river, which dries out by the time it passes Delhi, the national capital. This people's campaign is now in progress.

I invite scientists to estimate the impact of such large community actions on carbon sequestration and fixation. I already know by experience its impact on the rural population. Most farmers who left Alwar for lack of sustenance have returned to their villages.

(*Reproduced by the author with permission from the Central Ground Water Board*).

Making Science Fun

Arvind Gupta
Inter-University Centre for Astronomy and Astrophysics (IUCAA), Pune University Campus, Pune 411 007, India

The 1970s was a period of great political turmoil throughout the world. The anti-Vietnam war, the upcoming ecology and feminist movements imbued people with a rare sensitivity. The students were out on the streets of Paris challenging authority. Whenever there is a political churning of society, it unleashes a lot of social energy. Socially conscious scientists—instead of designing missiles and bombs—searched for a more 'meaningful' role for themselves. Instead of being associated with research leading to death and destruction in the name of 'nation' or 'religion' they wanted to do something more worthwhile.

During the early 1970s Dr Anil Sadgopal, a Ph.D. from California Institute of Technology (Caltech), left Tata Institute of Fundamental Research (TIFR), Mumbai to initiate the famous Hoshangabad Science Teaching Programme (HSTP) in Madhya Pradesh, India. The HSTP aimed at revitalising the learning of science in resource-starved village schools. Prof. Yashpal—a Ph.D. from the Massachusetts Institute of Technology (MIT) and architect of the Satellite Instructional Television Experiment (SITE) programme that aimed at beaming educational videos to remote villages via satellites—arrived as the first teacher trainer.

In most village schools there are no laboratories. All science is learnt by rote. Often elite schools in cities have laboratories full of glass burettes, beakers and test tubes, but all this 'fancy stuff' is often kept locked in school cupboards and gathers dust. These expensive apparatus are meant for the inspectors—not for the children. And all the while children mug up science and spit it out in the examinations!

This was not how good science was learnt. The HSTP started with 16 government schools and worked for 5 years to develop a suitable, contextual, hands-on curriculum relevant for village children. Many good teachers from various universities and Indian Institutes of Technology (IITs) helped the HSTP. In 1978, I took a year off from my job at Tata Motors, Pune to intern myself with the HSTP.

Matchstick Models

Here my engineering background came in handy. In the very first month I designed the simple 'Matchstick Meccano' using pieces of thin rubber tubes that are used as bicycle tube valves (called *valve tubes*) as joints and matchsticks as members (Fig. 1). Notably, it used very ordinary things—matchsticks, found in every household and bicycle valve tube, which was easily available in small towns. Children could make many two- and three-dimensional shapes and understand a great deal about angles and structures in a concrete way. Children could easily see that the triangle was the only rigid polygon and that is why it was used to build roofs and bridges. Children also understood the rigidity of the tetrahedron. Using simple, readily available stuff they could make physical models of methane and other molecular structures.

In 1986, I wrote my first book *Matchstick Models and Other Science Experiments*. Soon it was translated by various popular science groups into 12 Indian languages.

http://www.youtube.com/watch?v=E0rLC_WY02Q

Fig. 1 Arvind Gupta explaining matchstick models. *Source* Author's work

Simple Electric Motor

The electric motor is very much part of every high school curriculum. Electric motors are widely used everywhere. In a typical home, the electric fan, air cooler, refrigerator, washing machine, kitchen mixer and water pump, all use an electric motor. Modern life would come to a standstill without the electric motor. Despite being widely used at home very few children ever construct an electric motor. This is a scathing reflection on the way science is taught in our schools. Science is learnt by rote, seldom through experiments.

An electric motor is simple to make. If you have a 1.5-volt torch battery it takes just INR 5 more to make it. You need an ordinary ferrite magnet, two big safety pins, two wide rubber bands from an old bicycle tube and a 1-metre long piece of insulated copper wire used for motor rewinding. The coil is wound using the battery itself as a former. One end of the coil is completely scraped of the insulation exposing the shining copper below (Fig. 2). The other end is scraped only on three sides—the little remaining insulation on this end acts as an on–off switch. This improvised switch is the heart of the motor. This switch acts as the 'brush' or the 'commutator' of a DC motor.

Over the years we have made over 50,000 of these electric motor models with children and teachers. Every time a child makes an electric motor you can see a gleam in her eyes and a smile on her face. This has been our greatest reward. Watch this video on YouTube:

http://www.youtube.com/watch?v=vSPFwibREUg

Levitating Pencil

At the heart of the levitating pencil are six ferrite ring magnets. Four magnets are embedded in a rubber base. The magnet on the pencil near its pointed end attracts the two magnets below it (Fig. 3). The magnet near the pencil base repels the two

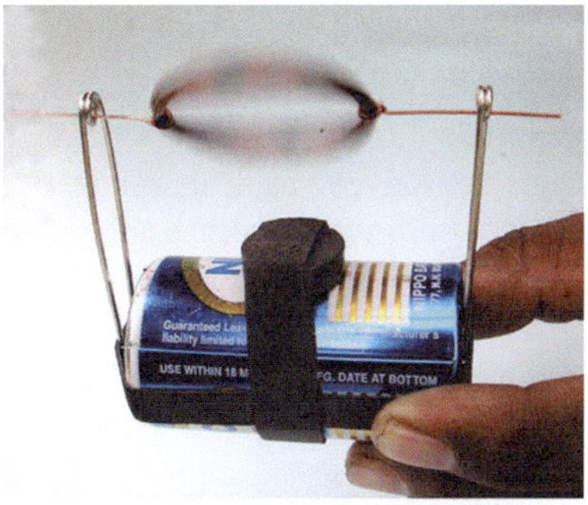

Fig. 2 Simple electric motor
Source Author's work

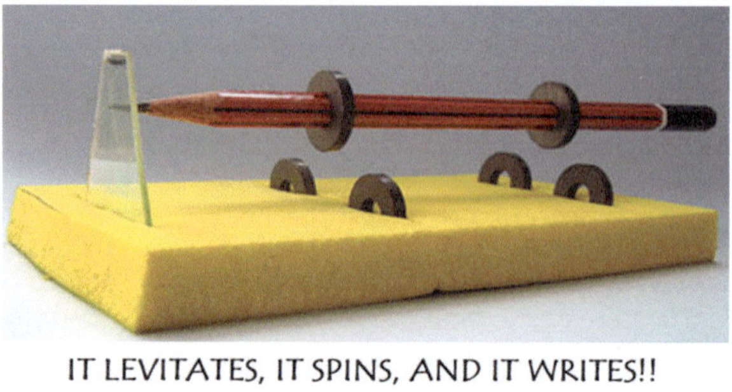

Fig. 3 Levitating pencil. *Source* Author's work

magnets below it. This makes the pencil levitate. As only the sharpened tip of the pencil touches and spins on a piece cut out of a discarded compact disc (CD), there is very little friction. Therefore, if you twirl the pencil once it keeps spinning for a very long time. Countries like China, Japan and Germany have installed magnetic levitation trains for mass rapid transport. Our children can at least get a glimpse of this phenomenon with this low-cost levitating pencil.

Few years ago Hamsa Padmanabhan, a student of 10th class at the Kendriya Vidyalaya (*Central School*), Ganeshkhind, Pune, wrote a short scientific article explaining the science behind the levitating pencil designed by us. This model costs less than INR 20 and can be made in 15 minutes. This won Hamsa the second Intel International Award of US $2500. Five years later a minor planet was named after Hamsa. This shows the innate potential of our children; they can compete with the best in the world. Watch a detailed video on how to make the levitating-pencil model at: http://www.youtube.com/watch?v=33e8bfUqK2o.

Touching Slate

The 'Touching Slate' is based on a simple idea, that wool sticks to Velcro. The slate is made by sticking Velcro strips on a stiff cardboard. The drawing pen is essentially a 'fishing line' that dispenses wool (Fig. 4). Visually impaired children can draw their artwork on the Velcro slate. Later they can literally 'feel' the drawing with their fingertips. The only thing to be bought is the Velcro. The rest of the materials can be easily collected from the surroundings. This low-cost slate has been extensively used in schools for the visually impaired throughout India. India has 13 million visually impaired people, the largest number in the whole world. This Touching Slate has been a real boon to them. With the help of our website a special-needs school in Thailand made these slates for their visually challenged children. Watch a video on making the 'Touching Slate' at: http://www.youtube.com/watch?v=LHXlKkS8La8.

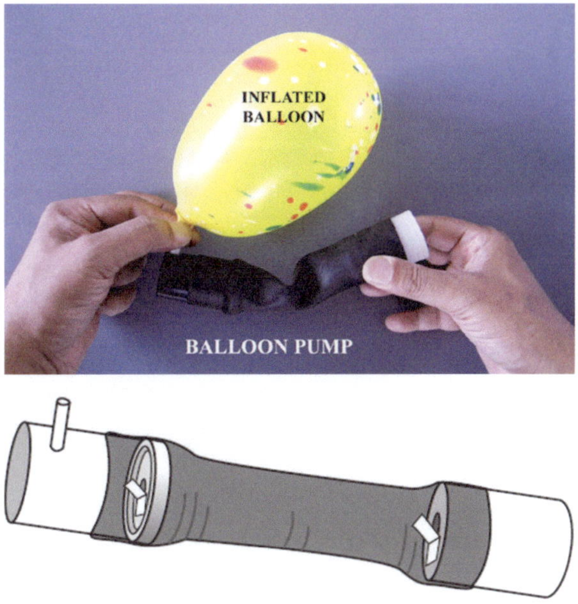

Fig. 4 Balloon pump. *Source* Author's work

Balloon Pump

We have designed over 30 different kinds of pumps but this one is stupendous. With this homemade pump you can inflate a balloon and pop it too! All the parts of this pump can be taken apart and assembled again. Two plastic film cans are fitted snugly at the two ends of a 25-centimetre-long old bicycle tube (Fig. 5). The left can lid has the delivery valve and the right film can base has a suction valve. These valves are simple 'flaps' that open and close and allow air to flow in only one direction. To demonstrate the working of this pump, tightly place a balloon on the delivery pipe and move the bellows to inflate the balloon. This superb pump very efficiently blows up a balloon. This pump certainly deserves a certificate of ISI standard! Children love playing with this pump. In the process of making it they learn a great deal of science.

Reduce, Reuse, Recycle

There is a very strong component of recycling in what we do. Our consumerist society produces a lot of junk. Every product comes packed in multiple layers. Tetra Pak containers, plastic bottles and cardboard boxes can be found littered everywhere. We encourage children to pick this waste and make wonderful toys and teaching aids out of them. This imbues children with eco-sensitivity. They feel happy about doing more with less and at the same time being good earth citizens.

Fig. 5 Velcro-basedtouching Slate for visuallyimpaired. *Source* Author's work

Take for instance the Tetra Pak packages—multilayered containers used for packaging of popular soft drinks, vegetable oils and so on. Tetra Pak containers are lethal for the environment. They are made by fusing several layers of polyethylene, aluminium foil and card sheets into a multiwalled laminate. It is very difficult and energy intensive to separate these layers; therefore, the discarded Tetra Pak containers can lie buried in landfills for hundreds of years. Nature is totally helpless against this manmade material. We have made over 40 science-teaching aids and toys using Tetra Pak materials. Here is an example of a very useful purse, which can be made from a discarded 1-litre fruit juice Tetra Pak container (Fig. 6). There are two pockets to keep money. The flap with the Velcro locks the purse. This material is water proof, so your money will remain safe in this wallet.

Fig. 6 Tetra Pak purse. *Source* Author's work

In the USA, a billion plastic bottles are thrown away every two days. We are learning the worst of the West. Plastic bottles can be found littered everywhere and have become an environmental menace in our cities. We have designed over 100 simple science experiments using plastic bottles, straws, lids and so on. Through these low-cost experiments children can learn a great deal about air and water pressure apart from making some delightful toys, which do not cost anything.

Educational Resources

Since 2003, I have been working with the Children's Science Centre located at the Inter-University Centre for Astronomy and Astrophysics (IUCAA) at Pune. Here a small, focused team of competent and compassionate people tries to make science accessible and fun for the poorest of our children.

Twice a week, 50 children from a local school visit our science centre. Here they see possibilities of doing science using the humblest materials—rubber slippers, cycle tubes, plastic bottles, newspapers, broomsticks, straws, buttons and old CDs. In the course of four hours children see numerous examples of low-cost science models. They also make a dozen things with their own hands—a newspaper cricket cap, an electric motor, a flying fish, helicopter, straw flute, straw sprinkler and a broomstick spinner. It is a unique rejuvenating experience, which the children never forget.

We have documented over 1000 hands-on experiments with sequential photographs and simple instructions. We have also collated them in activity books and made short-duration films titled *Toys from Trash*. As of 2016 we have over 7700 short videos in eighteen languages uploaded on the YouTube. Our films in various languages—Marathi, Hindi, Kannada, Telugu, Malayalam, Tamil, English,

Gujarati, Spanish, Japanese, French and Russian—have been viewed 30 million times. Every day over 40,000 children and teachers across the world view them (http://arvindguptatoys.com).

There are over 4000 passionate books on education, environment, science, mathematics, antiwar/peace topics and great children's books on the author's website in English, Hindi, Marathi and several other Indian languages. That over 15,000 books are downloaded every single day shows that our teachers and children have a great hunger for knowledge and an appetite for learning. Therein lies the hope.

References

Bagla, P., & Menon, S. (2008). *Destination moon: India's quest for the moon, mars, and beyond*. India: HarperCollins.

Janesick, J. R. (2001). *Scientific charge-coupled devices*. SPIE Press.

Nelkin, D. (1987). *Selling science: How the press covers science and technology*. New York: W.H. Freeman & Company.

Nyhan, B., Reifler, J., Richey, S., & Freed, G. L. (2014). Effective messages in vaccine promotion: A randomized trial. *Pediatrics, 133*, e835–e842.

Sankar, U. (2007). *The economics of India's space programme*. New Delhi: Oxford University Press.

Watve, M. (2013). *Dove, diplomats and diabetes: A Darwinian interpretation of type 2 diabetes and related Disorders*. New York: Springer Science+Business Media.

MIX
Papier aus verantwortungsvollen Quellen
Paper from responsible sources
FSC® C105338

If you have any concerns about our products,
you can contact us on
ProductSafety@springernature.com

In case Publisher is established outside the EU,
the EU authorized representative is:
Springer Nature Customer Service Center GmbH
Europaplatz 3, 69115 Heidelberg, Germany

Printed by Libri Plureos GmbH
in Hamburg, Germany